Learning Guide for Tortora and Anagnostakos:
Principles of Anatomy and Physiology

Learning Guide

for
Tortora and Anagnostakos: Principles of Anatomy and Physiology
Fifth Edition

Kathleen Schmidt Prezbindowski

College of Mount St. Joseph
and
The University of Cincinnati

Gerard J. Tortora

Bergen Community College

1817

HARPER & ROW, PUBLISHERS, New York
Cambridge, Philadelphia, San Francisco, Washington,
London, Mexico City, São Paulo, Singapore, Sydney

TO AMY AND LAURIE

Sponsoring Editor: Claudia M. Wilson
Development Editor: Robert Ginsberg
Project Editor: David Nickol
Cover Design: Ron Gross
Cover Art: Composite photograph of x ray of skull with brain by Bill Longcore/Longcore Maciel
   Studio. Anteroposterior projection of the right scapula and proximal end of the humerus. Courtesy of
   Garret J. Pinke, Sr., R.T. Scanning electron micrograph of fibrin threads and red cells. Courtesy of
   Fisher Scientific Company and S.T.E.M. Laboratories, Inc. Copyright 1975.
Text Art: Vantage Art, Inc.
Production: Kewal K. Sharma
Compositor: ComCom Division of Haddon Craftsmen, Inc.
Printer and Binder: The Murray Printing Company

LEARNING GUIDE FOR TORTORA AND ANAGNOSTAKOS:
PRINCIPLES OF ANATOMY AND PHYSIOLOGY, Fifth Edition

Library of Congress Cataloging in Publication Data

Prezbindowski, Kathleen Schmidt.
   Learning guide for Tortora and Anagnostakos'
Principles of anatomy and physiology, fifth edition.

   1. Human physiology—Problems, exercises, etc.
2. Anatomy, Human—Problems, exercises, etc.   I. Tortora,
Gerard J.   II. Tortora, Gerard J.   Principles of
anatomy and physiology   III. Title.   [DNLM:
1. Anatomy—problems.   2. Homeostasis—problems.
3. Physiology—problems.   QS 4 T712P Suppl.]
QP34.5.T67   1987 Suppl.   612′.0076   86-26962
ISBN 0-06-045274-9

86 87 88 89 9 8 7 6 5 4 3 2 1

Contents

Preface

This *Learning Guide* is intended to be used in conjunction with Tortora and Anagnostakos' *Principles of Anatomy and Physiology,* Fifth Edition. The emphasis is on active learning, not passive reading, and the student examines each concept through a variety of activities and exercises. By approaching each concept several times from different points of view, the student sees how the ideas of the text apply to real, clinical situations. In this way the student may be helped not simply to study but to *learn.*

The twenty-nine chapters of the *Learning Guide* parallel those of the Tortora and Anagnostakos text. Each chapter begins with an Introduction that presents an overview of the material and a Topics Summary that outlines the major divisions of the chapter. Chapter Objectives, repeated exactly from the text, are included for convenience in both previewing and reviewing. Learning Activities follow the exact sequence of topics in the chapter, with cross-references to pages in the text. This format makes the *Learning Guide* "user-friendly" or "student-friendly" and enhances the effectiveness of the excellent text.

This edition of the *Learning Guide* offers new opportunities for self-assessment of learning. The number of Learning Activities for which answers are provided has been increased by 50 to 100 percent in most chapters. Twenty-one new figures have been added. A number of new *For Extra Review* questions provide enrichment or assistance with especially difficult topics. The number of *Clinical Challenges* has been increased to 50. *Clinical Challenges* are exercises based on my own practice as an R.N. and on discussions with colleagues. They reflect our actual experience with patients in clinical settings. These challenging exercises enable students to apply abstract concepts to real situations. They also emphasize the relevance of anatomy and physiology to students' current or future clinical experiences.

ACTive Learning exercises are a new feature of the Fifth Edition. With a study partner or study group, or in a lab/classroom setting, students role play interrelated structures or functions involving a complex series of events. The book provides guidance for these activities and encourages student creativity and group interactions. Examples of the 10 *ACTive Learning* exercises in the *Learning Guide* are: role-playing life within cells, such as protein synthesis (page 52) and transport mechanisms (page 50), and acting out blood clotting (page 411), bile regulation (page 554), and the menstrual cycle (page 651).

A number of persons have contributed to the revision of this book. I offer particular thanks to the students and faculty at the College of Mount St. Joseph, especially to my colleague A. Kay Clifton, for endless discussions on teaching/learning strategies. I extend thanks to Annette Muckerheide, S.C., for her critique and valued suggestions regarding Chapter 22. For her unmatched organizational skills and patience, I am indebted to Marlene Pohlman. I thank my family, Laurie and Amy Prezbindowski and Harry, Norine, and Maureen Schmidt, for their constant encouragement and caring. I am grateful to the editorial and production staff at Harper & Row for their support and guidance. Special thanks to Robert Ginsberg, development editor, David Nickol, project editor, and Claudia Wilson, sponsoring editor.

To the Student

This *Learning Guide* is designed to help you do exactly what its title indicates —*learn*. It will serve as a step-by-step aid to help you bridge the gap between goals (objectives) and accomplishment (learning). The twenty-nine chapters of the *Learning Guide* parallel the twenty-nine chapters of Tortora and Anagnostakos' *Principles of Anatomy and Physiology,* Fifth Edition. Each chapter consists of the following parts: *Introduction, Topics Summary, Objectives, Learning Activities, Answers to Numbered Questions in Learning Activities,* and *Mastery Test.* Take a moment to turn to Chapter 1 and look at the major headings in this *Learning Guide.* Now consider each one.

Introduction.   Before you begin your study of each chapter, read this brief overview. It will present the scope of the chapter and will correlate each main topic with related groups of objectives. It will also help to integrate new material with what you have learned in previous chapters.

Topics summary.   This section provides a concise outline of chapter content.

Chapter Objectives.   The objectives listed at the beginning of each textbook chapter are included here for convenient reference. Use the objectives to organize the material of the chapter into manageable facts. In this way, you will find that the chapter as a whole will not overwhelm you. (Your instructor may wish to modify the list of objectives according to the goals and time frame of your course.) The list of objectives also serves as a useful review. At the completion of each chapter, return to the objectives and check off each one that you have studied and that you feel you understand. If any are not clear to you, look over the *Learning Guide* material again.

Learning Activities.   As you move to each new chapter topic, thoroughly read the corresponding text pages (noted in the *Learning Guide*). Then carry out the related activities in the *Learning Guide.* These thought-provoking exercises are designed to increase your depth of understanding of the text and to challenge and verify your knowledge of key concepts. In short, they will assist you, not only in *studying,* but in *truly learning.*

The *Learning Guide* includes a variety of learning activities. You will focus on the specific facts of the chapter by doing exercises that require you to provide definitions and comparisons, and you will fill in the blanks in paragraphs with key terms. You will also draw, complete, or color figures. (Colored felt-tipped pens or pencils will be useful.) Your understanding will be tested with matching exercises, multiple-choice questions, and identifications of parts of figures. The number of activities has been increased in this edition to aid you more during the learning process.

Two special types of learning activities are *For Extra Review* and *A Clinical*

Challenge. For Extra Review provides additional help for difficult topics or refers you by page number to specific activities earlier in the *Learning Guide* so that you may better integrate related information. Applications for enrichment and interest in areas particularly relevant to nursing or health sciences students are found in selected *Clinical Challenge* exercises. *ACTive Learning* exercises are a new feature of this edition. These provide guidance for you and a study group or lab group to *act* or portray parts of the body, such as molecules or organs. For example, you may act the part of chemicals in blood that interact to form a clot (Chapter 19), the parts of a cell that contribute to the health of the whole cell (Chapter 3), nerve transmitters and drugs that can block nerve transmission (Chapter 16), and hormones effecting changes of the menstrual cycle (Chapter 28). My own students have greatly benefitted from such role-playing. They say that they learn better as a result of the total involvement—and they enjoy it. (Be a ribosome or a gallbladder soon!)

When you have difficulty with an exercise, refer to the related pages (listed in the margin of the *Learning Guide*) of *Principles of Anatomy and Physiology* before proceeding further. You will find specific references by page number for all text figures and exhibits.

Answers to Numbered Learning Exercises.   As you glance through the activities, you will notice that a number of questions are marked with boxed numbers (such as $\boxed{14}$). Answers to such questions are given at the end of each chapter, in order to provide you with some immediate feedback about your progress. Incorrect answers alert you to review related objectives in the *Learning Guide* and corresponding text pages.

Activities without boxed numbers and answers are also purposely included in this book. The intention is to encourage you to verify some answers independently by consulting your textbook and also to stimulate discussion with students or instructors.

Mastery Test.   This twenty-five-question self-test provides an opportunity for a final review of the chapter as a whole. Its format will assist you in preparing for standardized tests or course exams that are objective in nature, since the first twenty questions are multiple choice, true-false, and arrangement questions. The final five questions of each test help you to evaluate your learning with fill-in and short answer questions. Answers for all twenty-nine mastery tests are placed together at the end of the book.

I wish you success and enjoyment in *learning* concepts and relevant applications of anatomy and physiology.

Kathleen Schmidt Prezbindowski

UNIT I
Organization of the Human Body

An Introduction to the Human Body

As you begin your study of *anatomy* and *physiology,* you will first define these sciences and recognize their subdivisions (Objective 1). You will then learn about structure-function relationships at different levels of organization within the body (2–3). You will identify important life processes of humans (4). In order to avoid getting lost during your travels through the human body, you will familiarize yourself with the structural plan, directional terms, planes, and sections of the body (6–8). You will also identify major body cavities, regions, and locations of organs (9–10). You will also learn about techniques used to study internal body structure, particularly those for diagnosis of disease (11–12). This chapter will also serve as an introduction to *homeostasis.* You will consider its central role in physiology, and you will begin to study some mechanisms which enable the human body to maintain this state of dynamic equilibrium (13–17).

Topics Summary

A. Anatomy and physiology defined
B. Levels of structural organization
C. Life processes
D. Structural plan, terms, planes, sections
E. Body cavities, regions
F. Radiographic anatomy
G. Homeostasis
H. Measuring the human body

Objectives

1. Define anatomy, with its subdivisions, and physiology.

2. Define each of the following levels of structural organization that make up the human body: chemical, cellular, tissue, organ, system, and organismic.

3. Identify the principal systems of the human body, list the representative organs of each system, and describe the function of each system.

4. List and define several important life processes of humans.

5. Describe the general anatomical characteristics of the structural plan of the human body.

6. Define the anatomical position and compare common and anatomical terms used to describe various regions of the human body.

7. Define several directional terms used in association with the human body.

8. Define the common anatomical planes that may be passed through the human body and distinguish a cross section, frontal section, and midsagittal section.

9. List by name and location the principal body cavities and the organs contained within them.

10. Explain how the abdominopelvic cavity is divided into nine regions and into quadrants.

11. Contrast the principles employed in conventional radiography and in computed tomography (CT) scanning.

12. Explain how the dynamic spatial reconstructor (DSR) operates and describe its diagnostic importance.

13. Define homeostasis and explain why homeostasis is a state that results in normal body activities and why the inability to achieve homeostasis leads to disorders.

14. Define a stress and identify the effects of stress on homeostasis.

15. Describe the interrelationships of body systems in maintaining homeostasis.

16. Contrast the homeostasis of blood pressure (BP) through nervous control and of blood sugar (BS) level through hormonal control.

17. Define a feedback system and explain its role in homeostasis.

4   **Learning Activities**

(page 6)   **A.** Anatomy and physiology defined

   1. Contrast each of the following pairs of terms.
      a. Anatomy/physiology

      b. Gross anatomy/histology

      c. Systemic anatomy/regional anatomy

      d. Developmental anatomy/embryology

   2. Look at the two items in each of the following pairs. Tell how their structural differences explain their functional differences.
      a. Spoon/fork

      b. Hand/foot

      c. Chair/bed

      d. Incisors (front teeth)/molars

(pages 6–7)   **B.** Levels of structural organization

   [1]   1. Rearrange these terms in order from highest to lowest level of organization, using lines provided.

Term	Level of organization	5
Organism	(highest) _____	
Cell	_____	
Tissue	_____ Organ _____	
Organ	_____	
Chemical	_____	
System	(lowest) _____	

2. Complete the following table describing systems of the body. Name two or more organs in each system. Then list one or more functions of each system.

System	Organs	Functions
a.	Skin, hair, nails	
b. Skeletal		
c. Muscular		
d.		Regulates body by nerve impulses
e.	Glands that produce hormones	
f.	Blood, heart, blood vessels	
g. Lymphatic		
h.		Supplies oxygen, removes carbon dioxide, regulates acid–base balance
i.		Breaks down food and eliminates solid wastes
j.	Kidney, ureters, urinary bladder, urethra	
k. Reproductive		

(pages 8–9) **C.** Life processes

1. Complete the list and briefly define each of seven characteristics which distinguish you (or other living organisms) from nonliving things. One is done for you.

   a. _____

   _____

   b. _____

   _____

   c. _____

   _____

   d. Contractility: ability of cells (especially muscle cells) to shorten or contract.

   e. _____

   _____

   f. _____

   _____

   g. _____

| 2 |

2. Refer to the list of terms at right, all related to the life process of metabolism. Demonstrate your understanding of these terms by selecting the term that best fits each description provided below.

   _____ a. Taking in of foods

   _____ b. Breakdown of foods into forms that can be absorbed and used by cells

   _____ c. The sum of all chemical processes in the body

   _____ d. Synthesis of body's structural and functional components; uses energy

   _____ e. Production and release of useful substances by cells

   A. Anabolism
   B. Catabolism
   D. Digestion
   E. Excretion
   I. Ingestion
   M. Metabolism
   S. Secretion

(pages 9–13) **D.** Structural plan, terms, planes, sections

1. Complete this exercise about the structural plan of the human body.
   a. Explain the meaning of the anatomical arrangement of a *tube within a tube*.

b. State one advantage of the following two characteristics of the human body:

vertebral column (backbone) of 26 separate bones

bilateral symmetry

2. Be sure that you understand the meaning of anatomical position by assuming that position yourself.
3. Complete the table relating common terms to anatomical terms. (See Figure 1-2, page 10, in your text to check your answers.) For extra practice use common and anatomical terms to identify each region of your own body and that of a study partner.

Common Term	Anatomical Term
a.	Axillary
b. Fingers	
c. Arm	
d.	Popliteal
e.	Cephalic
f. Mouth	
g.	Inguinal
h. Chest	
i.	Cervical
j.	Antebrachial
k. Buttock	
l.	Calcaneal

4. Using your own body, a skeleton, a torso, or Figure 1-3, (page 11) in your text, determine relationships among body parts. Write the correct directional term(s) to complete each of these statements.

3

   a. The liver is _____ to the diaphragm.

   b. Fingers (phalanges) are located _____ to wrist bones (carpals).

   c. The skin on the dorsal surface of your body can also be said to be located on your _____ surface.

   d. The great (big) toe is _____ to the little toe.

   e. The little toe is _____ to the great toe.

   f. The skin on your leg is _____ to muscle tissue in your leg.

   g. Muscles of your arm are _____ to skin on your arm.
   h. When you lie face down in a pool, as if to do the "deadman's float," you are lying on your _____ surface.

   i. The lungs and heart are located _____ to the abdominal organs.
   j. Since the stomach and the spleen are both located on the left side of the abdomen, they could be described as _____-lateral.

   k. The _____ pleura covers the external surface of the lungs.

5. Match each of the following planes with the phrase telling how the body would be divided by such a plane.

4

_____ a. Into superior and inferior portions   F. Frontal
_____ b. Into equal right and left portions   H. Horizontal
_____ c. Into anterior and posterior por-   M. Midsagittal (median)
          tions   S. Sagittal
_____ d. Into right and left portions

(pages 13–16) E. Body cavities, regions

5

1. After you have studied Figures 1-6, 1-7, and 1-8 (pages 13-16), complete this exercise about body cavities. Circle the correct answer in each statement.
   a. The *(dorsal? ventral?)* cavity consists of the cranial cavity and the vertebral canal.
   b. The viscera, including such structures as the heart, lungs, and intestines, are all located in the *(dorsal? ventral?)* cavity.
   c. Of the two body cavities, the *(dorsal? ventral?)* appears to be better protected.
   d. Pleural, mediastinal, and pericardial are terms that refer to regions of the *(thorax? abdominopelvis?)*.
   e. The *(heart? lungs? esophagus and trachea?)* are located in the pleural cavities.

f. The division between the abdomen and the pelvis is marked by *(the diaphragm? an imaginary line from the symphysis pubis to the superior border of the sacrum?).*

g. The stomach, pancreas, small intestine, and most of the large intestine are located in the *(abdomen? pelvis?).*

h. The urinary bladder, rectum, and internal reproductive organs are located in the *(abdominal? pelvic?)* cavity.

2. Complete Figure LG 1-1 according to directions below.

   a. Draw and label thoracic organs.

   b. Draw lines dividing the abdomen into nine regions. After carefully studying Figure 1-8 (pages 14–16) in your text, try to draw and label from memory the following organs in their correct locations: *stomach, liver, colon of large intestine, appendix,* and *small intestine.* Then draw and label the *kidneys* in contrasting color (since these organs are posterior to other abdominal viscera).

3. Using the names of the nine abdominal regions, complete these statements. Refer to the figure you just completed and to Figure 1-8 (pages 14–16) in your text. ⬜6

   a. From superior to inferior, the three abdominal regions on the right

      side are _____, _____, and

      _____.

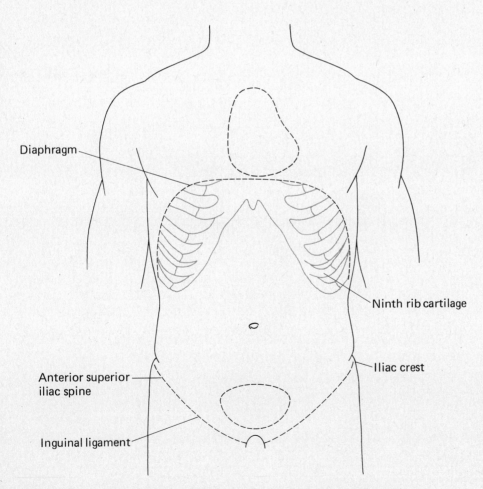

Figure LG 1-1 Regions of the ventral body cavity. Complete as directed.

b. The stomach is located primarily in the two regions named

_____ and _____ .

c. The navel is located in the _____ region.

d. The region immediately superior to the urinary bladder is named the

_____ region.

e. If the abdomen were divided into four quadrants, the left lower quadrant

(LLQ) would include all of the _____ region and

parts of the _____ , _____

_____ , and _____ regions.

4. A *clinical challenge*. Complete this exercise about the procedure known as an autopsy.

1. List three reasons why autopsies may be performed.

2. Briefly describe the three phases of an autopsy.

(page 17) **F. Radiographic anatomy**

1. Match the radiographic anatomy techniques listed at right with the correct descriptions.

| 7 |

_____ a. A recent development in radiographic anatomy, this technique produces moving, three-dimensional images of body organs.

_____ b. This method provides a cross-sectional picture of an area of the body by means of an x-ray source moving in an arc. Results are processed by a computer and displayed on a console.

_____ c. This technique produces a two-dimensional image (roentgenogram) via a single barrage of x-rays. Images of organs overlap, making diagnosis difficult.

CR. Conventional radiograph

CT. Computed tomography

DSR. Dynamic spatial reconstruction

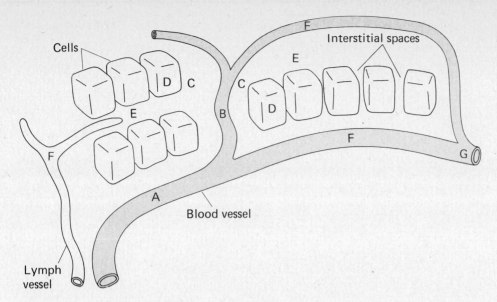

Figure LG 1-2 Internal environment of the body: types of fluid. Complete as directed. Areas labeled *A* to *G* refer to Learning Activity ⑨ .

G. Homeostasis

(pages 17–21)

1. Describe the types of fluid in the body and their relationships to homeostasis by completing these statements and Figure LG 1-2. ⑧

   a. Fluid inside of cells is known as _____ fluid. Color areas containing this type of fluid yellow.

   b. Fluid in spaces between cells is called _____. It surrounds and bathes cells and is one form of *(intracellular? extracellular?)* fluid. Color the spaces containing this fluid light green.

   c. Another form of extracellular fluid is that located in _____

   _____ vessels and _____ vessels. Color these areas dark green.

   d. The body's "internal environment" (that is, surrounding cells) is

   _____-cellular fluid (all green areas in your figure). The condition of maintaining ECF in relative constancy is known as

   _____.

2. Refer again to Figure LG 1-2. Show the pattern of circulation of body fluids ⑨ by drawing arrows connecting letters in the figure in alphabetical order (*A → B → C*, etc.). Now fill in the blanks below describing this lettered pathway:
   A (arteries and arterioles) → B (capillaries)

   → *C* (_____) → *D* (_____)

   → *E* (_____) → *F* (_____)

   → *G* (_____).

3. List six qualities of ECF which are maintained under optimal conditions when the body is in homeostasis. ⑩

4. Consider how your internal environment is affected by your external environment. Which of the six qualities of your ECF listed above would be altered if you were subjected to the following conditions? Compare your answers with those of a study partner.
   a. Taking a two-day hike at high altitude without appropriate food or clothing

   b. Invasion of your body by an infectious microorganism that causes diarrhea over a prolonged period of time

5. Suppose you decide to run a quarter mile at your top speed. Tell how your body would respond to this stress created by exercise. List according to system the changes your body would probably make in order to maintain homeostasis.
   a. Integumentary

   b. Muscular

   c. Cardiovascular

   d. Respiratory

   e. Digestive

   f. Urinary

   g. Nervous

   h. Endocrine

6. Consider an example of a feedback system. Ten workers on an assembly line are producing hand-made shoes. As shoes come off the assembly line, they pile up around the last worker, Worker Ten. When this happens, Worker Ten calls out, "We have hundreds of shoes piled up here." This information

(input) is heard by (fed back to) Worker One who determines when more shoes should be begun. Worker One is therefore the ultimate controller of output. This controller could respond in either of two ways. In a *negative feedback system,* Worker One says, "We have an excess of unsold shoes. Let us slow down or stop production until the excess (stress) is relieved—until these shoes are sold." What might Worker One's response be if this were a *positive feedback system?* 11

7. Most feedback systems of the body are *(positive? negative?)*. Why do you think this might be advantageous and directed toward maintenance of homeostasis?

8. Refer to Figure 1-12 (page 20) in your text and match the answers at right with the statements at left describing steps in a homeostatic mechanism for controlling elevated blood pressure (BP). 12

_____ a. What serves as the *input?*

_____ b. What structures receive this input?

_____ c. What structure then receives instructions from the brain to decrease contractile strength and so eject less blood into major arteries?

_____ d. What structures receive brain messages causing them to dilate so that blood runs out of major arteries and into smaller vessels?

_____ e. What is the final *output* as a result of the presence of less blood in the major arteries?

A. Arterioles
H. Heart
HBP. High blood pressure
LBP. Lowered blood pressure
P. Pressure-sensitive nerve cell receptors located in major arteries which send messages to the brain

9. Consider how homeostasis is regulated in the following situation: The endocrine system prevents blood sugar from increasing too much after a person has eaten a large meal. Complete Figure LG 1-3, identifying the *input,* the desired *output,* and the steps which lead from input to output. 13

14

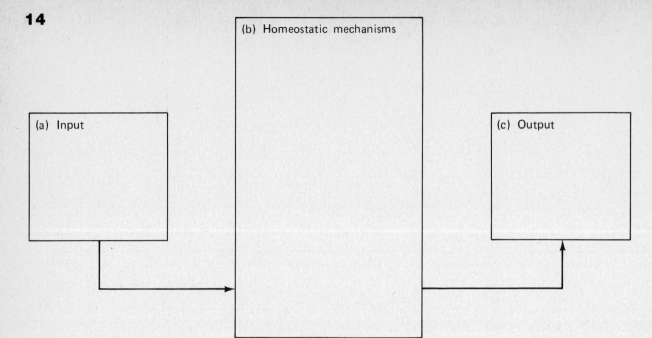

Figure LG 1-3 Complete as directed in Learning Activity 13

(page 21) **H. Measuring the human body**

If you have difficulty answering questions in this section, be sure to read Appendix A in the text.

1. Why are conversions in the metric system simpler than those in the American system of measurement?

14 2. Select ALL correct answers to each question.

a. When you drink all of the milk in a quart container, you are drinking ____ of milk.

A. Almost a liter   B. Almost half a liter   C. 2 liters
D. About 32 oz

b. One milliliter (ml) is equal to:

A. About 1 oz   B. $\frac{1}{100}$ liter   C. 0.1 liter   D. $\frac{1}{1000}$ liter
E. About 1 cubic centimeter (cc, or cm^3)

c. If a red blood cell (RBC) is about 7 micrometers (μm) in diameter, how many RBC's could fit side by side across a distance of 1 millimeter (mm)?

A. 0.15   B. 1.5   C. 15   D. 150   E. 1,500

d. One centimeter (cm) is equal to:

A. 0.1 mm   B. 1 mm   C. 10 mm   D. 100 mm   E. 0.1 meter (m)
F. 2 inches   G. 0.4 inch

e. If 20 ml of oxygen can be transported in 100 ml of your blood, how much oxygen can be transported in 1 liter of your blood?

   A. 0.2   B. 2   C. 20   D. 200   E. 2,000

f. Since there are 2.2 lb in 1 kilogram (kg), if you weigh 132 lb, you weigh about ____ kg.

   A. 60   B. 290   C. 2,112   D. 13.2   E. 132

g. A human liver weighs about 4 lb. This is about ____ in metric units.

   A. 8.8 kg   B. 1.8 kg   C. 180 grams (g)   D. 88 g   E. 1,800 g

h. A micrometer is equivalent to:

   A. 10 angstroms (Å)   B. 10,000 Å   C. 0.001 mm   D. 0.001 cm
   E. 1 millionth of a meter

i. The human body contains about 6 liters of blood. This is about ____ pints of blood.

   A. 0.2   B. 2   C. 6   D. 12   E. 36

j. If your body requires an intake of food that will provide 1 Calorie (Cal) of energy per minute, how many Calories do you require each day?

   A. 28,500   B. 1,440   C. 60   D. 720   E. 144

3. A *clinical challenge.* Complete this exercise applying metric system conversions to the clinical setting. Determine Elizabeth's weight and height in metric measurements. Elizabeth weighs 22 lb and is 30 in. tall. Weight: ____ kg; height: ____ cm. |15|

4. Another *clinical challenge.* Mr. Frederick is to receive 1 liter of 5% glucose solution intravenously (IV) over each 8-hour (hr) period. Answer the following questions. |16|
   a. How many milliliters (cc)/hr of the IV solution should Mr. Frederick receive?

   b. If Mr. Frederick receives no other food or fluid intake, what will be his total IV fluid intake each 24 hr?

c. Each 100 cc of a 5% glucose solution contains 5 g of glucose. Each gram of glucose provides about 4 Cal of energy to Mr. Frederick. How many Calories is this patient receiving each day?

17   5. A final metric *clinical challenge.* A serving of a certain food has been analyzed for its food content. It contains 1 oz of sugar, 1 oz protein, and 2 oz fat. Protein and sugar each provide 4 Cal/g, while fat provides 9 Cal/g. How many Calories are supplied by the serving described above? (*Hint:* How many grams = 1 oz?)

Answers to Numbered Questions in Learning Activities

1   From highest to lowest: organism, system, (organ), tissue, cell, chemical.

2   (a) I. (b) D. (c) M. (d) A. (e) S.

3   (a) Inferior (caudad). (b) Distal. (c) Posterior. (d) Medial. (e) Lateral. (f) Superficial. (g) Deep. (h) Anterior (ventral). (i) Superior (cephalad). (j) Ipsi-. (k) Visceral.

4   (a) H. (b) M. (c) F. (d) S.

5   (a) Dorsal. (b) Ventral. (c) Dorsal. (d) Thorax. (e) Lungs. (f) An imaginary line from the symphysis pubis to the superior border of the sacrum. (g) Abdomen. (h) Pelvic.

6   (a) Right hypochondriac, right lumbar, right iliac (inguinal). (b) Left hypochondriac, epigastric. (c) Umbilical. (d) Hypogastric (pubic). (e) Left iliac (inguinal), left lumbar, hypogastric (pubic), umbilical.

7   (a) DSR. (b) CT. (c) CR.

8   (a) Intracellular. (b) Interstitial (or intercellular or tissue) fluid, extracellular. (c) Blood, lymph. (d) Extra-, homeostasis.

9   C, interstitial (intercellular) fluid; D, intracellular fluid; E, interstitial (intercellular) fluid again; F, capillaries (again); G, venules and veins or lymph vessels.

10   Optimum concentration of gases, nutrients, ions, and water; optimal pressure; optimal temperature.

11   "Hurray! We have produced hundreds of shoes; let us go for thousands! Do not worry that the shoes are not selling. Step up production even more."

12 (a) HBP. (b) P. (c) H. (d) A. (e) LBP.

13 (a) Stress of high blood sugar following large meal. (b) High sugar level of blood flowing through pancreas triggers insulin production. This hormone facilitates transport of sugar out of blood and into body cells (so cells can use it for energy), and it also stimulates liver to convert the excess blood sugar to the storage form glycogen. (c) Blood sugar is consequently lowered to normal level.

14 (a) A, D. (b) D, E. (c) D. (d) C, G. (e) D. (f) A. (g) B, E. (h) B, C, E. (i) D. (j) B.

15 Weight: 10 kg = 22 lb × 1 kg/2.2 lb. Height: 75 cm = 30 in × 2.5 cm/in.

16 (a) 125 cc/hr = 1,000 cc/8 hr. (b) 3,000 cc/24 hr = 1,000 cc/8 hr. (c) 600 Cal = 150 g × 4 Cal/g. (*Note:* If 100 cc contains 5 g, then 3,000 cc contains 150 g.)

17 728 Cal = (28 g/oz × 4 Cal/g sugar) + (28 g/oz × 4 Cal/g protein) + (56 g/2 oz × 9 Cal/g fat).

MASTERY TEST: Chapter 1

Questions 1–10: Choose the one best answer to each question.

_____ 1. In a negative feedback system, when blood pressure rises slightly, the body will respond by causing a number of changes which tend to:

    A. Lower blood pressure   B. Raise blood pressure

_____ 2. The following structures are all located in the ventral cavity EXCEPT:

    A. Spinal cord   B. Urinary bladder   C. Heart
    D. Gallbladder   E. Esophagus

_____ 3. Which pair of common and anatomical terms is mismatched?

    A. Eye/ocular   B. Skull/cranial   C. Armpit/brachial
    D. Neck/cervical   E. Buttock/gluteal

_____ 4. Each answer below includes a correct example of a life process of humans EXCEPT:

    A. Absorption—uptake of sugar by a body cell
    B. Assimilation—the process by which a fat that has entered a cell becomes a part of that cell
    C. Excretion—production and elimination of urine or sweat
    D. Catabolism—synthesis of protein in muscle cell

_____ 5. Which is most inferiorly located?

    A. Abdomen   B. Pelvic cavity   C. Mediastinum
    D. Diaphragm   E. Pleural cavity

_____ 6. The spleen, tonsils, and thymus are all organs in which system?

    A. Nervous   B. Lymphatic   C. Cardiovascular
    D. Digestive   E. Endocrine

_____ 7. Which of the following structures is located totally outside of the upper right quadrant of the abdomen?

    A. Liver   B. Gallbladder   C. Transverse colon
    D. Spleen   E. Pancreas

_____ 8. Choose the FALSE statement. (It may help you to mark each statement _T_ for true or _F_ for false as you read it.)

    A. Stress disturbs homeostasis.
    B. Homeostasis is a condition in which the body's internal environment remains relatively constant.
    C. The body's internal environment is best described as extracellular fluid.
    D. Extracellular fluid consists of plasma and intracellular fluid.

_____ 9. The system responsible for providing support, protection, and leverage, for storage of minerals, and for production of blood cells is:

    A. Urinary   B. Integumentary   C. Muscular
    D. Reproductive   E. Skeletal

_____ 10. Which is most proximally located?

    A. Ankle   B. Hip   C. Knee   D. Toe

Questions 11–20: Circle T (true) or F (false). If the statement is false, change the underlined word or phrase so that the statement is correct.

T F   11. Anterior and ventral are synonyms.

T F   12. In order to assume the anatomical position, you should lie down with arms at your sides and palms turned backwards.

T F   13. The appendix is usually located in the left iliac region of the abdomen.

T F   14. Anatomy is the study of how structures function.

T F   15. One milliliter (ml) is equivalent to 1 cubic centimeter (cc).

T F   16. Tissue is one organizational level higher than organ.

T F   17. The right kidney is located mostly in the right iliac region of the abdomen.

T F   18. A single rentgenogram provides a three-dimensional image of the body which can be used for diagnostic purposes.

T F   19. A sagittal section is parallel to and lateral to a midsagittal section.

T F   20. The human body is bilaterally symmetrical.

Questions 21–25: fill-ins. Write the word which best fits the description.

_____ 21. Region of thorax located between the lungs. Contains heart, thymus, esophagus.

_____ 22. System of the body which produces movement and heat and maintains posture.

_____ 23. Name of fluid located within cells.

_____ 24. Life process in which unspecialized cells as in early embryo change into specialized cells such as nerve or bone.

_____ 25. Name of the organ which fills the right hypo-chondriac region.

The Chemical Level of Organization

The human body is composed of chemicals which serve as the first level of organization—the basis for all structure and function of the body. In this chapter you will examine the structure and kinds of atoms (elements) (Objectives 1–2) and the manner in which they combine to form important components of the human body (4–7, 9). You will study recently developed diagnostic techniques focusing on body chemistry (3, 8). Next, you will distinguish inorganic from organic compounds (11) and contrast roles of each. You will consider vital functions of water (12), as well as relationships of acids, bases, and salts to pH and buffers (13–15). Lastly, you will begin to look at the complex organic compounds which serve as building blocks and as regulators of the body (10, 16–18).

Topics Summary

A. Introduction to basic chemistry: atoms, elements, molecules, chemical bonds, radioisotopes
B. Chemical reactions
C. Inorganic compounds, water
D. Acids, bases, salts; pH, buffers
E. Organic compounds: carbohydrates, lipids, proteins
F. Organic compounds: nucleic acids (DNA and RNA)

Objectives

1. Identify by name and symbol the principal chemical elements of the human body.
2. Explain, by diagraming, the structure of an atom.
3. Describe the principle of magnetic resonance imaging (MRI) and its diagnostic importance.
4. Define a chemical reaction as a function of electrons in incomplete outer energy levels.
5. Describe ionic bond formation in a molecule of sodium chloride (NaCl).
6. Discuss covalent bond formation as the sharing of outer energy level electrons.
7. Explain the nature and importance of hydrogen bonding.
8. Define a radioisotope and explain the principle of positron emission tomography (PET) scanning and its diagnostic importance.
9. Explain the basic differences among synthesis, decomposition, exchange, and reversible chemical reactions.
10. Discuss how chemical reactions are related to metabolism.
11. Define and distinguish between inorganic and organic compounds.
12. Discuss the functions of water as a solvent, suspending medium, chemical reactant, heat absorber, and lubricant.
13. List and compare the properties of acids, bases, and salts.
14. Define pH as the degree of acidity or alkalinity of a solution.
15. Explain the role of a buffer system as a homeostatic mechanism that maintains the pH of a body fluid.
16. Compare the structure and functions of carbohydrates, lipids, and proteins.
17. Contrast the structure of deoxyribonucleic acid (DNA) and ribonucleic acid (RNA).
18. Identify the function and importance of adenosine triphosphate (ATP) and cyclic adenosine-3′,5′-monophosphate (cyclic AMP).

22   **Learning Activities**

(page 25)   **A.** Introduction to basic chemistry: atoms, elements, molecules, chemical bonds, radioisotopes

1. Define *matter* and list the three states in which matter may exist.

2. Contrast an *atom* with an *element.*

1   3. Refer to Exhibit LG 2-1, and do the following exercise.
   a. Color yellow the four boxes containing the elements which make up about 96 percent of the weight of the body. These four elements are relatively *(simple? complex?)* in atomic structure among the total 103 elements in the periodic table.
   b. Color light green the boxes containing calcium and phosphorus. These

   two elements together contribute about _____ percent of body weight.

4. Refer to Exhibit 2-2 (page 34) in your text. Match the elements listed at right with descriptions of their functions in the body. *For extra review.* Locate each of these elements in the periodic table (Exhibit LG 2-1) and write the
2   atomic number of each element after its description.

_____ a. Found in every organic molecule   C. Carbon
      _____   Ca. Calcium
                                                       Fe. Iron
_____ b. Component of all protein mole-   I. Iodine
         cules and DNA and RNA   N. Nitrogen
                                                       O. Oxygen
      _____
_____ c. Vital to normal thyroid gland

         function _____
_____ d. Essential component of hemoglo-

         bin _____
_____ e. Constituent of bone and teeth; re-
         quired for blood clotting and for

         muscle contraction _____
_____ f. Constituent of water; functions in

         cellular respiration _____
3   5. Complete this exercise about atomic structure.

   a. An atom consists of two main parts: _____ and

      _____.

   b. Within the nucleus are positively charged particles called

      _____ and uncharged (neutral) particles called

      _____. Together these are known as _____

      _____.

   c. Electrons bear *(positive? negative?)* charges. The number of electrons spinning around the nucleus is *(greater than? equal to? smaller than?)* the number of protons in the atomic nucleus.

Periodic Table

Key:

2	Atomic number
He	Chemical symbol
4.00	Atomic weight

Periods	1A	2A	3B	4B	5B	6B	7B	8	8	8	1B	2B	3A	4A	5A	6A	7A	Noble gases 0
1	1 H 1.0079																	2 He 4.00260
2	3 Li 6.941	4 Be 9.01218											5 B 10.81	6 C 12.011	7 N 14.0067	8 O -5.9994	9 F 18.998,403	10 Ne 20.179
3	11 Na 22.98977	12 Mg 24.305											13 Al 26.98154	14 Si 28.0855	15 P 30.97376	16 S 32.06	17 Cl 35.453	18 Ar 39.948
4	19 K 39.0983	20 Ca 40.08	21 Sc 44.9559	22 Ti 47.90	23 V 50.9415	24 Cr 51.996	25 Mn 54.9380	26 Fe 55.847	27 Co 58.9332	28 Ni 58.70	29 Cu 63.546	30 Zn 65.38	31 Ga 69.72	32 Ge 72.59	33 As 74.9216	34 Se 78.96	35 Br 79.904	36 Kr 83.80
5	37 Rb 85.4678	38 Sr 87.62	39 Y 88.9059	40 Zr 91.22	41 Nb 92.9064	42 Mo 95.94	43 Tc (98)	44 Ru 101.07	45 Rh 102.9055	46 Pd 106.4	47 Ag 107.868	48 Cd 112.41	49 In 114.82	50 Sn 118.69	51 Sb 121.75	52 Te 127.60	53 I 126.9045	54 Xe 131.30
6	55 Cs 132.9054	56 Ba 137.33	57 *La 138.9055	72 Hf 178.49	73 Ta 180.9479	74 W 183.85	75 Re 186.207	76 Os 190.2	77 Ir 192.22	78 Pt 195.09	79 Au 196.9665	80 Hg 200.59	81 Tl 204.37	82 Pb 207.2	83 Bi 208.9804	84 Po (209)	85 At (210)	86 Rn (222)
7	87 Fr (223)	88 Ra 226.0254	89 †Ac 227.0278	104 Unq (261)	105 Unp (262)	106 Unh (263)												

*Lanthanide Series

58 Ce 140.12	59 Pr 140.9077	60 Nd 144.24	61 Pm (145)	62 Sm 150.4	63 Eu 151.96	64 Gd 157.25	65 Tb 158.9254	66 Dy 162.50	67 Ho 164.9304	68 Er 167.26	69 Tm 168.9342	70 Yb 173.04	71 Lu 174.967

†Actinide Series

90 Th 232.0381	91 Pa 231.0359	92 U 238.029	93 Np 237.0482	94 Pu (244)	95 Am (243)	96 Cm (247)	97 Bk (247)	98 Cf (251)	99 Es (254)	100 Fm (257)	101 Md (258)	102 No (259)	103 Lr (260)

Exhibit LG 2-1 Periodic Table

   d. Electrons are arranged in orbits. The inner orbit has a maximum capacity of _____ electrons, while a maximum of _____ electrons may be present in the next orbit.

4   6. Do the following exercise about MRI.

   a. MRI stands for the diagnostic technique known as _____

_____. It focuses on the *(nucleus? electrons?)* of a single element.

   b. Most studies utilizing MRI to date have examined the atom

_____ since it is abundant in the body as part of water molecules. Body tissues are exposed to an external force, such as

_____. Energy changes in the nuclei of the hydrogen atoms in the tissue are ultimately reconstructed into a three-dimensional

_____.

   c. MRI *(is? is not?)* advantageous diagnostically and *(does? does not?)* utilize ionizing radiation.

  7. Refer to Exhibit LG 2-1 and Figure 2-2 (page 26) in your text and complete
5   this exercise about atomic structure and chemical bonding.

   a. The atomic number of potassium (K) is _____. Locate this number above the K on Exhibit LG 2-1. This means that a potassium atom has

_____ protons and _____ electrons.

   b. Identify the number located immediately under the K in Exhibit LG 2-1.

This number is _____, and is the atomic weight of the potassium atom. This indicates the total number of nucleons in the potassium nucleus. If you subtract the atomic number from the atomic weight, you will identify the number of neutrons in an atom. A typical potassium atom has

_____ neutrons.

   c. The type of chemical bonding that occurs depends on the number of electrons in the outer orbit. Write the chemical symbols of two atoms in the periodic table that have just one electron in the outer orbit: _____

_____. Write the symbol of one atom that is missing only one electron

from an otherwise complete outer orbit: _____.

   d. The type of bonding that tends to occur between potassium and chlorine is *(covalent? ionic?)*. In such a bond potassium *(gains? gives up?)* an electron (which is a negative entity) and so becomes more *(positive? negative?)*. The K^+ ion is a(n) *(anion? cation?)*.

   e. As potassium chloride is formed, chlorine *(gains? gives up?)* an electron and so becomes the *(anion? cation?)* Cl^-.

6   8. Match the following description with the types of bonds listed at right.

_____ a. Atoms lose electrons if they have just one or two electrons in their outer orbits; they gain electrons if they need just one or two electrons to complete the outer orbit.	C. Covalent H. Hydrogen I. Ionic
_____ b. This bond is a bridgelike, weak link between a hydrogen atom and another atom such as oxygen or nitrogen.	

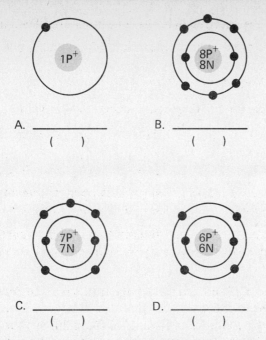

A. _____
( )

B. _____
( )

C. _____
( )

D. _____
( )

Figure LG 2-1 Atomic structure of four common atoms in the human body. Complete as directed in Learning Activity ⌐7⌐.

_____ c. This type of bond is most easily formed by carbon (C) since its outer orbit is half-filled.

_____ d. These bonds are only about 5 percent as strong as covalent bonds and so are easily formed and broken. They are vital for holding large molecules like protein or DNA in proper configurations.

9. Refer to Figure LG 2-1 and Exhibit LG 2-1 and complete this exercise. |7|
   a. Write the atomic number of each atom in Figure LG 2-1 in the parentheses ( ) under each atom. (*Hint*: Count the number of electrons or protons = atomic number.)
   b. Write the name or symbol for each atom on the line under that atom. You may identify these by atomic number in Exhibit LG 2-1.
   c. Based on the number of electrons which each of these atoms is missing from a complete outer orbit, draw the number of covalent bonds that each of the atoms can be expected to form. Note that this is the valence, or combining capacity, of these atoms. One is done for you.

   H   O   ⟍N⟋   C
                          |

d. *For extra review,* draw the atomic structure of an ammonia molecule (NH_3). Note that this is a covalently bonded compound.

| 8 | 10. Write C next to the molecules that are classified as compounds.

    _____ H   _____ Glucose
    _____ H_2 _____ N_2
    _____ H_2O _____ A molecule that contains two or more different kinds of atoms

| 9 | 11. Describe radioisotopes by completing this exercise. Consult Exhibit LG 2-1.

a. The isotope ^{14}C differs from ^{12}C in that ^{14}C has _____.

b. Isotopes that are unstable (their nuclear structure decays) are known as

_____. Radioactive iodine is used for studies of the

_____ gland.

c. The average atomic weight for iron (Fe) is ____. ^{59}Fe is used for study

of _____.

d. Short-lived isotopes, such as ^{11}C or ^{15}O, when injected into the body,

emit positively charged electrons known as_____.

These may be used diagnostically by means of the technique known as

PET, or _____.

(pages 31–33)   **B. Chemical reactions**

| 10 | 1. Identify the kind of chemical reaction described in each statement below.

_____ a. The end product can revert to the original combining molecules.
                D. Decomposition reaction
                E. Exchange reaction

_____ b. Two reactants combine to form an end product, for example, many glucose molecules bond to form glycogen.
                R. Reversible reaction
                S. Synthesis reaction

_____ c. Such a reaction is partly synthesis and partly decomposition (as in Figure LG 2-2b, page LG 30).

_____ d. All chemical reactions involve making or breaking of bonds. This type of reaction involves only breaking of bonds.

_____ e. This type of reaction is catabolic, as in digestion of foods, such as starch digestion to glucose.

_____ f. This type of reaction is most likely to release heat.

2. Define _metabolism._

3. Briefly describe in terms of chemical reactions what _high metabolism_ (or _high metabolic rate_) means.

   Explain why a person with such metabolism might be energetic and also feel hot.

4. Write a paragraph describing the collision theory explanation for chemical reactions. Include three factors which determine whether a collision will lead to a chemical reaction.

5. Contrast energy forms by completing this exercise. $\boxed{11}$

   a. _____ is the capacity to do work. _(Kinetic? Potential?)_ energy is energy of motion, while _____ energy is stored energy.

   b. In anabolic (or synthesis) reactions, chemical energy is _(required? released?)_ as chemical bonds are formed. The opposite type of metabolic process, called _____, involves release of energy as when chemical bonds of foods are broken.

   c. _____ energy is the energy of sunlight that warms your skin on a hot day. Another source of radiant energy is

   _____ .

   d. Conduction of nerve impulses involves _(mechanical? electrical?)_ energy.

(page 33)   **C.** Inorganic compounds, water

1. Write *inorganic* after phrases that describe inorganic compounds and *organic* after those that describe organic compounds.

|12| a. Held together almost entirely by covalent bonds: _____

_____

b. Tend to be very large molecules which serve as good building blocks for body structures: _____

c. The class that includes water, the most abundant compound in the body: _____

d. The class that includes carbohydrates, proteins, fats, and nucleic acids: _____

|13| 2. Answer these questions about the functions of water.

a. Water normally is a *(solute? solvent?)*.

b. A gas that is dissolved in water within body fluids is _____.

c. Water requires a *(large? small?)* amount of heat in order to change it into a gas. As a result, for each small amount of perspiration (sweat) evaporated during heavy exercise, a *(large? small?)* amount of heat can be released. This is fortunate since dehydration is less likely to occur.

d. Name a location where water's function as a lubricant is essential.

(pages 34–37)   **D.** Acids, bases, salts; pH, buffers

|14| 1. Match the descriptions below with the answers at right.

_____ a. A substance that dissociates into hydroxyl ions (OH^-) and one or more cations. Example: NaOH

_____ b. A substance that dissociates into hydrogen ions (H^+) and one or more anions. Example: H_2SO_4

_____ c. A substance that dissolves in water forming cations and anions neither of which is H^+ or OH^-. Example: $CaCl_2$

A. Acid
B. Base
S. Salt

2. In the space below draw the outline of a cell. Write inside of it symbols for ions concentrated in intracellular fluid. Write outside of it symbols for ions

|15| concentrated in extracellular fluid.

3. Draw a diagram showing the pH scale. Label the scale from 0 to 14. Indicate by arrows increasing acidity (H^+ concentration) and increasing alkalinity (OH^- concentration).

4. Choose the correct answers regarding pH.

   16

   a. Which pH is most acid?

      A. 4 B. 7 C. 10

   b. Which pH has the highest concentration of OH^- ions?

      A. 4 B. 7 C. 10

   c. Which solution has pH closest to neutral?

      A. Limewater B. Blood C. Lemon juice D. Gastric juice (digestive juice of the stomach)

   d. A solution with pH 8 has 10 times *(more? fewer?)* H^+ ions than a solution with pH 7.

   e. A solution with pH 5 has *(10? 20? 100?)* times *(more? fewer?)* H^+ ions than a solution with pH 7.

5. Contrast:
   a. Strong acid/weak acid

   b. Strong base/weak base

6. What are *buffers?* State their essential function in maintaining homeostasis.

(a) Buffer pair: weak acid (H·HCO$_3$) and its
related weak base (Na·HCO$_3$)

| NaOH | | H$_2$CO$_3$ | | H$_2$O | | NaHCO$_3$ |
| (strong base) | + | (weak acid) | | (very weak base) | + | (weak base) |

(b) Buffering strong base

| HCl | | NaHCO$_3$ | | (salt) | + | (weak acid) |
| (strong acid) | + | (weak base) | | | | |

(c) Buffering strong acid

Figure LG 2-2 Buffering action of carbonic acid-sodium bicarbonate buffer system. Complete as directed in Learning Activity | 17 |.

| 17 | 7. Refer to Figure LG 2-2 and complete this exercise about buffers.

a. The _____ system is the most important buffer system in the extracellular fluid. Like all buffers, this system consists of a pair of chemicals: a weak _____

_____ and the related weak _____

_____. (See Figure LG 2-2a.)

b. _____ is the weak acid of this buffer pair. Its chemical formula is _____. It serves as a proton (H$^+$) (receptor? donor?) in the presence of a strong (acid? base?) Add H$^+$ and Na$^+$ to finish Figure LG 2-2b.

c. When a strong acid threatens pH of body fluids, the weak base partner,

_____, exchanges its cation (Na$^+$) for the proton (H$^+$) of the strong acid (HCl). Consequently the strong acid is changed into a weak one, one that has (more? less?) of a tendency to ionize. Show this by completing Figure LG 2-2c. Label the products.

d. When the body continues to take in or produce excess strong acid, the concentration of the _____ member of this buffer pair will decrease as it is used up in the attempt to maintain homeostasis of pH.

e. Though buffer systems provide rapid response to acid–base imbalance, they are limited since one member of the buffer pair can be used up. And they can only convert strong acids or bases to weak ones; they cannot eliminate them. Two systems of the body that can actually eliminate acid or basic substances are _____ and

_____.

(pages 37–40)

E. Organic compounds: carbohydrates, lipids, proteins

1. Organic compounds all contain carbon (C). State two reasons why carbon is an ideal element to serve as the primary structural component for living systems.

2. Carbohydrates carry out important functions in your body.
   a. State one primary function of carbohydrates.

   b. State two secondary roles of carbohydrates.

3. Complete these statements about carbohydrates. 　　　　　　　　　　　　18

   a. The ratio of hydrogen to oxygen in all carbohydrates is _____ : _____. The general formula for carbohydrates is _____.

   b. The chemical formula for a *hexose* is _____. Two common hexoses are _____ and _____.

   c. One common pentose sugar found in DNA is_____ _____. Its name and its formula ($C_5H_{10}O_4$) indicate that it lacks one oxygen atom from the typical pentose formula which is

   _____.

   d. When two glucose molecules ($C_6H_{12}O_6$) combine, a disaccharide is formed. Its formula is _____, indicating that in a synthesis reaction such as this, a molecule of water is *(added? removed?)*. Refer to Figure LG 2-3. Identify which reaction *(1? 2?)* on that figure demonstrates synthesis.

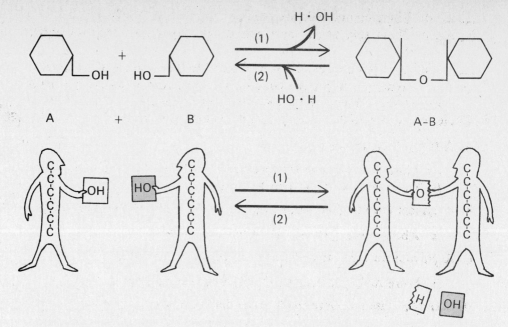

Figure LG 2-3 Dehydration synthesis and hydrolysis reactions. Refer to Learning Activity [18]

e. Continued dehydration synthesis leads to enormous carbohydrates called _____. One such carbohydrate is _____ _____.

f. When a disaccharide such as sucrose is broken, water is introduced to split the bond linking the two monosaccharides. Such a decomposition reaction is termed _____, which literally means "splitting using water." (See Figure 2-10, page 38 in your text.) On Figure LG 2-1, hydrolysis is shown by reaction *(1? 2?)*.

g. For extra review, role play dehydration synthesis and hydrolysis with study partners or in a classroom demonstration. Two students (A and B in Figure LG 2-3) play roles of glucoses. Each holds a large paper marked with large letters O H. These represent hydroxyl groups (OH) which are part of the $C_6H_{12}O_6$ structure of glucose. A third person plays the role of a helper enzyme (or hydrolase) who catalyzes the joining of glucose A and glucose B, much as a matchmaker. (Note that the enzyme is not shown in Figure LG 2-3.) The enzyme tears off H from the OH paper held by glucose A, and also removes the entire OH paper from glucose B. The enzyme has therefore removed water (H·OH) from the two glucoses. Glucose B is left literally "empty-handed" so must form a bond at the carbon site (or "hand") which had held an OH. So glucose A and B now chemically bond to each other by sharing (covalently bonding to) the oxygen (O) which glucose A had retained. Thus a larger compound, the disaccharide maltose (A—B) is formed by removal of water. Try the reverse process (hydrolysis). Have some tape handy to reassemble glucose A's OH.

4. Contrast carbohydrates with lipids in this exercise. Write C next to any statement true of carbohydrates, and write L next to any statement true of lipids.

[19]

_____ a. These compounds are insoluble in water. (*Hint:* Water is ineffective for washing such compounds out of clothes.)

_____ b. These compounds are organic.

_____ c. These substances have a hydrogen to oxygen ratio of about 2:1.

_____ d. Very few oxygen atoms (compared to the numbers of carbon and hydrogen atoms) are contained in these compounds. (Look carefully at Figure 2-11, page 39, in your text.)

5. Circle the answer that correctly completes each statement. | 20 |

a. A triglyceride consists of:

    A. Three glucoses bonded together
    B. Three fatty acids bonded to a glycerol

b. A typical fatty acid contains (see Figure 2-11, page 39 in your text):

    A. About three carbons   B. About 18 carbons

c. A polyunsaturated fat contains _____ than a saturated fat.

    A. More hydrogen atoms   B. Fewer hydrogen atoms

d. Saturated fats are more likely to be derived from:

    A. Plants, for example, corn oil or peanut oil
    B. Animals, as in beef or cheese

6. Match the following types of lipids with their functions. | 21 |

_____ a. Present in egg yolk and carrots; leads to formation of vitamin A, a chemical necessary for good vision

_____ b. The major lipid in cell membranes

_____ c. Present in all cells, but infamous for its relationship to "hardening of the arteries"

_____ d. Substances that break up (emulsify) fats before their digestion

_____ e. Steroid sex hormones produced in large quantities by females

_____ f. Important regulatory compounds with many functions, including stimulation of smooth muscle and alteration of blood pressure

B. Bile salts
Car. Carotene
Cho. Cholesterol
E. Estrogens
Pho. Phospholipids
Pro. Prostaglandins

7. Complete this exercise about fat-soluble vitamins. | 22 |

a. Name four fat-soluble vitamins (consult Exhibit 2-4, page 38 in your text, for help): _____ _____ _____ _____

b. Since these four vitamins are absorbed into the body along with fats, any problem with fat absorptions is likely to lead to:

c. The fat-soluble vitamin necessary for normal bone growth is vitamin

_____.

d. Vitamin _____ may be administered to a patient before surgery in order to prevent excessive bleeding since this vitamin helps in the clotting process.

    e. Normal vision is associated with an adequate amount of vitamin _____.

    f. Vitamin _____ has a variety of possible functions including the promotion of wound healing and prevention of scarring.

8. Contrast proteins with the other organic compounds you have studied so far by completing this exercise.

    a. Carbohydrates, lipids, and proteins all contain carbon, hydrogen, and oxygen. A fourth element, _____, makes up a substantial portion (16 percent) of proteins.

    b. Just as large carbohydrates are composed of repeating units (the _____), proteins are composed of building blocks called _____.

    c. As in the synthesis of carbohydrates or fats, when two amino acids bond together, a water molecule must be *(added? removed?)*. This is another example of *(hydrolysis? dehydration?)* synthesis.

    d. The product of such a reaction is called a _____-peptide. (See Figure 2-12, page 40 in your text.) When many amino acids are linked together in this way, a _____-peptide results. Several polypeptides form a _____.

9. There are at least ____ different kinds of amino acids, much like a 20-letter alphabet that forms protein "words." It is the specific sequence of amino acids that determines the nature of the protein formed. In order to see the significance of this fact, do the following activity. Arrange the five letters listed below in several ways so that you form different five-letter words. Use all five letters in each word.

a e m s t

__ __ __ __ __   __ __ __ __ __

__ __ __ __ __   __ __ __ __ __

Note that although all of your words contain the same letters, the sequence of letters differs in each case. The resulting words have different meanings, or functions. Such is the case with proteins too, except that protein "words" may be thousands of "letters" (amino acids) long, involving all or most of the 20-amino acid "alphabet." Proteins must be synthesized with accuracy. Think of the drastic consequences of omission or misplacement of even a single amino acid. (How different are *mates, meats, steam, tames, teams*!)

10. Match the four levels of structural organization of proteins with the description below.

_____ a. Specific sequence of amino acids   P. Primary structure

_____ b. Zigzag or coiled arrangement (as in spirals or pleated sheets)   Q. Quaternary structure

_____ c. Three-dimensional shape due to bending or folding of proteins   S. Secondary structure

_____ d. Two or more tertiary patterns bonded together   T. Tertiary structure

11. List six principal classes of proteins. **Consider** how you would describe to a friend how each of these six **kinds of proteins** helped your friend today.

F. Organic compounds: nucleic acids (DNA and RNA)   **(pages 40–43)**

1. Complete Figure LG 2-4 as follows: 25
   a. Circle one nucleotide.
   b. Draw in a complementary set of nucleotides on the right side of the figure to produce the double-stranded structure of DNA. (For the sake of clarity, DNA is shown straightened—not in a helix or twisted ladder.)

      Bases are paired _____ to _____ and _____ to _____. Bond base pairs by drawing red dotted lines to represent hydrogen bonding to connect the two sides into a double strand.
   c. DNA structure, like that of proteins, depends on sequence. Proteins differ

      from one another by their specific sequence of_____

      _____, whereas DNA molecules (for example, DNA determining blood type and DNA controlling eye color) differ from one another by the sequence of _____. (Note again, as with protein structure, the dire consequences of an omission or substitution in this "four-letter alphabet": A C G T.)
   d. How does a strand of RNA differ from DNA?

2. *For extra review.* Cover Figure LG 2-4 and (without glancing at it) extend the length of the DNA polynucleotide by adding three nucleotides on the left side only. Check your drawing and correct where necessary. Then complete the right side.

3. Describe the structure and significance of ATP by completing these statements.   26
   a. ATP stands for adenosine triphosphate. Adenosine consists of a base

      that is a component of DNA, that is, _____, along

      with the five-carbon sugar named _____.

   b. The "TP" of ATP stands for _____. The final *(one? two? three?)* phosphates are bonded to the molecule by high energy bonds.

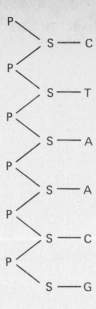

Figure LG 2-4 Structure of DNA. P, phosphate; S, sugar (deoxyribose); A, adenine; T, thymine; G, guanine; C, cytosine. Complete as directed in Learning Activity 25 .

    c. When the terminal phosphate is broken, a great deal of energy is released as ATP is split into _____ + _____ _____ .

    d. ATP is constantly reformed by the reverse of this reaction as energy is made available from foods you eat. Write this reversible reaction.

4. ATP serves as the body's primary energy-storing molecule by means of its high-energy bonds. Of what advantage is the fact that the body stores energy in ATP rather than in glucose?

5. A compound similar to ATP, but with only one phosphate, is named _____ . A cyclic form of this molecule can be made with the help of the enzyme _____ . Cyclic AMP has important regulatory functions and will be discussed in Chapter 18.

Answers to Numbered Questions in Learning Activities

1 (a) O (65 percent), C (18 percent), H (10 percent), N (3 percent); simple. (b) 3

2 (a) C, 6. (b) N, 7. (c) I, 53. (d) Fe, 26. (e) Ca, 20. (f) O, 8.

3 (a) Nucleus, electrons. (b) Protons, neutrons, nucleons. (c) Negative, equal to. (d) 2, 8.

4 (a) Magnetic resonance imaging, nucleus. (b) Hydrogen, magnetism (magnetic field), image. (c) Is, does not.

5 (a) 19, 19, 19. (b) 39, 20 (39 − 19). (c) Examples: H, Na, K; examples: F, Cl, Br, I. (d) Ionic, gives up, positive, cation. (e) Gains, anion.

6 (a) I. (b) H. (c) C. (d) H.

7 (a) A, 1; B, 8; C, 7; D, 6. (b) A, hydrogen (H); B, oxygen (O); C, nitrogen (N); D, carbon (C).

(c) H−   O−   (N),   −C−   (d)

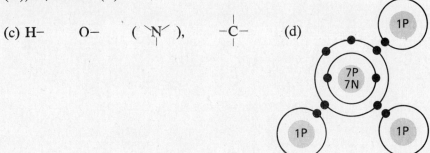

8 H_2O, glucose, a molecule that contains two or more different kinds of atoms.

9 (a) Two extra neutrons (eight neutrons). (b) Radioisotopes, thyroid. (c) 55.8 (or 56), red blood cell production. (d) Positrons, positron emission tomography.

10 (a) R. (b) S. (c) E. (d) D. (e) D. (f) D.

11 (a) Energy, kinetic, potential. (b) Required, catabolism or decomposition. (c) Radiant, light. (d) Electrical.

12 (a) Organic. (b) Organic. (c) Inorganic. (d) Organic.

13 (a) Solvent. (b) Oxygen (O_2) or carbon dioxide (CO_2). (c) Large, large. (d) The mucous lining of digestive or respiratory tract or the serous fluid of pleura, pericardium or peritoneum.

14 (a) B. (b) A. (c) S.

15

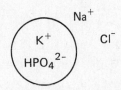

16 (a) A. (b) C. (c) B. (d) Fewer. (e) 100, more.

17 (a) Carbonic acid–bicarbonate buffer, acid, base. (b) Carbonic acid, H_2CO_3, donor, base. (c) $NaHCO_3$, less, products: $NaCl + H_2CO_3$. (d) $NaHCO_3$, or base. (e) Respiratory (lungs), urinary (kidneys).

18 (a) 2:1, $(CH_2O)_n$. (b) $C_6H_{12}O_6$; glucose, fructose, galactose. (c) Deoxyribose, $C_5H_{10}O_5$. (d) $C_{12}H_{22}O_{11}$, removed, 1. (e) Polysaccharides; starch, glycogen, cellulose. (f) Hydrolysis, 2.

19 (a) L. (b) C L. (c) C. (d) L.

20 (a) B. (b) B. (c) B. (d) B.

21 (a) Car. (b) Pho. (c) Cho. (d) B. (e) E. (f) Pro.

22 (a) A, D, E, K. (b) Symptoms of deficiencies of fat-soluble vitamins. (c) D. (d) K. (e) A. (f) E.

23 (a) Nitrogen. (b) Monosaccharides, amino acids. (c) Removed, dehydration. (d) Di, poly, protein.

24 (a) P. (b) S. (c) T. (d) Q.

25 (a) Phosphate-sugar-base (such as P-S-A in Figure LG 2-4). (b) A-T, C-G. (c) Amino acids, bases (A, T, C, G). (d) RNA is single-stranded, its sugar is ribose, and it contains the base uracil (U), not thymine (T).

26 (a) Adenine, ribose. (b) Triphosphate, two. (c) ADP + phosphate (P). (d) ATP ⟷ ADP + P + energy.

MASTERY TEST: Chapter 2

Questions 1–10: Choose the one best answer to each question.

_____ 1. When two monosaccharides join to form a disaccharide (or two amino acids form a dipeptide), the reaction is called a _____ reaction.

    A. Hydrolysis   B. Dehydration synthesis
    C. Isomerization

_____ 2. Metabolism is best described by which phrase?

    A. Ability to adjust to a variety of changes in internal environment

    B. Process of making foods from CO_2 and H_2O

    C. Ability to make copies of oneself

    D. Entire complex of anabolic and catabolic reactions occurring in an organism

_____ 3. In the formation of the ionically bonded salt NaCl, Na^+ has:

    A. Gained an electron from Cl^- B. Lost an electron to Cl^-
    C. Shared an electron with Cl^- D. Formed an isotope of Na^+

_____ 4. Over 99 percent of living cells consist of just six elements. Choose the element that is NOT one of these six.

    A. Calcium   B. Hydrogen   C. Carbon   D. Iodine
    E. Nitrogen   F. Oxygen   G. Phosphorus

_____ 5. Which of the following describes the structure of a nucleotide?

    A. Base-base   B. Phosphate-sugar-base   C. Enzyme
    D. Dipeptide   E. Adenine-ribose

_____ 6. Which of the following groups of chemicals includes only polysaccharides?

    A. Glycogen, starch   B. Glycogen, glucose, galactose
    C. Glucose, fructose   D. RNA, DNA
    E. Sucrose, polypeptide

_____ 7. $C_6H_{12}O_6$ is most likely the chemical formula for:

A. Amino acid   B. Fatty acid   C. Hexose
D. Polysaccharide   E. Ribose

_____ 8. Which of the following is an artificial sweetener used as a replacement for saccharin?

A. Aspartame   B. Fructose   C. Carotene
D. Deoxyribose   E. Glycogen

_____ 9. Which of the following substances are used mainly for structure and regulatory functions and are not normally used as energy sources?

A. Lipids   B. Proteins   C. Carbohydrates

_____ 10. Which is a component of DNA, but not of RNA?

A. Adenine   B. Phosphate   C. Guanine   D. Ribose
E. Thymine

Questions 11–20: Circle T (true) or F (false). If the statement is false, change the underlined word or phrase so that the statement is correct.

T   F   11. About 65 to 75 percent of living matter consists of <u>organic</u> compounds.

T   F   12. Oxygen can form <u>three bonds since it requires three electrons</u> to fill its outer energy level.

T   F   13. <u>Oxygen, water, NaCl, and glucose</u> are inorganic compounds.

T   F   14. There are about <u>four</u> different kinds of amino acids found in human proteins.

T   F   15. The number of protons always equals the number of <u>neutrons</u> in an atom.

T   F   16. <u>K^+ and Cl^-</u> are both cations.

T   F   17. A strong acid has <u>more</u> of a tendency to contribute H^+ to a solution than a weak acid has.

T   F   18. <u>$NaHCO_3$ is the weak base</u> of the carbonic acid-bicarbonate buffer system.

T   F   19. ATP contains <u>more</u> energy than ADP.

T   F   20. <u>Prostaglandins, cyclic AMP, steroids, and fats</u> are all classified as lipids.

Questions 21–25: fill-ins. Identify each organic compound below. Write names of compounds on lines provided.

21. _____

22. _____

23. _____

24. _____

25. _____

The Cellular Level of Organization

In this chapter you will move to the second level of organization and examine structure and function at the cellular level. You will look first at the general organizational plan of these living building blocks (Objective 1). Then you will study the manner in which materials enter or leave cells (2–3). You will look closely at the cell's interior and see the role of each organelle (5–16). You will learn how cells control their own activities by protein synthesis (17–18) and how cells divide (19–20). Finally, you will contrast the healthy, homeostatic cell with those experiencing aging or pathological disorders such as cancer (21–23).

Topics Summary

A. Generalized animal cell, plasma (cell) membrane
B. Movement of materials across plasma membranes
C. Cytoplasm, organelles
D. Cell inclusions, extracellular materials
E. Gene action: protein synthesis
F. Normal cell division: mitosis, meiosis
G. Abnormal cell division: cancer
H. Cells and aging
I. Medical terminology

Objectives

1. Define a cell and list its generalized parts.
2. Explain the chemistry, structure, and functions of the plasma membrane.
3. Describe how materials move across plasma membranes by diffusion, facilitated diffusion, osmosis, filtration, dialysis, active transport, and endocytosis.
4. Describe the chemical composition and functions of cytoplasm.
5. Describe the structure and functions of a cell nucleus.
6. Define the structure and function of ribosomes.
7. Distinguish between agranular and granular endoplasmic reticulum (ER) with regard to structure and function.
8. Describe the structure and functions of the Golgi complex.
9. Discuss the structure and function of mitochondria as ''powerhouses of the cell.''
10. Explain the structure and function of lysosomes.
11. Discuss the structure and role of peroxisomes.
12. Distinguish between the structure and function of microfilaments and microtubules as components of the cytoskeleton.
13. Discuss the structure and function of centrioles in cellular reproduction.
14. Differentiate between cilia and flagella.
15. Define a cell inclusion and give several examples.
16. Define extracellular material and give several examples.
17. Define a gene and explain the sequence of events involved in protein synthesis.
18. Describe the principle and importance of genetic engineering.
19. Describe the principal events of interphase, the stage between cell divisions.
20. Discuss the stages, events, and significance of somatic and reproductive cell division.
21. Explain the relationship of aging to cells.
22. Describe cancer as a homeostatic imbalance of cells.
23. Define medical terminology associated with cells.

Learning Activities

(pages 46–48)   **A.** Generalized animal cell, plasma (cell) membrane

   1. Define:
      a. Cell

      b. Cytology

      c. Microtomography

   2. List the four principal parts of a generalized animal cell.

☐1   3. Refer to Figure 3-2b (page 48) in your text and complete this exercise.

      a. The new concept in plasma membrane structure shown in the figure is

         known as the _____ model. The membrane
         is composed of two main chemical components: a bilayer of

         _____ embedded with _____
         molecules.

      b. Phospholipids appear to have "heads" and "tails." The *(heads? tails?)*
         face outward in the membrane. These *(do? do not?)* mix well with the
         watery environment surrounding the membrane. The phospholipid
         "tails" facing inward are not soluble in water. Of what advantage is this?

      c. Two categories of protein contribute to the plasma membrane structure.
         Embedded deeply in the phospholipid bilayer are *(integral? peripheral?)*
         proteins. State two functions of such proteins.

      d. Peripheral proteins are *(firmly? loosely?)* bound to membrane surfaces.
         What roles may these serve?

4. From the complex and varied structure of membranes, it is clear that membranes do more than just "sit there" around cells. State four important functions of plasma membranes.

5. Describe the permeability of the plasma membrane by answering these questions.

   a. Most proteins *(do? do not?)* pass through the plasma membrane. Explain.

   b. Substances that dissolve in *(lipid? water?)* can more readily pass across plasma membranes. Explain.

   c. Which part of the plasma membrane is highly permeable to carbon dioxide and oxygen?

   d. Which part is most permeable to water?

   e. Plasma membranes have special molecules that help to transport substances across membranes; these molecules are appropriately named

   _____. They are *(integral? peripheral?)* proteins.

B. Movement of materials across plasma membranes   **(pages 48–53)**

1. Why must substances move across your plasma membranes in order for you to survive and maintain homeostasis?

2. Complete parts a–e of Table LG 3-1.

Table LG 3-1.
Summary of mechanisms for movement of materials across membranes

Name of Process	What Moves (Solute or Solvent?)	Where (Direction, e.g., High to Low Concentration)?	Membrane Necessary (Yes or No)?	Carrier Necessary (Yes or No)?	Active or Passive?
a. Diffusion					
b. Facilitated diffusion					
c.	Solvent	From high to low concentration of water across semipermeable membrane			
d.		From high pressure area to low pressure area			
e. Dialysis				No	
f. Active transport		From low to high concentration of solute			
g. Phagocytosis			Yes		
h.	Solute	From outside cell to inside cell			
i. Receptor-mediated endocytosis					Active

3. Match the passive transport processes listed at right with the descriptive phrases.

_____ a. Net movement of any substance (such as cocoa powder in hot milk) from region of higher concentration to region of lower concentration; membrane not required

Dia. Dialysis
Dif. Diffusion
Fac. Facilitated diffusion
Fil. Filtration
O. Osmosis

_____ b. Same as (a) except movement across a semipermeable membrane with help of a carrier; ATP not required

_____ c. Net movement of water from region of high water concentration (such as 2 percent NaCl) to region of low water concentration (such as 10 percent NaCl) across semipermeable membrane; important in maintenance of normal cell size and shape

_____ d. Movement of molecules from high pressure zone to low pressure zone, for example, in response to force of blood pressure

_____ e. Movement of small molecules across a semipermeable membrane, leaving large molecules behind; the principle used in artificial kidney machines

4. A large part of the human diet is starch. When starch is digested, it is broken down to glucose. Answer these questions related to absorption of glucose.
   a. Explain why glucose cannot readily pass through plasma membranes by simple diffusion.

   b. By what process does it cross plasma membranes as it is absorbed into cells lining the digestive tract?

   c. List three factors that would speed up this rate of movement of glucose.

5. Complete the following exercise about osmosis in blood.

   a. Human red blood cells (RBCs) contain about _____ NaCl.

      A. 2.0 percent   B. 0.85 percent   C. 0 percent (pure water)

   b. A solution that is hypertonic to RBCs contains *(more? fewer?)* solute particles and *(more? fewer?)* water molecules than blood.

   c. Which of these solutions is hypertonic to RBCs?

      A. 2.0 percent NaCl B. 0.85 percent NaCl C. pure water

   d. If RBCs are surrounded by hypertonic solution, water will tend to move *(into? out of?)* them, so they will *(crenate? hemolyze?)*.

   e. A solution that is _____-tonic to RBCs will maintain the shape and size of the RBC. An example of such a solution is

   _____.

   f. Which solution will cause RBCs to hemolyze?

      A. 2.0 percent NaCl B. 0.85 percent NaCl C. pure water

   g. Which solution has the highest osmotic pressure?

      A. 2.0 percent NaCl B. 0.85 percent NaCl C. pure water

6 6. When blood is dialyzed in an artificial kidney, why must nutrients and certain ions be added to the blood?

   7. Contrast *active* with *passive transport* processes.

7 8. Complete parts f–i of Table LG 3-1.

8 9. Describe active transport in the following exercise.
   a. "Pumping" of certain ions occurs by active transport so that the ion

   _____ is greatly concentrated inside of cells, and the

   ion _____ is concentrated in extracellular fluid.
   b. Another chemical compound that may be actively transported is

   _____ which is moved from areas of *(high? low?)* concentration to areas of *(high? low?)* concentration, as from digestive tract cells into bloodstream.

   c. Active transport utilizes an _____ protein within the

   membrane, as well as an energy source, _____. In

   fact, a typical body may use up to _____ percent of its ATP for active transport.

9 10. Indicate your understanding of other active processes by completing this exercise.
   a. Phagocytosis is the process of cell *(eating? drinking?)*. Cell drinking is

   the process known as _____.
   b. In cell-mediated endocytosis, large molecules known as _____

   _____ bind to _____ on cell

   membranes. Invagination leads to formation of vesicles which fuse to form *(endosomes? lysosomes?)* What event later occurs in structures called CURLs? _____ Eventually ligands are digested by enzymes in _____ while receptors are recycled.

11. *ACTive learning.* Role-play transport mechanisms. Examples:
   a. Demonstrate diffusion of molecules by concentrating class members in one part of the room. As a result of high kinetic energy and rapid collisions, student "molecules" diffuse out to less concentrated areas.
   b. Role-play facilitated diffusion of student glucose molecule across the cell membrane of an intestinal cell. Other students can portray carrier molecules. The classroom wall with an open doorway can serve as the cell membrane with the hallway outside the room equated to extracellular spaces. Show the three factors which can alter the rate of facilitated diffusion.
   c. Show how increased blood pressure alters the rate of filtration of blood in the kidneys.
   d. Role-play the Na^+–K^+ "pump."
   e. Demonstrate the sequence of events in receptor-mediated endocytosis. A line of students can portray the cell membrane, invaginating to form vesicles, and fusing with other vesicles to form endosomes and CURLs.
   f. Contrast energy sources of passive versus active transport mechanisms.

C. Cytoplasm, organelles

(pages 53–61)

1. Describe the composition of cytoplasm using these key words: *fluid, tubules* and *filaments, solutes,* and *colloids.*

2. State briefly four main functions of cytoplasm.

3. The cell is compartmentalized by the presence of organelles. Of what advantage is this?

4. Study Figure 3-1 (page 47) in your text. Then on Figure LG 3-1 label the following parts of a generalized animal cell: *plasma membrane, cytoplasm, nucleus, nuclear membrane, nucleolus, chromatin, endoplasmic reticulum (rough* and *smooth), ribosomes, Golgi complex, mitochondria, lysosomes,* and *centrosome.*

47

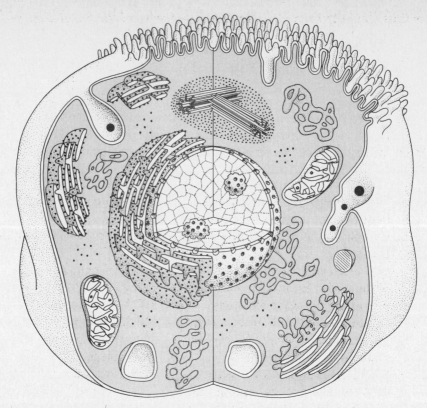

Figure LG 3-1 Generalized animal cell. Label as directed.

10 5. Describe parts of the nucleus by completing this exercise.

    a. The gel-like fluid in the nucleus is called _____.

    b. Nuclear storage sites of RNA used in protein synthesis are known as

    _____.

    c. During cell reproduction chromatin forms rod-shaped bodies called

    _____.

    d. Chromosomes consist of subunits known as _____
    which are composed of four types of proteins (called

    _____) plus DNA. A fifth type of histone helps to

    maintain adjacent nucleosomes in the configuration of a

    _____.

11 6. Describe the sequence of events in synthesis, secretion, and discharge of
    protein from the cell by numbering these five steps in correct order.

    _____ A. Proteins accumulate and are modified, sorted, and packaged in
             cisternae of Golgi complex.

    _____ B. Proteins pass through ER to Golgi complex.

    _____ C. Secretory granules move toward cell surface where they are dis-
             charged.

    _____ D. Proteins are synthesized at ribosomes on rough ER.

    _____ E. Vesicles containing protein pinch off of Golgi complex cisternae
             to form secretory granules.

7. Contrast roles of *cis, medial,* and *trans cisternae* of the Golgi complex.

8. Match the organelles listed at right with the following descriptions of their functions. $\boxed{12}$

_____ a. Site of direction of cellular activities by means of genes located here

_____ b. Cytoplasmic sites of protein synthesis; may occur attached to *ER* or scattered freely in cytoplasm

_____ c. System of membranous channels providing pathways for transport within the cell and surface areas for chemical reactions; may be granular or agranular

_____ d. Stacks of cisternae with vesicular ends; involved in packaging and secretion of proteins and lipids and synthesis of carbohydrates

_____ e. Called "suicide packets" since they may release enzymes which lead to autolysis of the cell

_____ f. Similar to lysosomes, but smaller; contain the enzyme catalase that breaks down hydrogen peroxide

_____ g. Cristae-containing structures, called "powerhouses of the cell" since ATP production occurs here

_____ h. Form part of cytoskeleton; involved with cell movement and contraction

_____ i. Part of cytoskeleton, proves support and gives shape to cell; form flagellae, cilia, centrioles, and spindle fibers; made of protein tubulin

_____ j. Help organize spindle fibers used in cell division

_____ k. Long, hairlike structures that help move entire cell, for example, sperm cell

_____ l. Short, hairlike structures that move particles over cell surface

Cen. Centrioles
Cil. Cilia
ER. Endoplasmic reticulum
F. Flagella
G. Golgi complex
L. Lysosomes
Mf. Microfilaments
Mt. Microtubules
Mit. Mitochondria
N. Nucleus
P. Peroxisomes
R. Ribosomes

9. *For extra review.* Considering the functions of organelles listed in the above exercise, choose the answers (organelles) that fit the following descriptions. $\boxed{13}$

_____ a. Abundant in white blood cells which use these organelles to help phagocytose bacteria

_____ b. Present in large numbers in mitochondria and liver cells which require much energy

_____ c. Present in liver cells to help metabolize H_2O_2

_____ d. Located on surface of cells of the respiratory tract; help to move mucus

_____ e. Absent in mature nerve cells, resulting in inability of these cells to produce if destroyed

10. *ACTive learning.* Role-play life within a cell. Study partners or groups of students can demonstrate activities occurring in different organelles designated to specific locations in a classroom/lab. Examples:

   a. Role-play sequence of events in synthesis, processing, packaging, and secretion of proteins.

   b. Role-play lysosomes of white blood cells digesting bacteria or lysosomes of injured or aging cell carrying out autolysis.

   c. Discuss and portray how life in a liver cell would differ from life in a white blood cell. Demonstrate which organelles would be more numerous and more active in each cell.

(pages 61–62) **D. Cell inclusions, extracellular materials**

1. Contrast *organelle* with *inclusion.*

2. List four examples of cell inclusions.

14 3. Name three examples of *extracellular materials.*

4. Complete the table about two major classes of matrix materials: *amorphous* and *fibrous.* Indicate to which category each of the substances listed below belongs. Then give an example of a location in the body where each type of matrix is found.

Matrix Material	Class	Location
a. Collagen		
b. Hyaluronic acid		
c. Chondroitin sulfate		
d. Reticulin		

1. State the significance of protein synthesis in the cell.

2. Describe the process of protein synthesis by doing this exercise.   15

   a. The first step in protein synthesis is called _____; it occurs in the *(nucleus? cytoplasm?)* of the cell.

   b. In transcription a portion of one side of a double stranded DNA serves as a mold or _____ and is called the *(sense? antisense?)* strand. Nucleotides of _____ line up in a complementary fashion next to DNA nucleotides (much as occurs between DNA nucleotides in replication of DNA prior to mitosis).

   c. The RNA formed by transcription is quite similar to DNA. RNA differs from DNA in that it is *(double? single?)*-stranded, contains the sugar _____ instead of deoxyribose, and has the nucleotide base _____ instead of thymine.

   d. Write in the complementary RNA strand that would be transcribed from the DNA template given.

   DNA   T A C A G T T A C A A T

   RNA   _ _ _ _ _ _ _ _ _ _ _ _

   e. Three types of RNA are formed by transcription: _____ _____ *(mRNA),* _____ *(tRNA),* and _____ *(rRNA).* These leave the nucleus and enter the cytoplasm.

   f. The second step of protein synthesis is known as _____ _____ since it involves translation of one "language" (the nucleotide base code of _____ RNA) into another "language" (the correct sequence of the 20 _____ to form a specific protein).

   g. Translation occurs in the *(nucleus? cytoplasm?),* specifically at a *(chromosome? lysosome? ribosome?)* to which mRNA attaches.

   h. Each amino acid is transported to this site by a particular _____-RNA characterized by a specific three-base unit or *(codon? anticodon?).* This portion of tRNA, for example, U A C, binds to a complementary portion of mRNA *(codon? anticodon?).*

   i. Write slash lines between codons in the mRNA strand below. Then write in the complementary tRNA anticodons that would bond to each codon. (The first one is done for you.)

   tRNA anticodon U A C _ _ _ _ _ _ _ _ _ _

   mRNA codon   A U G / U C A A U G U U A

This section of mRNA provides the "message" or code for correct placement of *(4? 12?)* amino acids carried by tRNA's.

j. As the ribosome moves along each mRNA codon, additional amino acids are transferred into place by _____-RNA. _____ bonds form between adjacent amino acids with the help of enzymes from the _____.

16 3. Summarize protein synthesis by completing this exercise.

a. A typical gene contains about _____ pairs of nucleotides. The mRNA strand transcribed from this gene would consist of about _____ nucleotides (containing _____ bases).

b. This 1,000-base mRNA strand contains about _____ mRNA codons (3 bases per codon). How many tRNA anticodons would pair with 333 mRNA codons? _____ (*Hint:* Refer to **15** i.)

c. Recall that each tRNA (with its 3-base anticodon) transfers *(1? 3?)* amino acid(s). Therefore the original 1,000 nucleotide DNA strand codes for formation of a protein containing about _____ amino acids. (*Note:* This summary is a simplification of numbers, omitting details such as termination codons.)

4. *For extra review.* Reread exercises **15** and **16** while carefully referring to Figure 3–14 (page 63) in your text.

5. *ACTive learning.* Role-play protein synthesis. Students carry large papers marked with letters A, C, G, T, and U (representing bases of DNA and RNA) or symbols for different amino acids. Student bases line up in complementary fashion in transcription. Each amino acid student attached to a three student tRNA anticodon is transferred into place during translation. Amino acid students bond (join hands) and separate from tRNA which is then recycled.

6. Define the "SOS response" for DNA repair.

7. Define and state the significance of *genetic engineering.*

8. List six therapeutic substances that have been produced by bacteria containing recombinant DNA.

1. Once you have reached adult size, do your cells continue to divide? *(Yes? No?)* Of what significance is this fact?

2. Contrast terms in each pair:

   a. Mitosis/cytokinesis

   b. Somatic cell division/reproductive cell division

3. Describe aspects of cell division by completing this exercise.   ⬜17

   a. Cell division includes two steps: division of the _____ _____ and division of the _____ _____ .

   b. In the formation of mature sperm and egg cells nuclear division is known as _____ . In the formation of all other body cells, that is, *somatic* cells, nuclear division is called _____ _____ . In *(meiosis? mitosis?)* the two daughter cells have the same hereditary material and genetic potential as the parent cell.

   c. Chromosomes consist of DNA present in the form of units called _____ bound by protein (histones). Each human chromosome contains about _____ genes.

   d. Prior to cell division DNA replicates. The helix partially uncoils and the double strand separates at the points where _____ are attached, so that new complementary nucleotides can attach here. (See Figure LG 2-4, page LG 36.) In this way the DNA of a single chromosome is doubled to form two identical _____ _____ during the _____-phase preceding a mitotic division.

4. Arrange the following time periods in correct sequence using lines below.   ⬜18 Then write a description of cell activity during each period.
   Cytokinesis, G1 period, G2 period, mitosis, S period

   ____

   ____

   ____

   ____

   ____

5. Carefully study the phases of mitosis shown in Figure 3-16 (page 66) in your text. Then check your understanding of major events of each phase by doing this matching exercise.

|19|

_____ a. Chromosomes are moved toward opposite poles of the cell.

_____ b. Nuclear membrane disappears; chromatin thickens into distinct chromosomes (chromatids). Centrioles move to opposite ends of cell, enabling microtubules to form mitotic spindles. The longest mitotic phase.

_____ c. The series of events is essentially the reverse of prophase; cytokinesis occurs.

_____ d. Chromatids line up on the equatorial plate and are attached to spindle fibers.

_____ e. Cell is not involved in cell division; it is in a metabolic phase which follows telophase.

A. Anaphase
I. Interphase
M. Metaphase
P. Prophase
T. Telophase

6. Describe the process of cytokinesis, including the cleavage furrow.

|20| 7. Examine Figure LG 3-2 and do the following exercise.

a. Meiosis is a special type of cell division that occurs only in

_____ and leads to the production of sex cells or

_____.

b. Cells that begin the process of meiosis are ovarian or testicular cells that have been undergoing mitosis up to this point. Each cell initially contains

_____ chromosomes, which is the *(diploid or 2n? haploid or n?)* number

for humans. These chromosomes consist of _____ homologous pairs. A simple description of these pairs in your own ovaries or testes is that for each maternal chromosome (black in Figure LG 3-2) you received in

your mother's egg, you received a comparable _____

_____ chromosome (white in Figure LG 3-2) from your father's sperm.

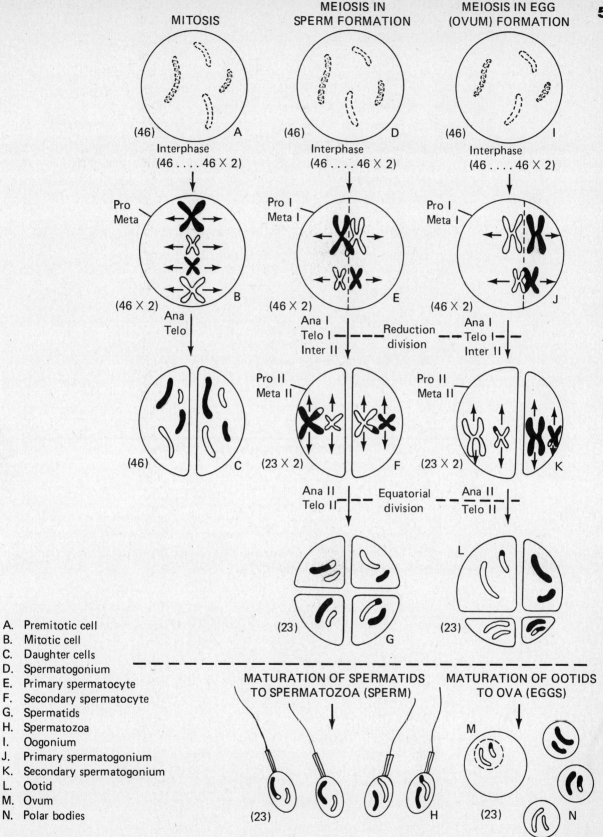

MITOSIS

MEIOSIS IN
SPERM FORMATION

MEIOSIS IN EGG
(OVUM) FORMATION

(46)   A

(46)   D

(46)   I

Interphase
(46 . . . . 46 × 2)

Interphase
(46 . . . . 46 × 2)

Interphase
(46 . . . . 46 × 2)

Pro
Meta

Pro I
Meta I

Pro I
Meta I

(46 × 2)   B

(46 × 2)   E

(46 × 2)   J

Ana
Telo

Ana I
Telo I   — — Reduction
Inter II   division

Ana I
— — Telo I — — ↓
Inter II

(46)   C

Pro II
Meta II

(23 × 2)   F

Pro II
Meta II

(23 × 2)   K

Ana II
Telo II   — — Equatorial
              division

— — Ana II
     Telo II

(23)   G

L

(23)   (23)   N

A.   Premitotic cell
B.   Mitotic cell
C.   Daughter cells
D.   Spermatogonium
E.   Primary spermatocyte
F.   Secondary spermatocyte
G.   Spermatids
H.   Spermatozoa
I.   Oogonium
J.   Primary spermatogonium
K.   Secondary spermatogonium
L.   Ootid
M.   Ovum
N.   Polar bodies

MATURATION OF SPERMATIDS
TO SPERMATOZOA (SPERM)

MATURATION OF OOTIDS
TO OVA (EGGS)

M

(23)   H

(23)   N

Figure LG 3-2 Comparison of mitosis with meiosis of sperm formation and meiosis of egg (ovum) formation. Maternal chromosomes are black; paternal ones are white. Chromosomes during interphase have dotted outlines indicating that they are threadlike and not visible under the microscope. Numbers of chromosomes (or chromatids) for each cell are in parentheses. Arrows within cells show directions in which chromosomes move. Maturation of gametes (after completion of meiosis) is shown below dotted line.

c. By the time that cells complete meiosis, they will contain *(46? 23?)* chromosomes, the *(2n? n?)* number.

d. Meiosis differs from mitosis in that meiosis involves *(one? two? three?)* divisions with replication of DNA only *(once? twice?)*. The first division is known as the _____ division, while the second division is the _____ division.

e. Replication of DNA occurs *(before? after?)* reduction division, resulting in doubling of chromosome number in humans from 46 chromosomes to _____ "chromatids." (Chromatid is a name given to a chromosome while the chromosome is attached to its replicated counterpart, that is, in the doubled state.) Doubled or paired chromatids are held together by means of a *(centromere? centrosome?)*.

f. A unique event that occurs in Prophase I is the lining up of chromosomes in homologous pairs, a process called _____. Note that this process *(does? does not?)* occur during mitosis. The four chromatids of the homologous pair are known as a _____. The proximity of chromatids permits the exchange or _____ of DNA between maternal and paternal chromatids. What is the advantage of crossing-over to sexual reproduction?

g. Another factor that increases possibilities for variety among sperm or eggs you can produce is the random (or chance) assortment of chromosomes as they line up on either side of the equator in Metaphase *(I? II?)*. Note that in cells shown in Figure LG 3-2, chromosomes aligned differently during the two cases of meiosis shown. Since there are actually 23 sets of tetrads lining up (only two tetrads, or doubled maternal/paternal pairs, are shown here), over 8 million possible sperm or egg types could be produced just on the basis of how chromosomes align (maternal on left, paternal on right, or vice versa) during this phase.

h. How does meiosis Metaphase I differ from mitosis metaphase? Refer to Figure LG 3-2.

i. Division I reduces the chromosome number to _____ _____ chromatids. Division II results in the separation of doubled chromatids by splitting of _____ so that resulting cells contain only ____ chromosomes.

j. Meiosis is only one part of the process of forming sperm or eggs. Each original diploid cell in a testis will lead to production of *(1? 2? 4?)* mature sperm, while each diploid ovary cell produces *(1? 2? 4?)* mature eggs (ova) plus three small cells known as _____.

1. Complete this exercise about cancer.

　　a. Certain types of cells of the body are not capable of reproduction after the early years of life. State three examples of such "post-mitotic" cells in your own body.

　　b. Other cells of the body can continue to undergo mitosis. State two examples of these.

　　c. Cancer cells can arise only from cells that *(can? cannot?)* undergo mitosis.

　　d. A term which means "tumor or abnormal growth" is _____

　　　_____. A cancerous growth is a *(benign? malignant?)* neoplasm, whereas a noncancerous tumor is a *(benign? malignant?)* neoplasm.

　　e. Which type of growth is more likely to spread and to possibly cause death? *(Benign? malignant?)* A term which means the "spread" of

　　　cancer cells is _____.

　　f. If kidney cancer metastasizes to the lungs, the lung cancer is known as the *(primary? secondary?)* tumor.

　　g. In the process of metastasis, cancer cells compete with normal cells for

　　　_____. They also migrate into space previously occupied by other cells since they do not follow rules of

　　　_____. By what routes do cancers reach distant parts of the body?

　　h. What may cause the pain associated with cancer?

22 2. Match terms at right with definitions below.

_____ a. General term for malignant tumor arising from any connective tissue

_____ b. Malignant tumor derived from bone (a connective tissue)

_____ c. General term for malignant tumor of epithelial tissue

_____ d. Malignant tissue derived from epithelial tissue of a gland

_____ e. Malignant tumor arising from cartilage (a connective tissue)

_____ f. A malignancy of tissue that forms blood

A. Adenocarcinoma
Ca. Carcinoma
Ch. Chondrosarcoma
M. Myeloma
O. Osteosarcoma
S. Sarcoma

3. Complete this exercise about causative factors associated with cancer.
   a. Define *carcinogen*.

   Name a carcinogen linked to cancer among cigarette smokers.

   b. Identify one or more types of cancer associated with the viruses named below.

   HTLV-1_____

   EBV_____

   HBV_____

   c. Contrast the roles of *oncogenes* with *proto-oncogenes* in causation of cancer.

   d. Do oncogenes appear to be involved in all cancers?
   (Yes? No?) Explain the apparent cause of the childhood eye cancer known as retinoblastoma.

4. *A clinical challenge.* Are all cancer cells alike within a single tumor? *(Yes? No?)* Of what significance is this in planning cancer treatment?

5. Make a statement about the role of the immune system in cancer.

H. Cells and aging   (pages 71–72)

1. Define *aging*.

2. Describe changes that normally occur in each of the following during the aging process.
   a. Number of cells in the body

   b. Collagen fibers

   c. Elastin

3. In a sentence or two, summarize the main points of each of these theories of aging.
   a. Free radical

b. Immune system

c. Pituitary gland hormone

(page 72) **I. Medical terminology**

23 1. Match the terms at right describing alterations in cells or tissues with descriptions below.

_____ a. Increase in size of tissue or organ by increase in size (not number) of cells, as growth of your biceps muscle with exercise

_____ b. Increase in size of tissue or organ due to increase in number of cells, such as a callus on your hand or breast tissue during pregnancy

_____ c. Change of one cell type to another normal cell type, such as change from single row of tall (columnar) cells lining airways to multilayers of cells as response to constant irritation of smoking

_____ d. Abnormal change in cells in a tissue as due to irritation or inflammation. May revert to normal if irritant removed, or may progress to neoplasia.

_____ e. Abnormal mass of cells which become increasingly different from original cells. Do not revert to normal, even if irritant removed. May be benign or malignant.

D. Dysplasia
HP. Hyperplasia
HT. Hypertrophy
M. Metaplasia
N. Neoplasia

24 2. In the above learning activity, alterations in cells and tissues are arranged such that c, d, and e are increasingly *(deranged or abnormal? normal?)* compared to those in a and b.

25 3. Match the terms at right with the definitions below.

_____ a. Death of a group of cells

_____ b. Decrease in size of cells with decrease in size of tissue or organ

_____ c. Study of tumors

_____ d. Study of use and effects of drugs in the treatment of disease

A. Atrophy
B. Biopsy
G. Geriatrics
N. Necrosis
O. Oncology
P. Pharmacology

_____ e. Branch of medicine devoted to medical problems and care of elderly persons

_____ f. Removal and examination of tissue from the living body for diagnosis

Answers to Numbered Questions in Learning Activities

1 (a) Fluid mosaic, phospholipid, protein. (b) Heads, do; the membrane forms a water-insoluble barrier between cell contents and extracellular fluid. (c) Integral; some serve as channels (for water or ions); others, in the form of glycoproteins, serve as receptors, for example, to "tell" hormones where to attach. (d) Loosely; may serve as enzymes or help change membrane shape.

2 (a) Do not; they are too large, that is, they are larger than the membrane pores. (b) Lipid; a major part of the plasma membrane, the bilayer, consists of lipid molecules. (c) Lipid bilayer. (d) Integral proteins (see channels in Figure 3-2b, page 48 in your text). (e) Carrier molecules; integral.

3 (a) –, Solute or solvent, from high to low, no, no, passive. (b) –, Solute (like glucose), from high to low, yes, yes, passive. (c) Osmosis, –, –, yes, no, passive. (d) Filtration, solute or solvent, –, yes (membrane or other barrier such as paper), no, passive. (e) –, Small solutes, from high to low, yes, –, passive.

4 (a) Dif. (b) Fac. (c) 0. (d) Fil. (e) Dia.

5 (a) B. (b) More, fewer. (c) A. (d) Out of, crenate. (e) Iso-, 0.85 percent NaCl (physiological saline) or 5.0 percent glucose. (f) C. (g) A.

6 Because small ions (even "good ones" needed by the body) are removed by dialysis.

7 (f) –, Solutes (like glucose, Na^+, K^+), –, yes, integral protein, active. (g) –, Large particles, from outside cell to inside cell, –, no, active. (h) Pinocytosis, –, –, yes, no, active. (i) –, Large molecules or particles, from outside cell to inside cell, yes, receptor, –.

8 (a) Potassium (K^+), Sodium (Na^+). (b) Glucose, low, high. (c) Integral, ATP, 40.

9 (a) Eating, pinocytosis. (b) Ligands, receptors, endosomes, receptors and ligands separate or uncouple, lysosomes.

10 (a) Karyolymph. (b) Nucleoli. (c) Chromosomes. (d) Nucleosomes, histones, helical coil.

11 3, 2, 5, 1, 4 (or D, B, A, E, C).

12 (a) N. (b) R. (c) ER. (d) G. (e) L. (f) P. (g) Mit. (h) Mf. (i) Mt. (j) Cen. (k) F. (l) Cil.

13 (a) L. (b) Mit. (c) P. (d) Cil. (e) Cen.

14 Body fluids such as plasma and interstitial fluid, mucus, and matrix surrounding cells.

15 (a) Transcription, nucleus. (b) Template, sense, RNA. (c) Single, ribose, uracil. (d) A U G U C A A U G U U A. (e) Messenger, transfer, ribosomal. (f) Translation, messenger, amino acids. (g) Cytoplasm, ribosome. (h) T, anticodon, codon, (i) tRNA: U A C/A G U/ U A C/ A A U; mRNA: A U G/ U C A/ A U G/ U U A, 4. (j) T, peptide, ribosome.

16 (a) 1,000, 1,000, 1,000. (b) 333, 333. (c) 1,333.

17 (a) Nucleus, cytoplasm. (b) Meiosis, mitosis, mitosis. (c) Genes, 20,000. (d) Nitrogen bases, chromatids, inter-.

18 G1 period, gap or growth period with production of substances necessary for cell division; S period, synthesis and replication of DNA; G2 period, additional gap or growth period; mitosis, nuclear division; cytokinesis, cytoplasmic division.

19 (a) A. (b) P. (c) T. (d) M. (e) I.

20 (a) Gonads (ovaries or testes), gametes. (b) 46, diploid or 2n, 23, paternal. (c) 23, n. (d) Two, once, reduction, equatorial. (e) Before, "46 × 2" (or 46 doubled), centromere. (f) Synapsis, does not, tetrad, crossing-over, permits exchange of DNA between chromatids and greatly increases possibilities for variety among sperm or eggs and thus offspring. (g) I. (h) In meiosis I, doubled chromatids line up in homologous pairs with centromeres equidistant from equator, whereas in mitosis the doubled chromatids are not in homologous pairs and they align with centromere directly on equator. (i) "23 × 2" (or 23 doubled chromatids), centromeres, 23. (j) 4, 1, polar bodies.

21 (a) Nerve tissue cells, as in brain or spinal cord, heart muscle, and striated (skeletal) muscle, as in your biceps muscle. (b) Skin, glands, liver, connective tissues such as bone, cartilage or blood. (c) Can. (d) Neoplasm, malignant, benign. (e) Malignant, metastasis. (f) Secondary. (g) Nutrients and space, contact inhibition, blood or lymph, or by detaching from an organ, such as liver, and "seeding" into a cavity (as abdominal cavity) to reach other organs which border on that cavity. (h) Pressure on nerve, obstruction of passageway, loss of function of a vital organ.

22 (a) S. (b) O. (c) Ca. (d) A. (e) Ch. (f) M.

23 (a) HT. (b) HP. (c) M. (d) D. (e) N.

24 Deranged or abnormal.

25 (a) N. (b) A. (c) O. (d) P. (e) G. (f) B.

MASTERY TEST: Chapter 3

Questions 1–10: Choose the one best answer to each question.

_____ 1. The organelle which carries out the process of autolysis, and is therefore called a "suicide packet" is:

A. Peroxisome   B. Beta-endorphin   C. Centrosome
D. Lysosome   E. Endoplasmic reticulum

_____ 2. Which statement about proteins is FALSE?

    A. Proteins are synthesized on ribosomes

    B. Proteins are so large that they will not pass out of blood that undergoes filtration in kidneys.

    C. Proteins are so large that they will not pass through the dialysis membrane used in artificial kidneys.

    D. Proteins are so large that they can be expected to create osmotic pressure important in maintaining fluid volume in blood.

    E. Proteins are small enough to pass through pores of typical plasma (cell) membranes.

_____ 3. Choose the FALSE statement about prophase of mitosis.

    A. It occurs just after metaphase.

    B. It is the longest phase of mitosis.

    C. Nucleoli become less visible.

    D. Centrioles move apart and become connected by spindle fibers.

_____ 4. Choose the FALSE statement about the Golgi complex.

    A. It is usually located near the nucleus.

    B. It consists of cisternae with expanded ends known as vesicles.

    C. It synthesizes glycoproteins.

    D. It is involved with lipid secretion.

    E. It pulls chromosomes toward poles of the cell during mitosis.

_____ 5. Choose the FALSE statement about genes.

    A. Genes contain DNA.

    B. Genes contain information which controls heredity.

    C. Each gene contains about 10 nucleotides.

    D. Genes are transcribed by messenger RNA during the first step of protein synthesis.

_____ 6. Choose the FALSE statement about protein synthesis.

    A. Translation occurs in the cytoplasm.

    B. Messenger RNA picks up and transports an amino acid during protein synthesis.

    C. Messenger RNA travels to a ribosome in the cytoplasm.

    D. Transfer RNA is attracted to mRNA due to their complementary bases.

_____ 7. Which of these structures is NOT located in the nucleus?

    A. Nuclcolus   B. Karyolymph   C. Chromatin
    D. Chromosomes   E. Centrosome

_____ 8. The fact that the aroma of cookies baking in the kitchen reaches you in the living room is due to:

    A. Active transport   B. Dialysis   C. Osmosis
    D. Diffusion   E. Phagocytosis

_____ 9. Bone cancer is an example of what type of cancer?

    A. Sarcoma   B. Myeloma   C. Carcinoma

64

_____ 10. All of the following are intracellular inclusions EXCEPT:

    A. Melanin   B. Hyaluronic acid   C. Mucus
    D. Glycogen   E. Lipid

Questions 11–20: Circle T (true) or F (false). If the statement is false, change the underlined word or phrase so that the statement is correct.

T F 11. Plasma membranes consist of a double layer of <u>carbohydrate molecules</u> with proteins embedded in the bilayer.

T F 12. Every cell of the body has <u>the same appearance</u> as the generalized animal cell.

T F 13. <u>Cristae</u> are folds of membrane in mitochondria.

T F 14. Sperm move by means of lashing of their long tails named <u>cilia.</u>

T F 15. Microtomography is a procedure which can produce high magnification, three-dimensional images of <u>living</u> cells.

T F 16. <u>Diffusion, osmosis, and filtration</u> are all passive transport processes.

T F 17. A 5 percent glucose solution is <u>hypotonic</u> to a 10 percent glucose solution.

T F 18. Most healthy cells <u>lack</u> the property of contact inhibition.

T F 19. The phase that follows metaphase of mitosis is <u>anaphase.</u>

T F 20. <u>Cytokinesis</u> is another name for mitosis.

Questions 21–25: fill-ins. Complete each sentence with the word which best fits.

_____ 21. The first meiotic division is known as the ____ division.

_____ 22. ____ proteins are loosely bound to the cell membrane and easily separated from it.

_____ 23. White blood cells engulf large solid particles by the process of ____.

_____ 24. The period of interphase during which chromosomes are replicated is known as the ____ period.

_____ 25. The sequence of bases of mRNA which would be complementary to DNA bases in the sequence A-T-T-C-A-C would be ____.

The Tissue Level of Organization

Cells of similar structure and function, along with intercellular substances they produce, are organized into tissues. In this chapter you will consider a variety of specialized tissue types which make up organs of the body (Objectives 1–2). You will study structure, functions, and locations of epithelium (3–6), including that composing glands (7–8). You will contrast different kinds of connective tissue (9–15), muscle, and nerve tissue (16–17). You will see the arrangement of a number of tissue types in membranes (19). You will also learn about the mechanisms of tissue repair (20–21).

Topics Summary

A. Types of tissue
B. Epithelial tissue
C. Connective tissue
D. Muscle tissue and nervous tissue
E. Membranes
F. Tissue repair

Objectives

1. Define a tissue.
2. Classify the tissues of the body into four major types and define each type.
3. Discuss the distinguishing characteristics of epithelial tissue.
4. Contrast the structural and functional differences between covering and lining epithelium and glandular epithelium.
5. Compare the layering arrangements and cell shapes of covering and lining epithelium.
6. List the structure, location, and function for the following types of epithelium: simple squamous, simple cuboidal, simple columnar (nonciliated and ciliated), stratified squamous, stratified cuboidal, stratified columnar, transitional, and pseudostratified.
7. Define a gland and distinguish between exocrine and endocrine glands.
8. Classify exocrine glands according to structural complexity and function and give an example of each.
9. Identify the distinguishing characteristics of connective tissue.
10. Contrast the structural and functional differences between embryonic and adult connective tissues.
11. Discuss the ground substance, fibers, and cells that constitute connective tissue.
12. List the structure, function, and location of loose (areolar) connective tissue, adipose tissue, and dense, elastic, and reticular connective tissue.
13. List the structure, function, and location of the three types of cartilage.
14. Distinguish between interstitial and appositional growth of cartilage.
15. Describe the structure and functions of osseous tissue (bone) and vascular tissue (blood).
16. Contrast the three types of muscle tissue with regard to structure and location.
17. Describe the structural features and functions of nervous tissue.
18. Define an epithelial membrane.
19. List the location and function of mucous, serous, cutaneous, and synovial membranes.
20. Describe the conditions necessary for tissue repair.
21. Explain the importance of nutrition, adequate circulation, and age to tissue repair.

66   Learning Activities

(page 77) **A. Types of tissue**

1. Define *tissue*.

Define *histology*.

2. Complete the table about the four main classes of tissues.

Tissue	General Functions
a. Epithelium	
b. Connective tissue	
c.	Movement
d.	Initiates and transmits nerve impulses which coordinate body activities

(pages 77–85) **B. Epithelial tissue**

1. State two general locations of epithelial tissue.

2. Write five structural characteristics of all epithelial tissue.

3. Complete these sentences describing *basement membrane.* $\boxed{1}$
   a. The structure which attaches epithelium to underlying connective tissue

      is called the _____. This membrane consists of *(cells? extracellular materials?).*
   b. The portion of this membrane adjacent to epithelium is secreted by *(epithelium? connective tissue?).* It consists of the protein named *(collagen? reticulin?)* along with glycoproteins, and is known as the *(reticular lamina? basal lamina?).*
   c. Many basement membranes consist of a second, underlying layer also. This is secreted by *(epithelium? connective tissue?),* and is called the

      _____ lamina.
4. Which tissues consist of cells that wear out constantly and are replaced by $\boxed{2}$ mitosis throughout life?

5. *A clinical challenge.* The fact that epithelial cells are sloughed off from tissues such as the cervix of uterus or lining of airways serves as the basis

   for the diagnostic test known as a _____. $\boxed{3}$
6. Contrast *simple* with *stratified* epithelium with regard to structure, function, and location.

7. Complete the table about epithelial types.

Type of Epithelium	Sketch of Tissue	Locations and Functions
a. Simple squamous		
b. Simple cuboidal		
c. Simple columnar (nonciliated)		

Type of Epithelium	Sketch of Tissue	Locations and Functions
d. Simple columnar (ciliated)		
e. Stratified squamous		
f. Stratified cuboidal		
g. Stratified columnar		
h. Stratified transitional		
i. Pseudostratified		

8. Check your understanding of the most common types of epithelium by writing the name of the type after the phrase which describes it.

4

a. Lines the inner surface of the stomach and intestine: _____

b. Lines urinary tract, as in bladder, permitting distension: _____

c. Lines mouth; present on outer surface of skin: _____

d. Single layer of cube-shaped cells; found in kidney tubules and ducts of some glands: _____

e. Lines air sacs of lungs where thin cells are required for diffusion of gases into blood: _____

Pseudostratified
Simple columnar
Simple cuboidal
Simple squamous
Stratified squamous
Stratified transitional

f. Not a true stratified; all cells on basement membrane, but some do not

reach surface of tissue: _____

g. Endothelium and mesothelium most

similar structurally to _____

9. Write a sentence describing each of these modifications of the epithelial
lining of the intestine.
a. Microvilli

b. Goblet cells

10. What is the function of *cilia* on cells lining the respiratory and reproductive ▢5
tracts?

11. Define *keratin* and state its function.

List three locations of nonkeratinized stratified squamous epithelium.

12. What are *glands?* Why are glands studied in this section on epithelium?

13. Write EXO before descriptions of *exocrine* glands, and ENDO before
descriptions of *endocrine* glands. (Endocrine glands will be studied further
in Chapter 18.) ▢6

_____ a. Their products are secreted into ducts that lead either directly or
indirectly to the outside of the body.
_____ b. Their products are secreted into the blood and so stay within the
body; they are ductless glands.
_____ c. Examples are glands that secrete sweat, oil, mucus, and digestive
enzymes.
_____ d. Examples are glands that secrete hormones.

7

14. Match the types of glands with the descriptions given.

_____ a. Tube-shaped   A. Acinar
_____ b. Flask-shaped   C. Compound
_____ c. One-celled   S. Simple
_____ d. Branched duct   T. Tubular
_____ e. Nonbranched duct   U. Unicellular

15. *For extra review.* On separate paper draw a diagram of each of the four
    types of glands listed below.
    a. Simple coiled tubular gland: _____

    b. Simple branched tubular gland: _____

    c. Simple branched acinar gland: _____

    d. Compound acinar gland: _____
16. Complete the table contrasting classes of glands.

Type of Gland	How Secretion Is Released	Example
a. Apocrine		
b.	Cell accumulates secretory product in cytoplasm; cell dies; cell and its contents are discharged as secretion.	
c.		Salivary glands

(pages 85–94) **C. Connective tissue**

1. Contrast connective tissue with epithelium according to location, blood
   supply, and amount of intercellular material.

2. What are the main functions of connective tissue?

3. The major factor that differentiates one type of connective tissue from
   another is *(appearance of cells? kind of intercellular substance?).*

8

4. Name several kinds of intercellular materials. (Note that *inter*cellular means "between cells" or "extracellular.")

5. The embryonic tissue from which all other connective tissues arise is called _____. Another kind of connective tissue of the embryo, called _____, is located only in the umbilical cord.

6. How can injection of hyaluronidase along with injected drugs help lessen pain associated with the procedure?

7. Name three types of fibers in connective tissue. Write a sentence describing each.

8. Identify characteristics of each of the cell types found in connective tissue by writing the name of the correct cell type after the correct description. |9|

a. Derived from lymphocyte, gives rise to antibodies, so helpful in defense: _____ _____

b. Phagocytic cell; engulfs bacteria and cleans up debris; important during infection: _____

c. Fat cell: _____ _____

d. Believed to form collagenous and elastic fibers in injured tissue: _____ _____

e. Abundant along walls of blood vessels; believed to produce heparin, an anticoagulant, as well as histamine, which dilates blood vessels: _____ _____

f. Pigment cell _____ _____

Adipocyte
Fibroblast
Macrophage
Mast cell
Melanocyte
Plasma cell

10

9. Match the types of mature connective tissue cells in the list at right with the cells at left which are believed to form them.

_____ a. Monocyte   F. Fibrocyte
_____ b. Fibroblast   Mac. Macrophage
_____ c. Lymphocyte B   Mas. Mast cell
_____ d. Basophil   P. Plasma cell

11

10. Which two types of tissue form subcutaneous tissue (superficial fascia)?

11. Explain what accounts for the "signet ring" (like a class ring with a stone) appearance of adipocytes.

12. State five locations where adipose tissue is normally found.

12

How is fat tissue helpful at those locations?

13. Describe the procedure called a *suction lipectomy*.

14. Match the common types of dense connective tissue with the descriptions given.

13

_____ a. Connects muscles to bones   A. Aponeurosis
_____ b. Holds bones together at joints   D. Deep fascia
_____ c. Flat band or sheet of tissue con-   L. Ligament
            necting muscles to each other or   P. Periosteum
            to bones   T. Tendon

_____ d. Sheet of connective tissue wrapped around muscle bundles, holding them in place

_____ e. Covering over bone

15. In which of the types of dense connective tissues in Learning Activity 13 are fibers arranged in an orderly, parallel, *regular arrangement?* | 14 |

16. List three locations of each of the following types of connective tissue.
    a. Elastic

    b. Reticular

17. In general, cartilage can endure *(more? less?)* stress than the connective tissues you have studied so far. | 15 |

18. Write *I* before descriptions of *interstitial* growth of cartilage, and *A* before descriptions of *appositional growth of cartilage.* | 16 |

_____ a. Occurs in childhood and adolescence

_____ b. Growth due to division of cells (chondrocytes) in existing cartilage

_____ c. Starts later in life and occurs throughout adulthood

_____ d. Growth due to division of cells in covering over cartilage (perichondrium)

19. Match the types of cartilage with the descriptions given. | 17 |

_____ a. Found where strength and rigidity are needed, as in discs between vertebrae and in symphysis pubis

_____ b. White, glossy cartilage covering ends of bones (articular), covering ends of ribs (costal), and giving strength to nose, larynx, and trachea

_____ c. Provides strength and flexibility, as in external part of ear.

E. Elastic
F. Fibrous
H. Hyaline

20. Complete this exercise about bone and blood tissue. | 18 |

   a. Bone tissue is also known as _____ tissue. Compact bone consists of concentric rings, or *(lamellae? canaliculi?)* with bone cells, called *(chondrocytes? osteocytes?),* located in tiny spaces called

   _____ . Nutrients in blood reach osteocytes by vessels

   located in the _____ canal and then via minute, radiating canals called _____ which extend out to lacunae.

74

   b. Blood, or _____ tissue, consists of a fluid called
   _____ containing three types of formed elements.
   Red blood cells, or _____, transport oxygen and
   carbon dioxide; white blood cells, also known as _____
   _____, provide defense for the body; thrombocytes
   assist in the function of blood _____.

21. Review connective tissue types in Exhibit 4-2 (pages 87–92) of your text.
   Then examine the diagrams in Figure LG 4-1 and label different cell types
   and structures characteristic of each tissue type. (Note that bone and blood,
   both classified as connective tissues, will be studied in greater detail in
   Chapters 6 and 19.)

(pages 94–95) **D. Muscle tissue and nervous tissue**

1. These tissues are more *(specific? generalized?)* in their structure and functions than epithelium or connective tissue. (They will be studied in greater detail in Chapters 10 and 12.)
2. Complete the table about kinds of muscle tissue.

Kind	Location	Striated or Smooth	Voluntary or Involuntary
a. Skeletal			Voluntary
b.		Smooth	
c.	Only in the heart		

3. Check your understanding of the three muscle types by selecting muscle types that best fit descriptions below.

_____ a. Tissue forming most of the wall of the heart.
_____ b. Attached to bones
_____ c. Spindle-shaped cells with ends tapering to points
_____ d. Contain intercalated discs
_____ e. Found in walls of intestine, urinary bladder, and blood vessels
_____ f. Cells are multinucleate

C. Cardiac
Sm. Smooth
Sk. Skeletal

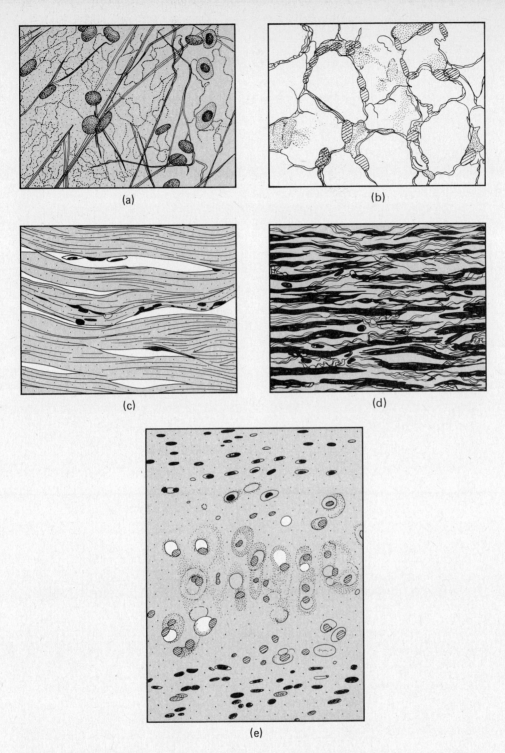

Figure LG 4-1 Types of connective tissue. (a) Loose (areolar). (b) Adipose. (c) Dense (collagenous). (d) Elastic. (e) Hyaline cartilage. Label as directed.

4. Write a sentence describing each of the following terms:

Sarcoplasm

Sarcolemma

Myofibrils

5. Constrast the functions of the two principal types of cells composing the nervous system: *neurons* and *neuroglia*.

6. Contrast the role of an *axon* and a *dendrite* of a neuron.

1. Complete the table about three types of membranes.

Type of Membrane	Location	Example of Specific Location	Function(s)
a.	Lines body cavities leading to exterior		
b. Serous			Allows organs to glide easily over each other
c.		Lines knee and hip joints	

2. Check your understanding of membrane types by doing this exercise. [20]
   a. The serous membrane covering the heart is known as the
   _____, whereas that covering the lungs is called the
   _____. The serous membrane over abdominal organs
   is the _____.
   b. The portion of serous membranes that covers organs (viscera) is called
   the _____ layer; that portion lining the cavity is
   named the _____ layer.
   c. The _____ layer of a mucous membrane binds epithelium to underlying muscle and serves as a route for oxygen and
   nutrients to the avascular epithelial layer of the membrane.
   d. Another name for skin is _____ membrane.
   e. A _____ membrane secretes a lubricating fluid
   known as synovial fluid. Such a membrane *(does? does not?)* contain
   epithelium, so *(is? is not?)* classified as an epithelial membrane.

F. Tissue repair (page 97)

1. Choose the correct term to complete each sentence. Write P for *parenchymal*
   or (S) for *stromal*. [21]

   _____ a. The part of an organ that consists of functioning cells, such as
   secreting epithelial cells lining the intestine, is composed of
   _____ tissue.

_____ b. The connective tissue cells that support the functional cells of the organ are called _____ cells.

_____ c. If only the _____ cells are involved in the repair process, the repair will be close to perfect.

_____ d. If _____ cells are involved in the repair, they will lay down fibrous tissue known as a fibrosis or a scar.

2. Define each of the following terms, and explain how these conditions occur.

Keloid scar

Adhesions

3. What is the function of *granulation tissue* in the process of tissue repair?

4. You have most likely experienced an injury that led to formation of a *scab*. What causes scabs to form?

5. List three factors that enhance tissue repair.

6. Which of these vitamins performs the following functions in the process of wound healing: A, B, C, D, E, or K?

22

_____ a. This vitamin is necessary for healing fractures since it enhances calcium absorption from foods in the intestine.

_____ b. This is important in repair of connective tissue and walls of blood vessels.

_____ c. Thiamine, riboflavin, and nicotinic acid are categorized as vitamins of this group. They enhance metabolic processes and therefore help to provide energy for wound repair.

_____ d. This vitamin is helpful in replacement of epithelial tissues, for example, in the lining of the respiratory tract.

_____ e. Believed to promote healing of injured tissues, this vitamin may also help to prevent scarring.

_____ f. This vitamin aids in blood clotting and so helps to prevent excessive blood loss from wounds.

Answers to Numbered Questions in Learning Activities

1 (a) Basement membrane, extracellular material. (b) Epithelium, collagen, basal lamina. (c) Connective tissue, reticular.

2 Epithelium of skin and gastrointestinal tract.

3 Pap smear (Papanicolaou test).

4 (a) Simple columnar. (b) Stratified transitional. (c) Stratified squamous. (d) Simple cuboidal. (e) Simple squamous. (f) Pseudostratified. (g) Simple squamous.

5 Cilia wave in unison to move mucus and foreign particles upward, away from the lungs and toward the throat where they can be swallowed. Cilia propel egg (ovum) from the ovary toward the uterus.

6 (a) EXO. (b) ENDO. (c) EXO. (d) ENDO.

7 (a) T. (b) A. (c) U. (d) C. (e) S.

8 Kind of intercellular substance.

9 (a) Plasma cell. (b) Macrophage. (c) Adipose. (d) Fibroblast. (e) Mast cell. (f) Melanocyte.

10 (a) Mac. (b) F. (c) P. (d) Mas.

11 Loose connective tissue and adipose tissue.

12 Since fat is a poor conductor of heat, it reduces heat loss from skin. It protects organs, such as kidneys and joints, and serves as an energy source.

13 (a) T. (b) L. (c) A. (d) D. (e) P.

14 A, L, T. Tissue is adapted for tension in one direction.

15 More.

16 (a) I. (b) I. (c) A. (d) A.

17 (a) F. (b) H. (c) E.

18 (a) Osseous, lamellae, osteocytes, lacunae, central (Haversian), canaliculi. (b) Vascular, plasma, erythrocytes, leukocytes, clotting.

19 (a) C. (b) Sk. (c) Sm. (d) C. (e) Sm. (f) Sk.

80

20 (a) Pericardium, pleura, peritoneum. (b) Visceral, parietal. (c) Lamina propria (connective tissue). (d) Cutaneous. (e) Synovial, does not, is not.

21 (a) P. (b) S. (c) P. (d) S.

22 (a) D. (b) C. (c) B. (d) A. (e) E. (f) K.

MASTERY TEST: Chapter 4

Questions 1–5: Choose the one best answer to each question.

_____ 1. A surgeon performing abdominal surgery will pass through the skin, then loose connective tissue (subcutaneous), then muscle, to reach the _____ membrane lining the inside wall of the abdomen.

   A. Parietal pleura   B. Parietal pericardium
   C. Parietal peritoneum   D. Visceral pleura
   E. Visceral pericardium   F. Visceral peritoneum

_____ 2. The type of tissue that covers body surfaces, lines body cavities, and forms glands is:

   A. Nervous   B. Muscular   C. Connective
   D. Epithelial

_____ 3. Endothelium lining the heart and blood vessels and mesothelium lining thoracic and abdominal cavities is most similar in structure to:

   A. Adipose tissue   B. Simple squamous epithelium
   C. Simple columnar epithelium   D. Elastic cartilage
   E. Loose connective tissue

_____ 4. Which statement about connective tissue is FALSE?

   A. Cells are very closely packed together.

   B. Connective tissue has an abundant blood supply.

   C. Intercellular substance is present in large amounts.

   D. It is the most abundant tissue in the body.

_____ 5. Modified columnar cells that are unicellular glands secreting mucus are known as:

   A. Cilia   B. Microvilli   C. Goblet cells
   D. Branched tubular glands

Questions 6–10: Match tissue types with correct descriptions.

6. Contains lacunae and chondrocytes:
_____

7. Forms fasciae, tendons, and dermis of skin: _____

8. Forms thick layer of skin on hands and feet, providing extra protection: _____
_____

9. Also called areolar connective tissue:
_____

10. Single layer of flat, scalelike cells: _____
_____

Adipose
Loose connective tissue
Cartilage
Dense connective tissue
Simple columnar epithelium
Simple squamous epithelium
Stratified squamous epithelium

Questions 11–20: Circle T (true) or F (false). If the statement is false, change the underlined word or phrase so that the statement is correct.

T   F   11. Stratified transitional epithelium lines the <u>urinary bladder</u>.

T   F   12. A person who has <u>good nutrition, has good blood supply, and is over 40 years of age</u> will tend to heal rapidly.

T   F   13. Hormones are classified as <u>exocrine</u> secretions.

T   F   14. Elastic connective tissue, since it provides both stretch and strength, is found in <u>walls of elastic arteries, in the vocal cords, and in some ligaments.</u>

T   F   15. Adipose tissue is a <u>good conductor of heat and therefore reduces heat loss (provides insulation).</u>

T   F   16. Another term for intercellular is <u>intracellular</u>.

T   F   17. The surface attachment between epithelium and connective tissue is called <u>basement membrane.</u>

T   F   18. <u>Simple</u> squamous epithelium is most likely to line the areas of the body that are subject to wear and tear.

T   F   19. <u>Loose connective tissue and adipose tissue</u> compose the subcutaneous layer that attaches skin to underlying tissue.

T   F   20. Appositional growth of cartilage <u>starts later in life than interstitial growth and continues throughout life.</u>

Questions 21–25: fill-ins. Complete each sentence with the word that best fits.

_____ 21. _____ is the study of tissue.

_____ 22. _____ is a dense connective tissue covering over bone.

_____ 23. The kind of tissue which lines alveoli (air sacs) of lungs is _____.

_____ 24. A term that means <u>without blood vessels</u> is _____.

_____ 25. All glands are formed of the tissue type named _____.

The Integumentary System

In this chapter you will see how cells, tissues, and organs are organized into a system—the integumentary system. You will learn about the largest organ in the body, the skin, as well as related special organs, such as hair and nails (Objective 1). You will study the layers of the skin (2–6) and its derivatives (8–10). You will consider the role of the skin in control of body temperature (11). You will examine the effects of aging on skin (12), as well as disorders and treatment of skin (13, 16). Special emphasis will be placed on the healing process and on burns (7, 14–15). You will also study development of skin (17).

Topics Summary

A. Skin
B. Skin wound healing
C. Epidermal derivatives
D. Homeostasis: temperature control
E. Effects of aging
F. Developmental anatomy
G. Disorders, medical terminology

Objectives

1. Define the integumentary system.
2. Describe the various functions of the skin.
3. List the various layers of the epidermis and describe their structure and functions.
4. Describe the composition and functions of the dermis.
5. Explain the basis for skin color.
6. Explain the basic pattern of epidermal ridges and grooves.
7. Outline the steps involved in epidermal wound healing and deep wound healing.
8. Describe the development, distribution, and structure of hair.
9. Compare the structure, distribution, and functions of sebaceous (oil), sudoriferous (sweat), and ceruminous glands.
10. List the parts of a nail and describe their composition.
11. Explain the role of the skin in helping to maintain the homeostasis of normal body temperature.
12. Describe the effects of aging on the integumentary system.
13. Describe the causes and effects for the following skin disorders: acne, systemic lupus erythematosus (SLE), psoriasis, decubitus ulcers, cellulitis, sunburn, and skin cancer.
14. Define a burn and list the systemic effects of a burn.
15. Classify burns into first, second, and third degrees and describe how to estimate the extent of a burn.
16. Define medical terminology associated with the integumentary system.
17. Describe the development of the epidermis, its derivatives, and the dermis.

(pages 101–104)   **A. Skin**

> [1]   1. Name the structures included in the integumentary system.

2. Describe some important aspects of skin by doing this exercise.
   a. Why is skin considered an organ?

> [2]   b. The skin is the largest organ in the body. It covers a surface area of 2 square meters, which is equivalent to about 21 square feet or _____ square inches.
   c. Which medical specialty deals with diagnosis and treatment of skin disorders?

   d. Skin may be one of the most underestimated organs in the body. What functions does your skin perform while it is "just lying there" covering your body? List six functions on the lines provided.

   _____   _____

   _____   _____

   _____   _____

> [3]   e. Write P next to the functions you listed above which in any way provide protection for the body.

> [4]   3. Answer these questions about the two portions of skin. Label them on Figure LG 5-1.

   a. The outer layer is named the _____. It is composed of *(connective tissue? epithelium?)*.

   b. The inner portion of skin, called the _____, is made of *(connective tissue? epithelium?)*. The dermis is *(thicker? thinner?)* than the epidermis.

> [5]   4. The tissue underlying skin is called *subcutaneous*, meaning _____ _____. This layer is also called _____.

   It consists of two types of tissue, _____ and _____. What functions does subcutaneous tissue serve?

> [6]   5. Epidermis contains four distinct cell types. Fill in the name of the cell type that fits each description.

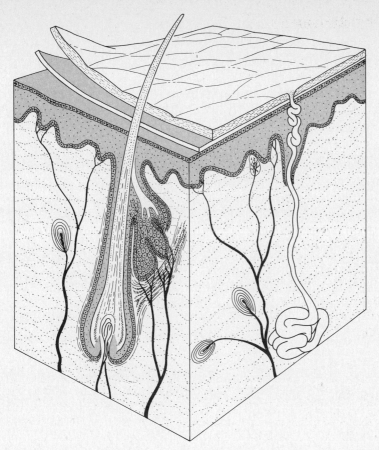

Figure LG 5-1 Structure of the skin. Label as directed in Learning Activity 4 .

a. Most numerous cell type, this cell produces keratin which helps to water-proof skin: _____ .

b. This type of cell produces the pigments which give skin its color:

_____ .

c. Two cell types that function in immunity: _____

and _____ .

6. Match the names of epidermal layers *(strata)* at right with the descriptions below.

7

_____ a. Deepest layer of epithelium, con-sisting of a single layer of cuboidal to columnar cells

_____ b. Eight to ten rows of polyhedral cells which may appear spiny mi-croscopically; together with stra-tum basale, forms stratum ger-minativum

_____ c. Third layer of epidermis, consist-ing of three to five rows of flat-tened, degenerating cells

B. Basale
C. Corneum
G. Granulosum
L. Lucidum
S. Spinosum

_____ d. Flat, dead cells clear due to presence of eleiden; normally found only in thick skin of palms and soles

_____ e. Most superficial layer of skin, consisting of 25 to 30 rows of flat, dead cells

7. Label each of the five epithelial layers in Figure LG 5-1. Check your labels with those on Figure 5-1a (page 102) in your text.

8. The superficial layers of epidermis are modified to prevent excess fluid loss. Write the names of compounds formed in the three layers shown below which gradually lead to such waterproofing.

8

a. _____ → b. _____ →

    (stratum granulosum)   (stratum lucidum)

c. _____

            (stratum corneum)

9

9. Describe the dermis in this exercise.

a. The outer one-fifth of the dermis is known as the *(papillary? reticular?)* region. It consists of *(loose? dense?)* connective tissue. Present in finger-

like projections known as dermal _____ are Meissner's corpuscles, sense receptors sensitive to *(pressure? touch?).*

b. The remainder of the dermis is known as the _____

_____ region, composed of *(loose? dense?)* connective tissue. Skin is strengthened by *(elastic? collagenous?)* fibers in the reticular layer. Skin is extensible and elastic due to _____ fibers.

c. Name three areas of the body in which skin is especially thick.

d. Thick skin is due primarily to the *(papillary? reticular?)* layer of the dermis.

10. Write a sentence defining each of these terms.

a. Striae

b. Pacinian corpuscles

c. Lines of cleavage

11. Explain what accounts for skin color by doing this exercise. ☐10☐
    a. Dark skin is due primarily to *(a larger number of melanocytes? greater melanin production per melanocyte?)*. Melanocytes are in greatest abundance in strata _____ and _____.
    b. Melanin is derived from the amino acid _____. The enzyme which converts tyrosine to melanin is _____. This enzyme is activated by _____ light. The hormone _____ plays a role in color production in skin of mammals.
    c. Melanin is produced by melanocytes which are located in the *(dermis? epidermis?)*. How do other epidermal cells pick up melanin?

    d. The yellowish color of skin of persons of Asian origin is due to the combination of the pigments _____ and _____.
    e. What accounts for the pink color of Caucasian skin, especially during blushing and as a cooling mechanism during exercise?

12. Explain how epidermal ridges are related to fingerprints.

13. *Clinical application:* explain how the following procedures may help to improve appearance of skin.
    a. Collagen implant

88   b. Chemical exfoliation (skin peel)

(pages 104–105)   **B. Skin wound healing**

11   1. Describe the process of epidermal wound healing in this exercise.
      a. State two examples of an epidermal wound.

      b. Usually the deepest part of the wound is in the *(central? peripheral?)* region.
      c. In the process of repair, epidermal cells of the stratum *(corneum? basale?)* break contact from the basement membrane. These are cells at the *(center? periphery?)* of the wound.
      d. These basal cells migrate toward the center of the wound, stopping when they meet other similar advancing cells. This cessation of migration is an

      example of the phenomenon known as _____. Cancer cells *(do? do not?)* exhibit this characteristic.
      e. Both the migrated cells and the remaining epithelial cells at the periphery

      undergo _____ to fill in the epithelium up to a normal (or close to normal) level.
   2. The process of deep wound healing involves four phases. List the three or four major events that occur in each phase.
      a. Inflammation

c. Proliferative

d. Maturation

3. *For extra review* of deep wound healing, match the phases at right with 12 descriptions below.

_____ a. Blood clot temporarily unites edges of wound; blood vessels dilate so neutrophils enter to clean up area.

_____ b. Clot forms a scab; epithelial cells migrate into scab; fibroblasts also migrate to start scar tissue; pink granulation tissue contains delicate new blood vessels.

_____ c. Epithelium and blood vessels grow; fibroblasts lay down many fibers.

_____ d. Scab sloughs off; epidermis grows to normal thickness; collagenous fibers give added strength to healing tissue; blood vessels are more normal.

I. Inflammation
Mat. Maturation
Mig. Migration
P. Proliferation

C. Epidermal derivatives

1. What is the main function of hair? _____ Give examples of how hair protects scalp, eyes, nose, and ears.

2. Explain how hair develops. Include the terms *lanugo, vellus (fleece),* and *terminal hairs* in your description.

3. Define *hirsutism* and state a cause for this condition.

4. Which parts of the body lack hairs?

5. On Figure LG 5-1 label the following parts of a hair and its associated structures: *hair shaft, root, bulb, internal root sheath, external root sheath,* and *connective tissue papilla.*

13 6. Complete this exercise about the structure of a hair and its follicle.
   a. Arrange the parts of a hair from superficial to deep:
      A. Shaft   B. Bulb   C. Root

   b. Arrange the layers of a hair from outermost to innermost:
      A. Medulla   B. Cuticle   C. Cortex

c. A hair is composed of *(cells? no cells, but only secretions of cells?).* What accounts for the fact that hairs are waterproofed?

d. Surrounding the root of a hair is the hair _____.

The follicle consists of two parts. The *(external? internal?)* root sheath is an extension of deeper layers of the *(epidermis? dermis?).* The internal

root sheath consists of cells derived from the _____

_____.

e. The _____ is the part of a hair follicle where cells undergo mitosis permitting growth of a new hair. What is the function of the papilla of the hair?

f. Which treatment destroys the bulb of a hair so that the hair cannot regrow? *(Use of depilatory? Electrolysis?)*

7. Describe the relationship between the terms in each pair of terms.
   a. Arrector pili muscles/"goosebumps"

   b. Sebaceous glands/"blackheads"

8. In which areas of the body are sebaceous glands most commonly found?

List three functions of these glands.

9. Contrast *apocrine sweat glands* and *eccrine sweat glands* according to distribution in the body and location of the gland and duct.

10. Describe the composition of perspiration (sweat) and state its functions.

14 11. Check your understanding of skin glands by stating whether the following descriptions refer to *sebaceous, sudoriferous,* or *ceruminous* glands.

a. Sweat glands: _____

b. Simple branched acinar glands leading directly to hair follicle; secrete

sebum which keeps hair and skin from drying out: _____.

c. Line the outer ear canal; secrete ear wax: _____.

12. Look at one of your own nails and Figure 5-6 (page 109). Identify these parts of your nail: *free edge, nail body, lunula, eponychium (cuticle), and hyponychium.*

13. Why does the nail body appear pink, yet the lunula and free edge appear white?

14. How do nails grow?

(page 109) **D. Homeostasis: temperature control**

1. Humans are homeotherms. What is the meaning of the term *homeotherm?*

2. Describe how human body temperature is maintained close to 37° C (98.6° F) with the help of sweat glands. Explain why this temperature control mechanism can be called a *negative feedback system.*

E. Effects of aging

(pages 109–110)

1. Complete the table relating observable changes in aging of the integument [15] to their causes.

Changes	Causes
a. Wrinkles; skin springs back less when gently pinched.	
b.	Macrophages become less efficient.
c.	Loss of subcutaneous fat
d. Dry, easily broken skin	
e.	Decrease in number and size of melanocytes

2. How is aging of skin related to sunlight?

F. Developmental anatomy

    1. Name the three primary germ layers.

16 2. Describe the development of the integument by completing this learning activity.

    a. The epidermis is derived from the *(ecto-? meso-? endo-?)* derm. All of its layers are formed by the *(second? fourth?)* month.

    b. The dermis arises from the _____-derm.

    c. Hair, nails, and glands all develop from _____-derm. However, the connective tissue and blood vessels of glands develop from

    _____-derm.

    d. At what point do nails begin to develop? _____ month. The nails reach the end of the digits during the *(sixth? ninth?)* month.

    e. When does lanugo form? _____ month. Lanugo *(is? is not?)* usually present on a full-term newborn.

    f. Which glands develop from the sides of hair follicles?

G. Disorders, medical terminology

    1. Write a description of these disorders. Include causes, parts of the body most often affected, and groups of people (by age or other factors) most often afflicted.

    a. Acne

    b. Decubitus ulcers

2. Match the name of the disorder with the description given. ☐17

_____ a. Bedsores
_____ b. An autoimmune disease with many symptoms, including "butterfly rash"
_____ c. Staphylococcal or streptococcal infection which may become epidemic in nurseries
_____ d. Inflammation of sebaceous glands especially in chin area, occurs under hormonal influence
_____ e. One kind of skin cancer
_____ f. Chronic disease of 6 to 8 million people in the United States, characterized by reddish plaques or papules, most severe among ages 10 to 50
_____ g. An inflammation that spreads through connective tissue, e.g., subcutaneous tissue. Typical signs of inflammation occur (such as pain, warmth, erythema). Most often caused by _Streptococcus pyogenes._

A. Acne
C. Cellulitis
D. Decubitus ulcers
I. Impetigo
L. Lupus (SLE)
M. Melanoma
P. Psoriasis

3. What reasons would you give if you were advising a person against constant overexposure to sun?

4. What is sunstroke?

5. Contrast systemic effects with local effects of burns.

[18] 6. Identify characteristics of the three different classes of burns by completing this exercise.
   a. In a first-degree burn, only the superficial layers of the *(dermis? epidermis?)* are involved. The tissue appears _____ in color. Blisters *(do? do not?)* form. Give one example of a first-degree burn.

   b. Which parts of the skin are injured in a second-degree burn?

   Blisters usually *(do? do not?)* form. Epidermal derivatives, such as hair follicles and glands, *(are? are not?)* injured. Healing usually occurs in about three to four *(days? weeks?)*. For most second-degree burns, grafting *(is? is not?)* required.
   c. Third-degree burns are called *(partial? full?)*-thickness burns. Such skin appears *(red and blistered? white, brown, or black and dry?)*. Such burned areas are usually *(painful? not painful?)* since nerve endings are destroyed. Grafting *(is? is not?)* required, and scarring *(does? does not?)* result from third-degree burns.
7. Explain how sonography can assist in determining the depth of a burn.

[19] 8. *Clinical challenge.* Answer these questions about the Lund–Browder method.
   a. What does the Lund–Browder method estimate? *(Depth of burn? Amount of body surface area burned?)* This method is based upon differences in body *(size? proportions?)* of age groups.
   b. If an adult and a one-year-old each experience burns over the entire

anterior surface of both legs, who is more burned? *(Adult? One-year-old?)* If both are burned over the entire anterior surface of the head, who is more burned? *(Adult? One-year-old?)*

c. The chest, buttocks, and arms of an adult account for *(a greater? a smaller? about the same?)* percentage of the total body surface compared with those areas of an infant.

9. Briefly describe treatment procedures for severely burned patients.

10. Contrast the following pairs of terms.
   a. Subcutaneous/intradermal

   b. Hypodermic/topical

   c. Eczema/pruritis

   d. Papule/nodule

20

11. Match the terms at right with the correct definition.

_____ a. Excessive redness of skin due to enlarged capillaries in skin

_____ b. Small, round elevation of skin containing pus

_____ c. Common, contagious, noncancerous growth of epithelium due to virus

_____ d. Boil due to infection of hair follicle

_____ e. Area of hardened, thickened skin, as in palms and soles, due to pressure or friction

_____ f. Mole

C. Callus
E. Erythema
F. Furuncle
N. Nevus
P. Pustule
W. Wart

Answers to Numbered Questions in Learning Activities

1 Skin, its derivatives (hair, nails, and glands), and sense receptors.

2 3,000.

3 All are protective either directly or indirectly. For example, temperature control prevents the destruction of cell enzymes by excessive heat; perception of cold or heat against the skin protects against injury; excretion permits removal of chemicals which would be toxic to the body; vitamin D synthesis assists in development and maintenance of strong bones which protect vital organs; immune factors protect the underlying tissues against invading organisms.

4 (a) Epidermis, epithelium. (b) Dermis, connective tissue, thicker.

5 Under the skin, superficial fascia or hypodermis, areolar and adipose, anchors skin to underlying tissues and organs.

6 (a) Keratinocyte. (b) Melanocyte. (c) Langerhan's cell and Granstein cell.

7 (a) B. (b) S. (c) G. (d) L. (e) C.

8 (a) Keratohyalin. (b) Eleiden. (c) Keratin.

9 (a) Papillary, loose, papillae, touch. (b) Reticular, dense, collagenous, elastic. (c) Palms, soles, and dorsal surface of the body. (d) Reticular.

10 (a) Greater melanin production per melanocyte, basale, spinosum. (b) Tyrosine, tyrosinase, ultraviolet (UV), melanocyte-stimulating hormone (MSH). (c) Epidermis, by phagocytosis of melanin from long extensions of melanocytes located between other epidermal cells. (d) Carotene and melanin. (e) Vasodilatation of blood vessels.

11 (a) Skinned knee, first- or second-degree burn. (b) Central. (c) Basale, periphery. (d) Contact inhibition, do not. (e) Mitosis.

12 (a) I. (b) Mig. (c) P. (d) Mat.

13 (a) A C B. (b) B C A. (c) Cells, keratin (especially in cuticle). (d) Follicle, external, epidermis, matrix. (e) Matrix (of bulb of hair follicle), contains blood vessels that nourish the hair. (f) Electrolysis.

14 (a) Sudoriferous. (b) Sebaceous. (c) Ceruminous.

15 (a) -, Elastic fibers thicken and fray; fibroblasts (which produce elastic fibers) decrease in number. (b) Skin is more susceptible to pathological conditions, -. (c) Person looks more hollowed, has less "padding," so more subject to decubiti, -. (d) -, Atrophy of sebaceous glands. (e) Gray hair, -.

16 (a) Ecto-, fourth. (b) Meso. (c) Ecto, meso. (d) Third, ninth. (e) Fifth or sixth, is not. (f) Sebaceous.

17 (a) D. (b) L. (c) I. (d) A. (e) M. (f) P. (g) G.

18 (a) Epidermis, redder, do not, typical sunburn. (b) All of epidermis and upper regions of dermis, do, are not, weeks, is not. (c) Full; white, brown or black and dry; not painful; is; does.

19 (a) Amount of surface burned, proportions. (b) Adult, One-year-old. (c) About the same.

20 (a) E. (b) P. (c) W. (d) F. (e) C. (f) N.

MASTERY TEST: Chapter 5

Questions 1–5: Choose the one best answer to each question.

_____ 1. "Goosebumps" occur as a result of:

    A. Contraction of arrector pili muscles
    B. Secretion of sebum
    C. Contraction of elastic fibers in the bulb of the hair follicle
    D. Contraction of papillae

_____ 2. Select the one FALSE statement about the stratum basale.

    A. It is the one layer of cells that can undergo cell division.
    B. It consists of a single layer of squamous epithelial cells.
    C. It is part of the stratum germinativum.
    D. It is the deepest layer of the epidermis.

_____ 3. Select the one FALSE statement.

    A. Epidermis is composed of epithelium.
    B. Dermis is composed of connective tissue.
    C. Pressure-sensitive Pacinian corpuscles are normally more superficial in location than Meissner's touch receptors.
    D. The amino acid tyrosine is necessary for production of the skin pigment melanin.

_____ 4. When older hairs are shed, the cells of the _____ produce new hairs by cell division.

    A. Arrector pili   B. Connective tissue papilla
    C. Internal sheath   D. External sheath
    E. Matrix in the base of the bulb

_____ 5. Which is derived from ectoderm?

    A. Epidermis   B. Dermis
    C. Blood vessels and connective tissue associated with glands

100

Questions 6–10: Arrange the answers in correct sequence.

____ ____ ____ 6. From most serious to least serious type of burn:
  A. First-degree
  B. Second-degree
  C. Third-degree

____ ____ ____ 7. Deep wound healing involves four phases. List in order the phases following the inflammatory phase.
  A. Migration
  B. Maturation
  C. Proliferation

____ ____ ____ 8. From outside of hair to inside of hair:
  A. Medulla
  B. Cortex
  C. Cuticle

____ ____ ____ 9. From most superficial to deepest:
  A. Dermis
  B. Epidermis
  C. Superficial fascia

____ ____ ____ 10. From most superficial to deepest:
  A. Stratum lucidum
  B. Stratum corneum
  C. Stratum germinativum

Questions 11–20: Circle T (true) or F (false). If the statement is false, change the underlined word or phrase so that the statement is correct.

T F 11. Hairs are <u>noncellular structures composed entirely of nonliving substances secreted by follicle cells.</u>

T F 12. The color of skin is due to a pigment named <u>keratin.</u>

T F 13. The <u>outermost layers of epidermis</u> are composed of dead cells.

T F 14. Eccrine sweat glands are <u>more</u> numerous than apocrine sweat glands, and are especially dense on <u>palms and soles.</u>

T F 15. The dermis consists of two regions; <u>the papillary region is most superficial, and the reticular region is deeper.</u>

T F 16. Temperature regulation is a <u>positive</u> feedback system.

T F 17. The <u>internal root sheath</u> is a downward continuation of the epidermis.

T F 18. <u>Both epidermis and dermis</u> contain blood vessels (are vascular).

T F 19. Hair, glands, and nails are all derived from the <u>dermis.</u>

T F 20. Pacinian corpuscles are <u>pressure</u>-sensitive nerve endings most abundant in the <u>subcutaneous tissue,</u> rather than in the <u>epidermis.</u>

Questions 21–25: fill-ins. Complete each sentence with the word or phrase that best fits.

_____ 21. Skin contains a chemical which, under the influence of ultraviolet radiation, leads to formation of vitamin ____.

_____ 22. The cells which are sloughed off as skin cells and undergo keratinization are those of the stratum ____.

_____ 23. Fingerprints are the result of a series of grooves called ____.

_____ 24. The oily glandular secretion which keeps skin and hairs from drying is called ____.

_____ 25. When the temperature of the body increases, nerve messages from brain to skin will decrease body temperature by ____.

UNIT II

Principles of Support and Movement

Skeletal Tissue

In Unit II you will consider the parts of the body that provide support and the ability to move—skeleton, joints, and muscles. In Chapter 6 you will take a brief look at the components and functions of the skeletal system (Objective 1). You will examine the gross and microscopic structure of bone tissue (2–3) and the manner in which bones form, continue to grow and age (4–9). You will study some common disorders and medical terminology related to the skeletal system, with particular emphasis on fractures (10–14).

Topics Summary

A. Functions
B. Histology
C. Ossification: bone formation
D. Bone growth and homeostasis
E. Aging and developmental anatomy of the skeletal system
F. Disorders, medical terminology

Objectives

1. Discuss the components and functions of the skeletal system.
2. List and describe the gross features of a long bone.
3. Describe the histological features of compact and spongy bone tissue.
4. Contrast the steps involved in intramembranous and endochondral ossification.
5. Identify the zones and growth pattern of the epiphyseal plate.
6. Describe the processes of bone construction and destruction involved in the homeostasis of bone remodeling.
7. Describe the conditions necessary for normal bone growth and replacement.
8. Explain the effects of aging on the skeletal system.
9. Describe the development of the skeletal system.
10. Define rickets and osteomalacia as vitamin deficiency disorders.
11. Contrast the causes and clinical symptoms associated with osteoporosis, Paget's disease, and osteomyelitis.
12. Define a fracture and describe several common kinds of fractures.
13. List the sequence of events involved in fracture repair.
14. Define medical terminology associated with the skeletal system.

106   **Learning Activities**

(page 120) **A.** Functions

    1. List and discuss briefly five functions of the skeletal system.

☐1 2. What other systems of the body depend on a healthy skeletal system? Explain why in each case.

    3. Name the branch of medical specialty concerned with the bones and joints.

(pages 120–123) **B.** Histology

☐2 1. Describe the components of bone by doing this exercise.

    a. Typical of all connective tissues, bone consists mainly of *(cells? intercellular material?).* The bone cells that are present are called _____

    _____-cytes.

    b. The intercellular substance of bone is unique among connective tissues. Collagenous fibers form about *(33? 67?)* of the weight of bone, whereas mineral salts account for about _____ percent of the weight of bone.

    c. The two main salts present in bone are _____ and

    _____. Together these salts are referred to as

    _____.

    2. On Figure LG 6-1 label the following structures: *diaphysis, epiphysis, articular cartilage, periosteum, medullary (marrow) cavity, endosteum,* and areas of *compact bone* and *spongy bone.*

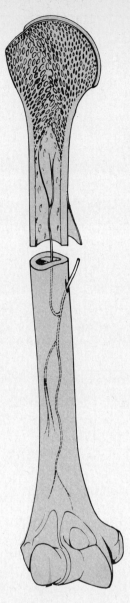

Figure LG 6-1 Diagram of a long bone that has been partially sectioned lengthwise. Label as directed.

3. Match the names of parts of a long bone listed at right with the description below.

3

_____ a. Thin layer of hyaline cartilage at end of long bone

_____ b. Region of mature bone where diaphysis joins epiphysis

_____ c. Outer layer of covering over bone into which ligaments and tendons attach

_____ d. Inner layer of covering over bone; osteoblasts here permit increase in diameter of bone

_____ e. Layer of osteoblasts and osteoclasts lining the marrow cavity

A. Articular cartilage
E. Endosteum
F. Fibrous periosteum
M. Metaphysis
O. Osteogenic periosteum

4. Describe some functional advantages provided by the following aspects of long bones.

   a. Long bones are hollow cylinders, not solid bone.

   b. Much of bone is spongy, porous bone, but a layer of compact bone is present on the outside of bones.

   c. Ends (epiphyses) of bones are bulbous, while shafts (diaphyses) are much narrower in diameter.

5. Contrast compact bone with spongy bone in terms of structure and location.

6. Refer to Figure LG 6-2 and complete this exercise about bone structure.

   a. Compact bone is arranged in concentric circle patterns known as _____ systems. Each individual concentric layer of bone is known as a _____, labeled with letter _____ in the figure.

   b. The Haversian pattern of compact bone permits blood vessels and nerves to supply bone cells trapped in hard bone tissue. Blood vessels and nerves penetrate bone from the periosteum, labeled with letter _____ in the figure. These structures then pass through horizontal canals, labeled _____, and known as _____ canals. These vessels and nerves finally pass into microscopic channels, labeled _____, in the center of each osteon (Haversian system). Color blood vessels red and blue in the figure.

...

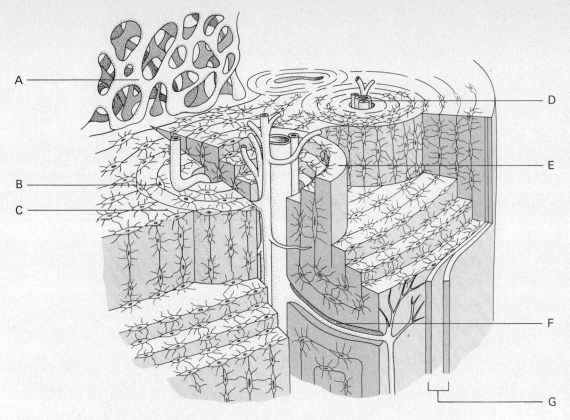

A

B

C

D

E

F

G

Figure LG 6-2 Osteons (haversian systems) of compact bone. Identify lettered structures and color as directed in Learning Activity ⬚5⬚.

   c. Mature bone cells, known as _____, are located relatively far apart in bone tissue. These are present in "little lakes," or

   _____, labeled _____ in the figure. Color ten lacunae green.

   d. Minute canals, known as _____ and labeled with

   letter _____, permit communication between lacunae and Haversian canals. These tiny channels permit diffusion of fluids, nutrients, wastes, and gases between blood vessels and bone cells trapped in lacunae.

7. What are interstitial lamellae?

C. Ossification: bone formation

1. What is mesenchyme? What does it have to do with the skeleton?

2. The two main kinds of tissue that compose the "skeleton" of a developing embryo or fetus are _____ and _____

   6

   _____.

3. Which bones of the body form by the process of intramembranous ossification?

4. Describe the process of intramembranous ossification. Include these terms in your paragraph: *mesenchymal cells, osteoblasts, center of ossification, collagenous fibers, calcium salts, trabeculae, spongy bone,* and *osteocytes periosteum.*

5. Write a sentence about each of the steps of endochondral ossification using these key terms in order:
   a. Perichondrium

   b. Collar of bone

   c. Primary ossification center

d. Calcification

e. Death of cartilage cells

f. Spaces, marrow

g. Secondary ossification centers

h. Articular cartilage

i. Epiphyseal plate

6. *For extra review.* Number in correct sequence the events of endochondral $\boxed{7}$ bone formation.

_____ a. Secondary ossification centers develop in epiphyses, forming spongy bone there.

_____ b. Meanwhile cartilage cells in the center of the diaphysis enlarge and burst.

_____ c. A cartilage model of future bone is laid down.

_____ d. Cartilage cells die since they are deprived of nutrients; in this way large spaces are formed.

_____ e. Blood vessels penetrate the perichondrium, stimulating periochondrial cells to form osteoblasts. A collar of bone forms and gradually thickens around the diaphysis.

_____ f. Release of alkaline chemicals from these cells cause calcification of cartilage.

_____ g. Blood vessels grow into spaces, forming the marrow cavity.

D. Bone growth and homeostasis

8 | 1. The epiphyseal plate consists of four zones. Arrange the letters representing these zones in order starting from the zone closest to the diaphysis.

C. Zone of calcified matrix
H. Zone of hypertrophic cartilage
P. Zone of proliferating cartilage
R. Zone of reserve cartilage

Diaphysis → _____ → _____ → _____ → _____

9 | 2. Now match the names of these zones with the correct description

_____ a. Cells do not function in bone growth, but anchor epiphyseal plate to bone of epiphysis.

_____ b. New cartilage cells are made by mitosis.

_____ c. Cartilage cells mature, surround themselves with calcium salts, then die due to these salts.

_____ d. Dead cartilage cells and broken up matrix are replaced by bone-forming cells.

C. Zone of calcified matrix
H. Zone of hypertrophic cartilage
P. Zone of proliferating cartilage
R. Zone of reserve cartilage

10 | 3. Complete this summary statement about bone growth at the epiphyseal plate.

Cartilage cells multiply on the *(epiphysis? diaphysis?)* side of the epiphyseal plate, providing temporary new tissue. But cartilage cells then die and are replaced by bone cells on the *(epiphysis? diaphysis?)* side of the epiphyseal plate.

4. Explain how bones grow in diameter. Contrast the roles of *osteoblasts* and *osteoclasts*.

5. Contrast the *epiphyseal plate* with the *epiphyseal line*.

6. What is osteogenic sarcoma? Which age group and which body parts is it most likely to affect?

7. Defend or dispute this statement: "Once a bone, such as your thighbone, is formed, the bone tissue is never replaced unless the bone is broken."

Explain the role of osteoclasts in the remodeling process.

Explain the roles of lysosomes and acids in osteoclastic activity. $\boxed{11}$

8. Describe the role of the skeleton in homeostasis of calcium distribution throughout the body.

9. Name factors necessary for normal bone growth:
   a. Two minerals

   b. Three vitamins

114

10. Match the hormones listed at right with their functions in bone growth and remodeling.

| 12 |

_____ a. Produced by the parathyroid gland, it increases osteoclast activity, causing bone destruction.

_____ b. Produced by the thyroid gland, it inhibits osteoclast activity.

_____ c. A pituitary hormone, it enhances bone growth in general; deficiency leads to dwarfism.

_____ d. These hormones stimulate osteoblasts to form new bone and so are responsible for growth spurt at puberty; they can also cause premature closure of epiphyseal plate (short stature) if puberty occurs early.

CT. Calcitonin
GH. Growth hormone
PTH. Parathyroid hormone
S. Sex hormones (testosterone, estrogens)

(pages 126–128)   **E. Aging and developmental anatomy of the skeletal system**

| 13 |

1. Describe the two major changes in the skeleton which occur during aging by completing this exercise.

   a. The amount of calcium in bones _____-creases with age. This change occurs at a younger age in *(men? women?)*. As a result, bones of the elderly are likely to be *(stronger? weaker?)* than in younger persons.
   b. Another component of bones which decreases with age is

   _____. What is the significance of this change?

| 14 |

2. Complete this learning activity about development of the skeleton.
   a. Bones and cartilage are formed from *(ecto-, meso-, endo-)* derm which later differentiates into the embryonic connective tissue called

   _____. Some of these cells become chondroblasts

   which eventually form *(bone? cartilage?)*, while _____

   _____-blasts develop into bone.
   b. Limb buds appear during the *(fifth? seventh? tenth?)* week of development. At this point the skeleton in these buds consists of *(bone? cartilage?)*. Bone begins to form during week _____.
   c. By week *(six? seven? eight?)* the upper extremity is evident, with defined shoulder, arm, elbow, forearm, wrist, and hand. The lower extremity is also developing at a slightly slower pace.
   d. The notochord develops in the region of the *(head? vertebral column? pelvis?)*. Most of the notochord eventually *(forms vertebrae? disappears?)*.

   Parts of it persist in the _____.

F. Disorders, medical terminology

1. Osteoporosis is associated with a loss of the hormone _____ 15

   _____, and so occurs especially in older *(men? women?)*. List three therapies which can help to prevent osteoporosis.

2. Suggest causes of the following symptoms that occur with each disorder listed below.
   a. Rickets: bowed legs in children

   b. Osteomalacia: bowed legs in adults

   c. Osteoporosis: loss of height, hunched back, and fractures

3. Check your understanding of bone disorders by matching the terms with related descriptions.   16

   a. Benign bone tumor: _____   Osteoma
                                            Osteomyelitis
   b. Malignant bone tumor: _____   Osteosarcoma
                                            Paget's disease
   c. Bone infection, for example, caused by   Rickets
      Staphylococcus: _____

   d. Bones thicken and soften so that extremities may bow; usually affects persons over 50 years: _____

   e. Caused by vitamin D deficiency which prevents normal calcium absorption; soft bones: _____

4. Contrast the terms related to types of fractures in each pair.
   a. Simple/compound

116

b. Partial/complete

c. Pott's/Colles'

5. What is the major difference between these two methods of setting a fracture: *closed reduction* and *open reduction?*

6. Describe the three main steps involved in repair of a fracture.

7. Define PEMFs and briefly describe how this treatment helps bone repair.

Answers to Numbered Questions in Learning Activities

1   Essentially all do. For example, muscles need intact bones for movement to occur; bones are site of blood formation; bones provide protection for viscera of nervous, digestive, urinary, reproductive, cardiovascular, respiratory, and endocrine systems; broken bones can injure integument.

2 (a) Intercellular material, osteo. (b) 33, 67. (c) Calcium phosphate, calcium carbonate, hydroxyapatites.

3 (a) A. (b) M. (c) F. (d) O. (e) E.

4 (a) Skeleton is lighter than it would be if it were solid bone. (b) Though skeleton is light, it is strong. (c) Joints are stable with broad area for muscle attachments.

5 (a) Haversian, lamella, *E.* (b) *G, F,* Volkmann's, *D.* (c) Osteocytes, lacunae, *B.* (d) Canaliculi, *C.*

6 Hyaline cartilage, fibrous membranes.

7 (a) 7. (b) 3. (c) 1. (d) 5. (e) 2. (f) 4. (g) 6.

8 C H P R.

9 (a) R. (b) P. (c) H. (d) C.

10 Epiphysis, diaphysis.

11 Lysosomes release enzymes that may digest the protein collagen while acids may cause minerals of bones to dissolve. Both contribute to bone resorption.

12 (a) PTH. (b) CT. (c) GH. (d) S.

13 (a) De, women, weaker. (b) Protein, bones are more brittle and vulnerable to fracture.

14 (a) Meso-, mesenchyme, cartilage, osteo. (b) Fifth, cartilage, six or seven. (c) Eight. (d) Vertebral column, disappears, intervertebral discs.

15 Estrogen, women; estrogen replacement therapy (ERT), calcium in diet and/or supplements, exercise.

16 (a) Osteoma. (b) Osteosarcoma. (c) Osteomyelitis. (d) Paget's disease. (e) Rickets.

MASTERY TEST: Chapter 6

Questions 1–16: Circle T (true) or F (false). If the statement is false, change the underlined word or phrase so that the statement is correct.

T F 1. Greenstick fractures occur only in underlined adults.

T F 2. Sunlight can help keep bones healthy since sunlight helps the body produce vitamin D.

T F 3. About two-thirds of the weight of bone is due to mineral salts, and one-third is due to bone cells.

T F 4. The epiphyseal plate appears earlier in life than the epiphyseal line.

T F 5. Osteons (haversian systems) are found in compact bone, but not in spongy bone.

T F 6. Another name for the epiphysis is the shaft of the bone.

T F 7. A compound fracture is defined as one in which the bone is broken into many pieces.

T F 8. Haversian canals run longitudinally (lengthwise) through bone, but Volkmann's canals run horizontally across bone.

T F 9. Canaliculi are tiny canals containing blood which nourishes bone cells in lacunae.

T F 10. Compact bone that is of intramembranous origin differs structurally from compact bone developed from cartilage.

T F 11. The metaphysis is the portion of a bone where growth occurs, that is, between diaphysis and epiphysis.

T F 12. In a long bone the primary ossification center is located in the diaphysis, whereas the secondary center of ossification is in the epiphysis.

T F 13. Osteoblasts are bone-destroying cells.

T F 14. Most bones start out in embryonic life as fibrous cartilage.

T F 15. The layer of compact bone is thicker in the diaphysis than in the cpiphysis.

T F 16. Lamellae are small spaces containing bone cells.

Questions 17–20: Arrange the answers in correct sequence.

_____ _____ _____ 17. From most superficial to deepest:

    A. Endosteum
    B. Periosteum
    C. Compact bone

_____ _____ _____ 18. Phases in repair of a fracture, in chronological order:

    A. Fracture hematoma formation
    B. Remodeling
    C. Callus formation

_____ _____ _____ 19. Phases in formation of bone in embryonic life, in chronological order:

    A. Mesenchyme cells
    B. Osteocytes
    C. Osteoblasts

_____ _____ _____ 20. Portions of the epiphyseal plate, from closest to epiphysis to closest to diaphysis:

    A. Zone of reserve cartilage
    B. Zone of proliferating cartilage
    C. Zone of calcified matrix

Questions 21–25: fill-ins. Write the word or phrase that best completes the statement.

_____ 21. _____ is a term which refers to the shaft of the bone.

_____ 22. _____ is a growth of new tissue in a fractured bone, and consists of cells derived from periosteum, endosteum, and marrow.

_____ 23. _____ is the connective tissue covering over cartilage in adults and also over embryonic cartilaginous skeleton.

_____ 24. The majority of bones formed by intramembranous ossification are located in the _____.

_____ 25. Vitamin _____ is a vitamin which is necessary for absorption of calcium from the gastrointestinal tract, and so it is important for bone growth and maintenance.

The Skeletal System: The Axial Skeleton

In this chapter you will see how the construction of a bone—its size, shape, and markings—relates to its functions (Objectives 1–3). You will distinguish the two principal classifications of bones, axial and appendicular, and their components (4). You will then examine bones of the axial skeleton: skull (5–8), vertebral column (9–10), and thorax (11). Finally, you will study some disorders associated with the axial skeleton (12).

Topics Summary

A. Types of bones
B. Surface markings
C. Divisions of the skeletal system
D. Skull
E. Hyoid bone
F. Vertebral column
G. Thorax
H. Disorders

Objectives

1. Define the four principal types of bones.
2. Describe the various markings on the surfaces of bones.
3. Relate the structure of a surface marking to its function.
4. List the components of the axial and appendicular skeleton.
5. Identify the bones of the skull and the major markings associated with each.
6. Identify the principal sutures and fontanels of the skull.
7. Identify the paranasal sinuses of the skull.
8. Identify the principal foramina of the skull.
9. Identify the bones of the vertebral column and their principal markings.
10. List the defining characteristics and curves of each region of the vertebral column.
11. Identify the bones of the thorax and their principal markings.
12. Contrast herniated (slipped) disc curvatures, spina bifida, and fractures of the vertebral column as disorders associated with the skeletal system.

122

Learning Activities

(page 133) **A. Types of bones**

1. Complete the table about the four major and two minor types of bones.

Type of Bone	Structural Features	Examples
a. Long	Slightly curved to absorb stress better	
b.		Wrist, ankle bones
c.	Composed of two thin plates of bone	
d. Irregular		
e. Sutural (or Wormian)		
f.	Small bones in tendons	

(page 133) **B. Surface markings**

1. In general, what is the purpose of surface markings of bones?

2. Contrast the bone markings in each of the following pairs.
   a. Tubercle/tuberosity

   b. Crest/line

   c. Fossa/foramen

   d. Condyle/epicondyle

3. Match the general descriptions of markings with the examples of specific bone markings. Take particular note of italicized terms.

a. Air-filled cavity within a bone, connected to nasal cavity: _____
_____

b. Narrow, cleftlike opening between adjacent parts of bone; passageway for blood vessels and nerves: _____

c. Rounded hole, passageway for blood vessels and nerves: _____
_____

d. Tubelike passageway through bone: _____

e. Large, rounded projection above constricted neck: _____
_____

f. Large, blunt projection on the femur: _____

g. Sharp slender projection: _____
_____

h. Smooth, flat surface: _____
_____

Articular *facet* on vertebra
External auditory *meatus*
Greater trochanter
Head of humerus
Maxillary sinus
Optic foramen
Styloid process
Superior orbital fissure

C. Divisions of the skeletal system

(page 133)

1. Describe the bones in the two principal divisions of the skeletal system by completing this exercise.

2

a. Bones that lie along the axis of the body are included in the *(axial? appendicular?)* skeleton.

b. The axial skeleton includes the following groups of bones. Indicate how many bones are in each category.

_____ Skull (cranium, face)   _____ Vertebrae

_____ Earbones   _____ Sternum

_____ Hyoid   _____ Ribs

c. The total number of bones in the axial skeleton is ____.

d. The appendicular skeleton consists of bones in which parts of the body?

e. Write the number of bones in each category. Note that you are counting bones on one side of the body only.

_____ Left shoulder girdle   _____ Left hipbone

_____ Left upper extremity (arm, forearm wrist, hand)   _____ Left lower extremity (thigh, kneecap, leg, foot)

f. The total number of bones in both (right and left) upper extremities and shoulder girdles is _____.

g. The total number of bones in both (right and left) hipbones and lower extremities is _____.

h. There are _____ bones in the appendicular skeleton.

i. In the entire human body there are _____ bones.

(pages 133–146) **D. Skull**

1. Try to locate the major bones of the skull by using these aids: Figures 7-2, 7-4, and 7-5 (pages 137–142) in your text, a mirror to examine your own facial contours, and a skull specimen (if available). At this time, do not concentrate on specific markings of bones; rather, try to identify bones by name and position relative to other bones of the skull. Be sure to find all skull bones in the following list. Color each bone on Figures LG 7-1 and LG 7-2. Use colors indicated in parentheses. (Code letters for each bone will be used in exercises below.)

List of Skull Bones

E. Ethmoid (blue)	O. Occipital (red)
F. Frontal (red)	Pal. Palatine (orange)
I. Inferior nasal concha (purple)	Par. Parietal (blue)
L. Lacrimal (yellow)	S. Sphenoid (yellow)
Man. Mandible (blue)	T. Temporal (orange)
Max. Maxilla (green)	V. Vomer (red)
N. Nasal (yellow)	Z. Zygomatic (purple)

2. Check your understanding of location and general functions of each skull bone by matching the correct bones with the following descriptions. Use the code letters from the list above.

3

_____ a. This bone forms the lower jaw, including the chin.

_____ b. These are the cheek bones; they also form lateral walls of the orbit of the eye.

_____ c. Tears pass through tiny foramina in these bones; they are the smallest bones in the face.

_____ d. The bridge of the nose is formed by these bones.

_____ e. Organs of hearing (internal part of ears) are located in and protected by these bones.

_____ f. This bone sits directly over the spinal column; it contains the hole through which the spinal cord connects to the brain.

_____ g. The name means "wall." The bones form most of the roof and much of the side walls of the skull.

_____ h. These bones form most of the roof of the mouth (hard palate) and contain the sockets into which upper teeth are set.

_____ i. L-shaped bones form the posterior parts of the hard palate and nose.

_____ j. Commonly called the forehead, it provides protection for the anterior portion of the brain.

_____ k. A light spongy bone, it forms much of the roof and internal structure of the nose.

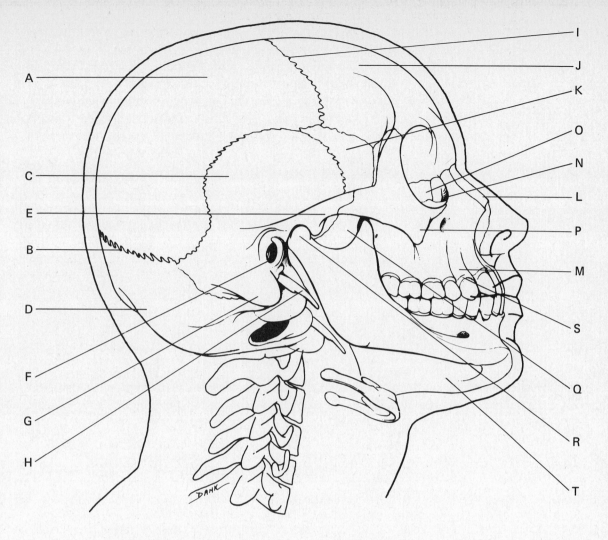

Figure LG 7-1 Right lateral view of the skull. Abbreviations in key: MN, mandible; MX, maxilla; S, sphenoid; T, temporal. Color as directed.

KEY

A.	PARIETAL BONE	K.	Greater wing (S)
B.	Lambdoidal suture	L.	NASAL BONE
C.	Squamous portion (T)	M.	MAXILLA
D.	OCCIPITAL BONE	N.	LACRIMAL BONE
E.	Zygomatic process (T)	O.	ETHMOID BONE
F.	External auditory meatus (T)	P.	ZYGOMATIC (MALAR) BONE
G.	Mastoid process (T)	Q.	Pterygoid process (S)
H.	Styloid process (T)	R.	Condyle or head (MN)
I.	Coronal suture	S.	Alveolar process (MX)
J.	FRONTAL BONE	T.	Body (MN)

_____ l. It serves as a "keystone," since it binds together many of the other bones of the skull. It is shaped like a bat, with the wings forming part of the sides of the skull and the legs at the back of the nose.

_____ m. This bone forms the inferior part of the septum dividing the nose into two nostrils.

_____ n. Two delicate bones form the lower parts of the side walls of the nose.

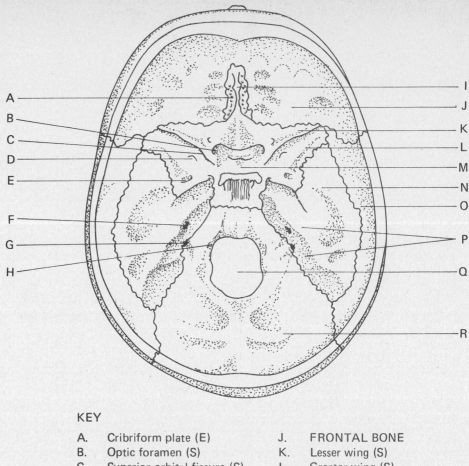

KEY

A.	Cribriform plate (E)	J.	FRONTAL BONE
B.	Optic foramen (S)	K.	Lesser wing (S)
C.	Superior orbital fissure (S)	L.	Greater wing (S)
D.	Foramen rotundum (S)	M.	Sella turcica (S)
E.	Foramen ovale (S)	N.	Squamous portion (T)
F.	Internal auditory meatus (T)	O.	PARIETAL BONE
G.	Jugular foramen (T–O)	P.	Petrous portion (T)
H.	Hypoglossal canal (O)	Q.	Foramen magnum (O)
I.	Crista galli (E)	R.	OCCIPITAL BONE

Figure LG 7-2 Floor of cranium. Abbreviations in key: E, ethmoid; O, occipital; S, sphenoid; T, temporal. Color as indicated.

3. What is the main function of the *cranium?*

Which bones are considered parts of the cranium, rather than parts of the face? Write code letters from the list of skull bones on LG page 124.

4

_____ _____ _____ _____ _____ _____

4. Which six of the bones of the skull are unpaired (that is, there is only one
5 bone of the name)?

5. What are sutures?

Label each of the following sutures in Figures LG 7-1 and LG 7-2 on pages LG 125 and 126: *coronal, sagittal, lambdoidal,* and *squamosal.*
6. Dcfinc *fontanel.*

Of what functional advantages are these "soft spots" of the skull during delivery of a baby?

7. Identify the location of the largest fontanel.

$\boxed{6}$

What shape is this fontanel? *(Round? Triangular? Diamond?)* What accounts for this shape?

8. Try to identify all of the labeled structures on Figures LG 7-1 and LG 7-2. Take particular note of markings and what bones they are parts of (indicated in parentheses in the key).
9. Complete the table describing major markings of the skull.

128

Marking	Bone	Function
a. Greater wings	Sphenoid	
b. External auditory meatus		
c.		Site of pituitary gland
d. Petrous portion		
e.		Largest hole in skull; passageway for spinal cord
f.		Passageway into skull for carotid artery
g. Occipital condyles		
h.		Site of only air sinuses that do not drain into nose
i. Pterygoid processes		
j.		Bony sockets for teeth

10. *For extra review.* The 12 pairs of nerves attached to the brain are called cranial nerves (Figure 14-4, page 312 of your text). Holes in the skull permit passage of these nerves to and from the brain. These nerves are numbered according to the order in which they attach to the brain (and leave the cranium) from I (most anterior) to XII (most posterior). To help you visualize their sequence, foramina for cranial nerves are labeled in order on the left side of Figure LG 7-2. Complete the table summarizing these foramina. The first one is done for you.

7

Number and Name of Cranial Nerve	Location of Opening for Nerve
a. I Olfactory	Cribriform plate of ethmoid bone
b. II Optic	
c. III Oculomotor IV Trochlear V Trigeminal (ophthalmic branch) VI Abducens	
d. V Trigeminal (maxillary branch)	
e. V Trigeminal (mandibular branch)	
f. VII Facial VIII Vestibulocochlear	
g. IX Glossopharyngeal X Vagus XI Accessory	
h. XII Hypoglossal	

11. Match the cranial fossa with the description that best fits. 8

_____ a. Formed by the greater wings of the sphenoid, the squamous portions of temporal bones, and small parts of the occipital bones. The temporal lobes of the brain rest here.

_____ b. Formed mostly by the occipital bones, it accommodates cerebellum and brainstem.

_____ c. The highest level of fossae, it is formed by orbital surface of frontal bone, lesser wings of sphenoid, and superior portions of ethmoid. Frontal lobes of the cerebrum are housed here.

A. Anterior
M. Middle
P. Posterior

130

9. 12. List three functions of paranasal sinuses.

10. 13. Name four bones that contain paranasal sinuses. Draw their approximate locations on Figure LG 7-3. Match your drawing to Figure 7-8, page 145 of the text. (Practice identifying their locations the next time you have a cold!)

11. 14. List several functions of the nasal conchae.

Identify and label two conchae on Figure LG 7-3.

12. 15. What structures form the septum of the nose?

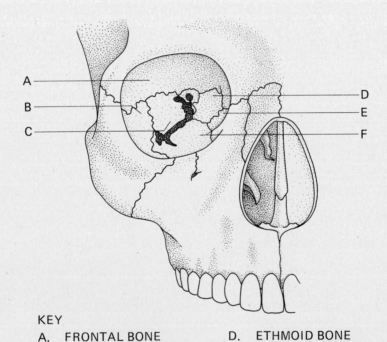

KEY
A. FRONTAL BONE   D. ETHMOID BONE
B. ZYGOMATIC BONE   E. LACRIMAL BONE
C. SPHENOID BONE   F. MAXILLA BONE

Figure LG 7-3 Bones of the right orbit. Color as directed.

Label the two bony portions of the septum shown on Figure LG 7-3.

16. A good way to test your ability to visualize locations of many important skull bones is to try to identify bones that form the orbit of the eye. See if you can determine the correct positions of the six major bones comprising the orbit shown in Figure LG 7-3. Using the color code in the list of skull bones (p. LG 124), color each bone.

E. Hyoid bone   (page 146)

1. In what way is the hyoid bone unique among the bones of the axial skeleton?

2. Describe the location and functions of the hyoid bone.

Now label the hyoid bone on Figure LG 7-1.

F. Vertebral column   (pages 146–152)

1. List several functions of the vertebral column.

2. Identify the five regions of the vertebral column by labeling regions *A–E* on [13] Figure LG 7-4. Color these vertebral regions according to this code:

   Cervical (red)   Sacral (green)
   Coccygeal (blue)   Thoracic (orange)
   Lumbar (yellow)

   Now write on the lines next to letters *A–E* the number of vertebrae located in each of these regions of an adult. Identify the location of each vertebral region on a study partner or on a skeleton.

3. Name structures which provide flexibility to the vertebral column: [14] _____ .

4. Note which regions of the vertebral column on Figure LG 7-4 normally [15] retain an anteriorly concave curvature in the adult: _____ and _____. These are considered *(primary? secondary?)* curvatures. This classification is based upon the fact that these curves *(were present originally during fetal life? are more important?)*.

5. On Figure LG 7-5, label structures *A–K*. [16]

6. Choose the terms that fit the descriptions of parts of vertebrae. (Not all terms will be used.) [17]

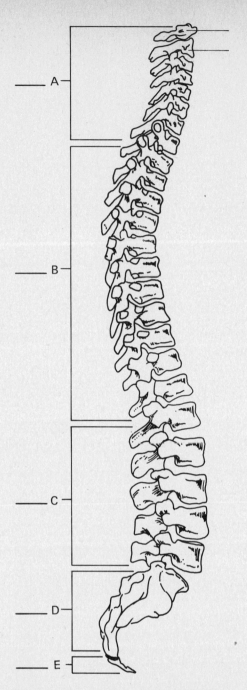

Figure LG 7-4 Lateral view of vertebral column. Label and color as directed in Learning Activity ⑬ .

a. Anterior portion of vertebral arch: _____

b. Drum-shaped portion designed to bear weight: _____

c. Opening through which spinal cord passes: _____

d. Forms joint with vertebra below: _____

Body
Inferior articular process
Intervertebral foramen
Lamina
Pedicle
Spinous process
Superior articular process
Transverse process
Vertebral foramen

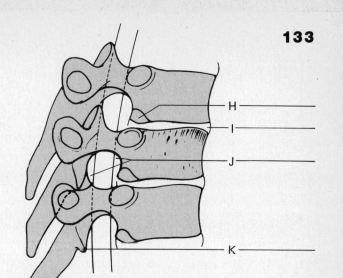

Figure LG 7-5 A typical thoracic vertebra. (a) Superior view. (b) Right lateral view. Label as directed in Learning Activity 16 .

e. Space between two vertebrae where spinal nerves exit: _____

f. Most posterior part of vertebra: _____

_____

7. Identify distinctive features of vertebrae in each region. 18

_____ a. Small body, foramina for vertebral blood vessels in transverse
processes
_____ b. Atlas and axis
_____ c. Five in the adult
_____ d. Only vertebrae that articulate
with ribs
_____ e. Massive body, blunt spinous process and articular processes directed medially or laterally
_____ f. Long spinous processes that point
inferiorly
_____ g. Most inferior part of vertebral column
_____ h. Articulate with the two hipbones

C. Cervical
Co. Coccyx
L. Lumbar
S. Sacral
T. Thoracic

8. Label the atlas and axis on Figure LG 7-4. Contrast them according to structure and function.

134

9. *Clinical challenge.* Identify the following sacral markings on a skeleton or on Figure 7-16, page 153 of your text. State the clinical significance of these markings.

   a. Sacral hiatus

   b. Auricular surface

19 c. Sacral promontory

(pages 152–155) **G. Thorax**

1. Name the structures that compose the thorax. (Why is it called a cage, the "thoracic cage"?)

2. Color the three parts of the sternum on Figure LG 7-6 as directed below. Note which structures form joints (articulations) with each part of the sternum.
   a. Manubrium/green

c. Xiphoid process/purple

3. Complete this exercise about ribs. Refer again to Figure LG 7-6.   20

   a. There are a total of _____ ribs (_____ pairs) in the human skeleton.
   b. Ribs slant in such a way that the anterior portion of the rib is *(superior? inferior?)* to the posterior end of the rib.

   c. Posteriorly, all ribs articulate with _____. Ribs 1 to 10 also pass *(anterior? posterior?)* to and articulate with transverse processes of vertebrae. (Check this on a skeleton or Figure 7-17, page 154, of the text.) Of what functional advantage is such an arrangement?

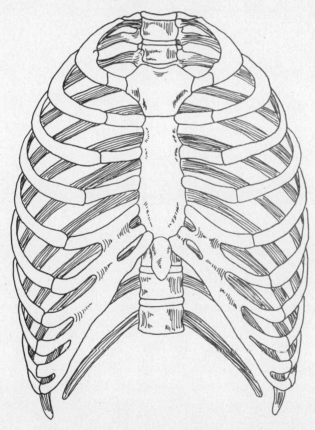

Figure LG 7-6 Anterior view of thorax. Label and color as directed.

d. Anteriorly, ribs numbered _____ to _____ attach to the sternum directly by means of strips of hyaline cartilage, called _____

_____ cartilage. These ribs are called *(true? false?)* ribs. Color them red on Figure LG 7-6, leaving the costal cartilages white.

e. Ribs 8 to 10 are called _____. Color them yellow, again leaving costal cartilages white. Do these ribs attach to the sternum? *(Yes? No?)* If so, in what manner?

f. Ribs _____ and _____ are called "floating ribs." Color them brown. Why are these ribs so named?

g. The head and neck of a rib are located at the *(anterior? posterior?)* end.
h. What function is served by the costal groove?

i. What occupies intercostal spaces?

21 4. At what point are ribs most commonly fractured?

(pages 155–156) **H. Disorders**

22 1. Complete this exercise about slipped discs.
a. The normal intervertebral disc consists of two parts: an outer ring of *(hyaline? elastic? fibrous?)* cartilage called _____

and a soft, elastic, inner portion called the _____.
b. Ligaments normally keep discs in alignment with vertebral bodies. What may happen if these ligaments weaken?

c. Why might pain result from a slipped disc?

d. In what part of the vertebral column are slipped discs most common? What symptoms may result from a slipped disc in this region?

2. Describe the following disorders and tell how they differ from one another.
   a. Scoliosis

   b. Kyphosis

   c. Lordosis

3. Imperfect union of the vertebral arches at the midline is the condition known as _____. Why is it crucial that the vertebral foramen be completely surrounded by bone? What problems may result from incomplete closure?

4 Fracture of the vertebral column occurs more frequently in the thoracolumbar region than in the cervical region. What structural difference between vertebrae in these two regions accounts for this fact? ⬚23

Answers to Numbered Questions in Learning Activities

⬚1 (a) Maxillary *sinus*. (b) Superior orbital *fissure*. (c) Optic *foramen*. (d) External auditory *meatus*. (e) *Head* of humerus. (f) Greater *trochanter*. (g) Styloid *process*. (h) Articular *facet* on vertebra.

2   (a) Axial. (b) 22 skull, 6 earbones (not studied as part of axial skeleton), 1 hyoid, 26 vertebrae, 1 sternum, 24 ribs. (c) 80. (d) Shoulder girdles, upper extremities, hipbones, lower extremities. (e) 2 left shoulder girdle, 30 left upper extremity, 1 left hipbone, 30 left lower extremity. (f) 64. (g) 62. (h) 126. (i) 206.

3   (a) Man. (b) Z. (c) L. (d) N. (e) T. (f) O. (g) Par. (h) Max. (i) Pal. (j) F. (k) E. (l) S. (m) V. (n) I.

4   F, Par, O, T, E, S.

5   Ethmoid, sphenoid, frontal, occipital, vomer, and mandible.

6   It is at the junction of the two parietal bones and the frontal bone. The fontanel is diamond-shaped since the frontal bone of the fetus still consists of two parts which will later fuse.

7   (b) Optic foramen of sphenoid bone. (c) Superior orbital fissure of sphenoid bone. (d) Foramen rotundum of sphenoid bone. (e) Foramen ovale of sphenoid bone. (f) Internal auditory meatus of temporal bone. (g) Jugular foramen between temporal and occipital bones. (h) Hypoglossal canal of occipital bone.

8   (a) M. (b) P. (c) A.

9   Lighten skull bones; serve as resonance chambers for speech and other sounds; warm and humidify air since sinuses are lined with mucous membrane.

10   Frontal, ethmoid, sphenoid, and maxillary.

11   Increase mucosal surface of nasal cavity, so warm, humidify, and increase circulation of air.

12   Bony portions: perpendicular plate of the ethmoid bone (superior portion) and vomer (inferior portion) (both shown on Figure LG 7-3). Also nasal cartilage (anterior portion).

13   7-Cervical *(A)*, 12-thoracic *(B)*, 5-lumbar *(C)*, 1-sacral *(D)*, 1-coccygeal *(E)*.

14   Intervertebral discs (and also ligaments).

15   Thoracic *(B)* and sacral *(D)*; primary; were present originally during fetal life.

16   Body *(A)*, vertebral foramen *(B)*, pedicle *(C)*, superior articular process *(D)*, transverse process *(E)*, lamina *(F)*, spinous process or spine *(G)*, demifacet for articulation with rib *(H)*, intervertebral disc *(I)*, intervertebral foramen *(J)*, inferior articular process *(K)*.

17   (a) Pedicle. (b) Body. (c) Vertebral foramen. (d) Inferior articular process. (e) Intervertebral foramen. (f) Spinous process.

18   (a) C. (b) C. (c) L. (d) T. (e) L. (f) T. (g) Co. (h) S.

19   It is an important obstetrical landmark, used for pelvic measurements.

20 (a) 24, 12. (b) Inferior. (c) Bodies of thoracic vertebrae, anterior; prevents ribs from slipping posteriorward. (d) 1, 7, costal, true. (e) False, yes; they attach to sternum indirectly via seventh costal cartilage. (f) 11, 12; they have no anterior attachment to sternum. (g) Posterior. (h) Provides protective channel for intercostal nerve, artery, and vein. (i) Intercostal muscles.

21 Just anterior to the costal angle, especially among the middle ribs.

22 (a) Fibrous, annulus fibrosis, nucleus pulposus. (b) Nucleus pulposus may protrude (herniate) through annulus fibrosus, a "slipped disc." (c) Disc tissue may press on spinal nerves. (d) L4 to L5 or L5 to sacrum; pain in posterior of leg(s) due to pressure on sciatic nerve which exits spinal cord at L4 to S3.

23 Articular surfaces are more horizontal in cervical region, allowing forward dislocation without fracture.

MASTERY TEST: Chapter 7

Questions 1–10: Choose the one best answer to each question.

_____ 1. All of these bones contain paranasal sinuses EXCEPT:

    A. Frontal   B. Maxilla   C. Nasal   D. Sphenoid
    E. Ethmoid

_____ 2. Choose the one FALSE statement.

    A. There are seven vertebrae in the cervical region.
    B. The cervical region normally exhibits a curve that is slightly concave anteriorly.
    C. The lumbar vertebrae are superior to the sacrum.
    D. Intervertebral discs are located between bodies of vertebrae.

_____ 3. The hard palate is composed of ____ bones.

    A. Two maxilla and two mandible
    B. Two maxilla and two palatine
    C. Two maxilla
    D. Two palatine
    E. Vomer, ethmoid, and two temporal

_____ 4. The lateral wall of the orbit is formed mostly by which two bones?

    A. Zygomatic and maxilla   B. Zygomatic and sphenoid
    C. Sphenoid and ethmoid   D. Lacrimal and ethmoid
    E. Zygomatic and ethmoid

_____ 5. Choose the one TRUE statement.

    A. All of the ribs articulate posteriorly with demifacets on transverse processes of vertebrae.
    B. Ribs 8 to 10 are called true ribs.
    C. There are 23 ribs in the male skeleton and 24 in the female skeleton.
    D. Rib 7 is larger than rib 3.
    E. Cartilage discs between vertebrae are called costal cartilages.

_____ 6. Which is the largest fontanel?

    A. Frontal   B. Occipital   C. Sphenoid   D. Mastoid

_____ 7. Immovable joints of the skull are called:

  A. Wormian bones   B. Sutures   C. Conchae
  D. Sinuses   E. Fontanels

_____ 8. _____ articulate with every bone of the face except the mandible.

  A. Lacrimal bones   B. Zygomatic bones   C. Maxillae
  D. Sphenoid bones   E. Ethmoid bones

_____ 9. All of these markings are parts of the sphenoid bone EXCEPT:

  A. Lesser wings   B. Optic foramen   C. Crista galli
  D. Sella turcica   E. Pterygoid processes

_____ 10. All of these bones are included in the axial skeleton EXCEPT:

  A. Rib   B. Sternum   C. Clavicle
  D. Hyoid   E. Ethmoid

Questions 11–15: Arrange the answers in correct sequence.

_____ _____ _____ 11. From anterior to posterior:

  A. Ethmoid bone
  B. Sphenoid bone
  C. Occipital bone

_____ _____ _____ 12. Parts of vertebra from anterior to posterior:

  A. Vertebral foramen
  B. Body
  C. Lamina

_____ _____ _____ 13. Vertebral regions, from superior to inferior:

  A. Lumbar
  B. Thoracic
  C. Cervical

_____ _____ _____ 14. From superior to inferior:

  A. Atlas
  B. Axis
  C. Occipital bone

_____ _____ _____ 15. From superior to inferior:

  A. Atlas
  B. Manubrium of sternum
  C. Hyoid

Questions 16–20: Circle T (true) or F (false). If the statement is false, change the underlined word or phrase so that the statement is correct.

T   F   16. The space between two ribs is called the costal groove.

T   F   17. The thoracic and sacral curves are called primary curves, meaning that they retain the original curve of the fetal vertebral column.

T   F   18. The jugular vein passes through the same foramen as cranial nerves V, VI, and VII.

T   F   19. In general, <u>foramina, meati, and fissures</u> serve as openings in the skull for nerves and blood vessels.

T   F   20. <u>Scoliosis</u> is another name for "hunchback."

Questions 21–25: fill-ins. Write the word or phrase that best completes the statement.

_____ 21. The dens is part of the _____ bone.

_____ 22. The ramus, angle, mental foramen, and alveolar processes are all markings on the _____ bone.

_____ 23. The squamous, petrous, and zygomatic portions are markings on the _____ bone.

_____ 24. The perpendicular plate, crista galli, and superior and middle conchae are markings found on the _____ ethmoid bone.

_____ 25. The sternum is often used for a marrow biopsy because _____.

8

The Skeletal System: The Appendicular Skeleton

In the previous chapter you learned about the bones comprising the axial skeleton. In this chapter you will study the remaining bones—those in the appendicular skeleton. You will identify bones of the shoulder girdles (Objective 1), upper extremities (2), pelvic girdle (3), and lower extremities (4–5). You will also consider structural differences between the male and female skeleton (6).

Topics Summary

A. Pectoral (shoulder) girdles
B. Upper extremities
C. Pelvic girdle
D. Lower extremities
E. Male and female skeletons

Objectives

1. Identify the bones of the pectoral (shoulder) girdle and their major markings.
2. Identify the upper extremity, its component bones, and their markings.
3. Identify the components of the pelvic (hip) girdle and their principal markings.
4. Identify the lower extremity, its component bones, and their markings.
5. Define the structural features and importance of the arches of the foot.
6. Compare the principal structural differences between female and male skeletons, especially those that pertain to the pelvis.

144   Learning Activities

(pages 159–160)   **A. Pectoral (shoulder) girdle**

1. Name the bones that form the pectoral (shoulder) girdle.

Identify these on Figure LG 8-1 and also on yourself or a study partner.

Do these bones articulate with vertebrae or ribs?

[1]

[2] 2. Identify the point on Figure LG 8-1 at which the shoulder girdle portion
of the appendicular skeleton articulates with the axial skeleton. This area is
marked by an asterisk (*). Name the two bones forming that joint:
_____ and _____.

3. How does the medial end of the clavicle differ in shape from the lateral end?

[3] 4. Study a scapula carefully, using a skeleton or Figure 8-3 (page 161) in your
text. Then match the markings at right with the descriptions given.

_____ a. Sharp ridge on the posterior sur-   A. Axillary border
face   C. Coracoid process
_____ b. Depression inferior to the spine;   I. Infraspinatus fossa
location of infraspinatus muscle   M. Medial border
_____ c. Edge closest to the vertebral col-   S. Spine
umn
_____ d. Thick edge closest to the arm
_____ e. Projection on the anterior surface;
used for muscular attachment

[4] 5. Write the names of bones that articulate with these areas of the scapula.
a. Acromion process

b. Glenoid cavity

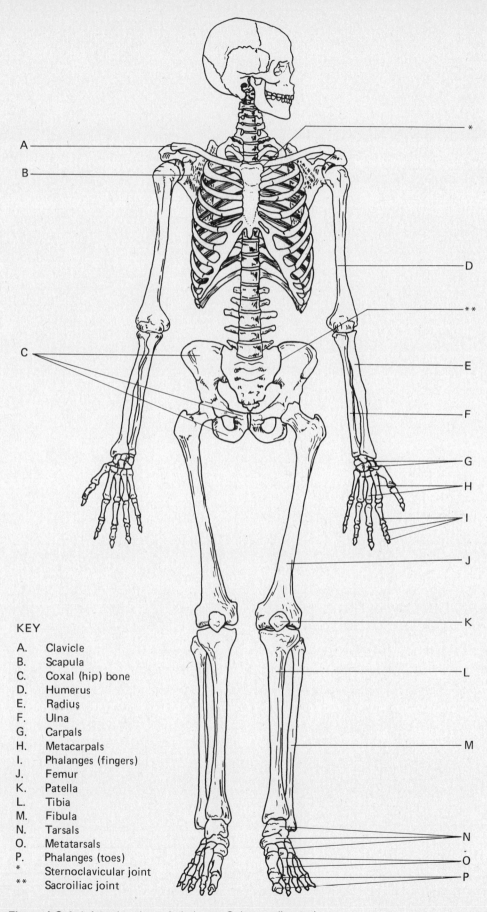

A
B
*
C
D
**
E
F
G
H
I
J
K
L
M
N
O
P

KEY

A. Clavicle
B. Scapula
C. Coxal (hip) bone
D. Humerus
E. Radius
F. Ulna
G. Carpals
H. Metacarpals
I. Phalanges (fingers)
J. Femur
K. Patella
L. Tibia
M. Fibula
N. Tarsals
O. Metatarsals
P. Phalanges (toes)
* Sternoclavicular joint
** Sacroiliac joint

Figure LG 8-1 Anterior view of skeleton. Color as directed.

B. Upper extremities

1. List the bones (or groups of bones) in the upper extremity from proximal to distal. Indicate how many of each bone there are. Two are done for you. Refer to Figure LG 8-1 to check your answers.

a. <u>Humerus</u>　(1)　d. _____ (　　)

b. _____ (　　)　e. <u>Metacarpals</u>　(5)

c. _____ (　　)　f. _____ (　　)

2. After carefully examining Figures 8-4 and 8-5 (pages 000–000) in your text, as well as bone specimens (if available), complete this exercise. For each marking write the name of the bone in which it is found, its location on the bone, and its function.

Marking	Bone	Location	Function
a. Head		Proximal end	
b. Intertubercular sulcus			Tendon of biceps muscle passes through here
c. Deltoid tuberosity			
d. Trochlea			
e. Trochlear (semilunar) notch	Ulna		
f. Coronoid process			
g. Coronoid fossa		Anterior of distal end	

Marking	Bone	Location	Function
h. Olecranon process			
i. Olecranon fossa			
j. Radial notch		Lateral side of proximal end	
k. Radial tuberosity			
l. Radial fossa			
m. Capitulum			Rounded knob that articulates with depressed head of radius

3. Circle all of the TRUE statements about the human arm and forearm in the anatomical position.

   6

   A. The radius and ulna are parallel to each other.
   B. The radius and ulna are crossed.
   C. The radius is lateral to the ulna.

4. Define *Colles' fracture*. Explain how it is likely to occur.

5. Answer these questions about wrist bones.

   7

   a. Wrist bones are called _____. There are *(5? 7? 8? 14?)* of them in each wrist.
   b. The wrist bone most subject to fracture is the bone named _____, located just distal to the *(radius? ulna?)*.

6. Trace an outline of your hand. Draw in and label all bones.

8 7. Referring to the figure you just drew, complete this exercise about the hand.
   a. The bones that constitute the palm of the hand are called
   _____ bones. There are _____ in each hand. The one
   on the thumb side is numbered *(I? V?).*

   b. Metacarpals articulate proximally with _____, laterally with _____, and distally with _____

   _____.

   c. Finger bones are called _____. Each digit, except for
   the thumb, has _____ bones; the thumb has _____ phalanges.

C. Pelvic girdle

1. Describe the bony pelvis by completing this Learning Activity.
   a. Name the bones that form the pelvic girdle.

   b. Which of these bones is/are part of the axial skeleton?

   c. Locate the point (**) on Figure LG 8-1 at which the pelvic girdle portion of the appendicular skeleton articulates with the axial skeleton. Name the bones involved in that joint: _____ and

   _____.

2. Contrast the two principal parts of the pelvis. Describe their location and name the structures which compose them.
   a. Greater (false) pelvis

   b. Lesser (true) pelvis

3. Look at the pelvic inlet on Figure 8-7 (page 165) of your text and Figure LG 8-2. Find the brim of the true pelvis *(pelvic brim)* which demarcates the opening known as the pelvic inlet. This line is very *(smooth? irregular?).*
4. Now consider the pelvic outlet. With your fingertip, trace it on a skeleton, if available. It is very *(smooth? irregular?).* Why is the pelvic outlet so named?

150

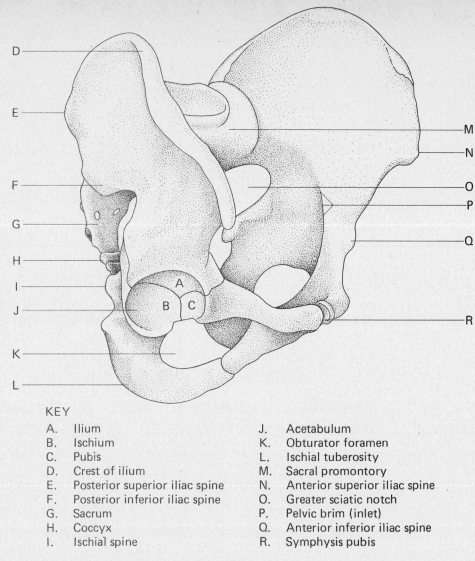

KEY

A. Ilium
B. Ischium
C. Pubis
D. Crest of ilium
E. Posterior superior iliac spine
F. Posterior inferior iliac spine
G. Sacrum
H. Coccyx
I. Ischial spine

J. Acetabulum
K. Obturator foramen
L. Ischial tuberosity
M. Sacral promontory
N. Anterior superior iliac spine
O. Greater sciatic notch
P. Pelvic brim (inlet)
Q. Anterior inferior iliac spine
R. Symphysis pubis

Figure LG 8-2 Pelvic girdle. Color as directed.

5. Define *pelvimetry.* For what purpose is it used?

6. Answer the following questions about coxal bones.

   a. The coxal bones are also known as _____ or

      _____.

   b. Each coxal bone originates as three bones which fuse early in life. These
      bones are the _____, _____, and

      _____.

   c. At what location do the three bones fuse?

   d. The largest of the three bones is the _____. A ridge
      along the superior border is called the iliac crest. Locate this on yourself.

   e. The iliac crest ends anteriorly as the _____ spine.
      The crest ends posteriorly as the _____

      _____. This marking causes a dimpling of the skin
      just lateral to the sacrum, which can be used as a landmark for administer-
      ing hip injections accurately.

7. Color Figure LG 8-2 according to the following code: *ilium,* yellow; *ischium,* red; *pubis,* blue; *sacrum,* green; *coccyx,* black.
8. Complete the table about markings of the coxal bones.

Marking	Location on coxal bone	Function
a. Greater sciatic notch		
b.		Supports most of body weight in sitting position
c. Auricular surface		
d.		Fibrocartilaginous joint between two coxal bones

Marking	Location on coxal bone	Function
e.		Socket for head of femur
f. Obturator foramen	Large foramen surrounded by pubic and ischial rami and acetabulum	

(pages 166–173) **D. Lower extremities**

1. Refer to Figure LG 8-1 and list the bones (or groups of bones) in the lower extremity from proximal to distal. Indicate how many of each bone there are. One is done for you.

$\boxed{13}$

    a.   Femur   (1)   e. _____ ( )

    b. _____ ( )   f. _____ ( )

    c. _____ ( )   g. _____ ( )

    d. _____ ( )

2. Contrast size, location, and names of bones of the upper and lower extremities by coloring the bones as follows. Color on one side of Figure LG 8-1 only.

    Humerus and femur (red)   Carpals and tarsals (blue)

    Patella (brown)   metacarpals and metatarsals (orange)

    Ulna and tibia (green)   phalanges (purple)

    Radius and fibula (yellow)

3. Circle the term that correctly indicates the location of these parts of the lower extremity.

$\boxed{14}$

    a. The head is the *(proximal? distal?)* epiphysis of the femur.

    b. The greater trochanter is *(lateral? medial?)* to the lesser trochanter.

    c. The intercondylar fossa is on the *(anterior? posterior?)* surface of the femur.

    d. The tibial condyles are more *(concave? convex?)* than the femoral condyles.

    e. The lateral condyle of the femur articulates with the *(fibula? lateral condyle of the tibia?).*

    f. The tibial tuberosity is *(superior? inferior?)* to the patella.

    g. The tibia is *(medial? lateral?)* to the fibula.

    h. The outer portion of the ankle is the *(lateral? medial?)* malleolus which is part of the *(tibia? fibula?).*

4. Define *shinsplints* and state probable causative factors.

5. The tarsal bone which is most superior in location (and which articulates with the tibia and fibula) is the _____. The largest and strongest of the tarsals is the _____. Name the other five tarsals: _____, _____, and three _____ bones.

6. Answer these questions about the arch of the foot.
   a. How is the foot maintained in an arched position?

   b. Locate each of these arches on your own foot. Refer to Figure 8-14 (page 172) of your text. (*For extra review:* List the bones that form each arch.)
   Longitudinal: medial side

   Longitudinal: lateral side

   Transverse

   c. What causes flatfoot?

7. Now that you have seen all of the bones of the appendicular skeleton, complete this table relating common and anatomical names of bones. 15

Common Name	Anatomical Name
a. Shoulder blade	
b.	Pollex
c. Collarbone	
d. Heel bone	
e.	Olecranon process
f. Kneecap	
g.	Tibial crest
h. Toes	
i. Palm of hand	
j. Wrist bones	

8. Contrast: flatfoot/clawfoot

E. Male and female skeletons

16
1. State several characteristics of the female pelvis which make it more suitable for childbirth than the male pelvis.

2. Identify specific differences in pelvic structure in the two sexes by placing M before characteristics of the male pelvis and F before structural descriptions of the female pelvis.

_____ a. Shallow greater pelvis
_____ b. Heart-shaped inlet
_____ c. Pubic arch greater than 90° angle
_____ d. Ischial spine turned inward
_____ e. Pelvic outlet comparatively small
_____ f. General structure light and thin

Answers to Numbered Questions in Learning Activities

1 No

2 Sternum (manubrium) and clavicles.

3 (a) S. (b) I. (c) M. (d) A. (e) C.

4 (a) Clavicle. (b) Humerus (head of).

5

Marking	Bone	Location	Function
a. Head	Humerus	Proximal end	Articulates with glenoid cavity of scapula
b. Intertubercular sulcus (bicipital groove)	Humerus	Proximal end	Tendon of biceps muscle passes through here
c. Deltoid tuberosity	Humerus	Midshaft, lateral side	Point of attachment for deltoid muscle
d. Trochlea	Humerus	Distal end	Articulates with ulna
e. Trochlear (semilunar) notch	Ulna	Proximal end	Notch into which trochlea of humerus fits
f. Coronoid process	Ulna	Anterior of proximal end	Fits into coronoid fossa of humerus when forearm flexed
g. Coronoid fossa	Humerus	Anterior of distal end	Receives coronoid process of ulna when forearm flexed
h. Olecranon process	Ulna	Posterior of proximal end	Fits into olecranon fossa of humerus when forearm extended
i. Olecranon fossa	Humerus	Posterior of distal end	Receives oelcranon process of ulna when forearm extended

Marking	Bone	Location	Function
j. Radial notch	Ulna	Lateral side of proximal end	Head of radius pivots in this notch
k. Radial tuberosity	Radius	On medial side toward proximal end	Point of attachment for biceps muscle
l. Radial fossa	Humerus	Distal end	Receives head of radius when forearm flexed
m. Capitulum	Humerus	Distal end	Rounded knob that articulates with depressed head of radius

6   A, C.

7   (a) Carpals, 8. (b) Scaphoid, radius.

8   (a) Metacarpal, 5, I. (b) Carpals, adjacent metacarpals, proximal phalanges. (c) Phalanges, 3, 2.

9   (a) Two coxal (hip) bones, the sacrum and coccyx. (b) Sacrum and coccyx. (c) Sacrum and iliac portion of coxal bone (sacroiliac joint).

10   Irregular. The outlet permits substances in the abdominopelvic cavity to exit from the body via openings in the muscular floor attached to the bony outlet. Feces, urine, semen, menstrual flow, and baby (at birth) pass through here.

11   (a) Hipbones, innominate bones. (b) Ilium, ischium, pubis. (c) Acetabulum. (d) Ilium. (e) Anterior superior iliac, posterior superior iliac spine.

12   (a) Ilium, space through which sciatic nerve passes from anterior of sacrum to posterior of thigh. (b) Ischial tuberosity, ischium. (c) Ilium, site of articulation with sacrum. (d) Symphysis pubis, cartilage between two pubic bones. (e) Acetabulum; point of fusion of ilium, ischium, and pubis. (f) Reduces weight of coxal bone and provides passageway for some blood vessels and nerves.

13   (a) Femur, 1. (b) Patella, 1. (c) Tibia, 1. (d) Fibula, 1. (e) Tarsals, 7. (f) Metatarsals, 5. (g) Phalanges, 14.

14   (a) Proximal. (b) Lateral. (c) Posterior. (d) Concave. (e) Lateral condyle of the tibia. (f) Inferior. (g) Medial. (h) Lateral, fibula.

15   (a) Scapula. (b) Thumb. (c) Clavicle. (d) Calcaneus. (e) Elbow. (f) Patella. (g) Shinbone. (h) Phalanges. (i) Metacarpals. (j) Carpals.

16   (a) F. (b) M. (c) F. (d) M. (e) M. (f) F.

MASTERY TEST: Chapter 8

Questions 1–5: Choose the one best answer to each question.

_____ 1. The point at which the upper part of the appendicular skeleton is joined to (articulates with) the axial skeleton is at the joint between:

     A. Sternum and ribs   B. Humerus and clavicle

     C. Scapula and clavicle   D. Scapula and humerus

     E. Sternum and clavicle

_____ 2. A "separated shoulder" means dislocation of the joint between:

    A. Acromion and clavicle   B. Sternum and clavicle

    C. Acromion and head of humerus

    D. Glenoid cavity and humerus

    E. Coracoid process and clavicle

_____ 3. Which structures are on the posterior surface of the upper extremity (in anatomical position)?

    A. Radial fossa and radial notch

    B. Trochlea and capitulum

    C. Coronoid process and coronoid fossa

    D. Olecranon process and olecranon fossa

_____ 4. The humerus articulates with all of these bones EXCEPT:

    A. Ulna   B. Radius   C. Clavicle   D. Scapula

_____ 5. Choose the FALSE statement.

    A. The capitulum articules with the head of the radius.

    B. The medial and lateral epicondyles are located at the distal ends of the tibia and fibula.

    C. The coronoid fossa articulates with the ulna when the ulna is flexed.

    D. The trochlea articulates with the trochlear notch of the ulna.

Questions 6–10: Arrange the answers in correct sequence.

_____ _____ _____ 6. According to size of the bones, from largest to smallest:

    A. Femur

    B. Ulna

    C. Humerus

_____ _____ _____ 7. Parts of the humerus, from proximal to distal:

    A. Anatomical neck

    B. Surgical neck

    C. Head

_____ _____ _____ 8. From proximal to distal:

    A. Phalanges

    B. Metacarpals

    C. Carpals

_____ _____ _____ 9. Parts of a metacarpal, from proximal to distal:

    A. Head

    B. Base

    C. Shaft

_____ _____ _____ 10. From superior to inferior:

    A. True pelvis

    B. False pelvis

    C. Pelvic brim

Questions 11–20: Circle T (true) or F (false). If the statement is false, change the underlined word or phrase so that the statement is correct.

T   F   11. The organs contained within the right iliac, hypogastric, and left iliac portions (ninths) of the abdomen are located in the <u>true</u> pelvis.

T F 12. Another name for the true pelvis is the <u>greater</u> pelvis.

T F 13. The scapulae <u>do</u> articulate with the vertebrae.

T F 14. The olecranon process is a marking on the <u>ulna, and the olecranon fossa is a marking on the humerus.</u>

T F 15. The female pelvis is <u>deeper and more heart-shaped</u> than the male pelvis.

T F 16. The greater tubercle of the humerus is <u>lateral</u> to the lesser tubercle.

T F 17. There are <u>14 phalanges in each hand and also in each foot.</u>

T F 18. The fibula articulates with the <u>femur, tibia, talus, and calcaneus.</u>

T F 19. The coracoid process articulates with <u>the clavicle.</u>

T F 20. The total number of bones in one upper extremity (including arm, forearm, wrist, hand, fingers, but excluding shoulder girdle) is <u>29.</u>

Questions 21–25: fill-ins. Write the word or phrase that best completes the statement.

_____ 21. In about three-fourths of all carpal bone fractures, only the ____ bone is involved.

_____ 22. A fracture of the distal end of the fibula with injury to the tibial articulation is known as a ____ fracture.

_____ 23. The ____ is the thinnest bone in the body compared to its length.

_____ 24. The point of fusion of the three bones forming the coxal bone is the ____.

_____ 25. The total number of phalanges in all of the fingers and toes is ____.

Articulations

In the past three chapters you have learned a great deal about the 206 bones in the body. Separated—or disarticulated—these bones would be no more than a disorganized pile. But bones are arranged in precise order and are held together in various ways so that they can function effectively. In this chapter you will look at different kinds of joints and the movements possible at each (Objectives 1–4). You will examine closely three selected joints: the shoulder, hip, and knee (5). You will also learn about a number of disorders involving joints (6), as well as medical terminology associated with articulations (7).

Topics Summary

A. Classification
B. Fibrous joints, cartilaginous joints
C. Synovial joints: structure
D. Synovial joints: movements
E. Synovial joints: types
F. Selected articulations of the body
G. Disorders, medical terminology

Objectives

1. Define an articulation and identify the factors that determine the degree of movement at a joint.
2. Contrast the structure, kind of movement, and location of fibrous, cartilaginous, and synovial joints.
3. Explain the principle of arthroscopy and its clinical importance.
4. Discuss and compare the movements possible at various synovial joints.
5. Describe selected articulations of the body with respect to the bones that enter into their formation, structural classification, and anatomical components.
6. Describe the causes and symptoms of common joint disorders, including rheumatism, rheumatoid arthritis (RA), osteoarthritis, gouty arthritis, bursitis, dislocation and sprain.
7. Define medical terminology associated with articulations.

160   **Learning Activities**

A. Classification

1. Define the term *articulation (joint)*.

|1| 2. Fill in the blanks below to name three classes of joints according to the amount of movement they permit.

   a. Synarthroses: _____

   b. _____: slightly movable

   c. _____: freely movable

|2| 3. Name three classes of joints based on structure.

(pages 176–177)   **B. Fibrous joints, cartilaginous joints**

|3| 1. Describe fibrous joints by completing this exercise.
   a. Fibrous joints *(have? lack?)* a joint cavity. They are held together by _____ connective tissue.

   b. One type of fibrous joint is a _____ found between skull bones. Some sutures are replaced by bone in the adult; an example is the site where two fetal _____ bones fuse to form one single "forehead" bone. Such a suture is classified as a _____ joint.

   c. The tibia/fibula and radius/ulna joints are fibrous joints which have *(more? less?)* flexibility than sutures. These joints are known as *(syndesmoses? gomphoses?)*.

   d. State one location of a *gomphosis* joint.

|4| 2. Contrast two types of cartilaginous joints in this exercise.
   a. Synchondroses involve *(hyaline? fibrous?)* cartilage between bone. An example is the _____ cartilage between diaphysis and epiphysis of a growing bone. This cartilage *(persists through life? is replaced by bone during adult life?)*. Synchondroses are *(somewhat movable? immovable?)*.

   b. Fibrocartilage is present in the type of joint known as _____ _____. These joints permit *(some? no?)* movement, so are called _____-arthrotic. Two locations of symphyses are _____ and _____.

C. Synovial joints: structure

1. What structural features of synovial joints make them more freely movable than fibrous or cartilaginous joints? ☐5

2. Complete the table about parts of a synovial joint. ☐6

Part of Joint	Structure	Function
a. Synovial cavity		
b.	Hyaline cartilage covering surfaces of articulating bones; not covered by synovial membrane	
c. Fibrous capsule		
d.		Secretes synovial fluid
e. Synovial fluid		
f. Ligaments		

3. Contrast *fibrous capsule* with *ligaments*.

4. On Figure LG 9-1 label parts *A-F* of a synovial joint: *articulating bone, articular cartilage, fibrous capsule, periosteum, synovial (joint) cavity,* and *synovial membrane.* ☐7

5. Structures *A* and *B* in Figure LG 9-1 together compose the _____. ☐8

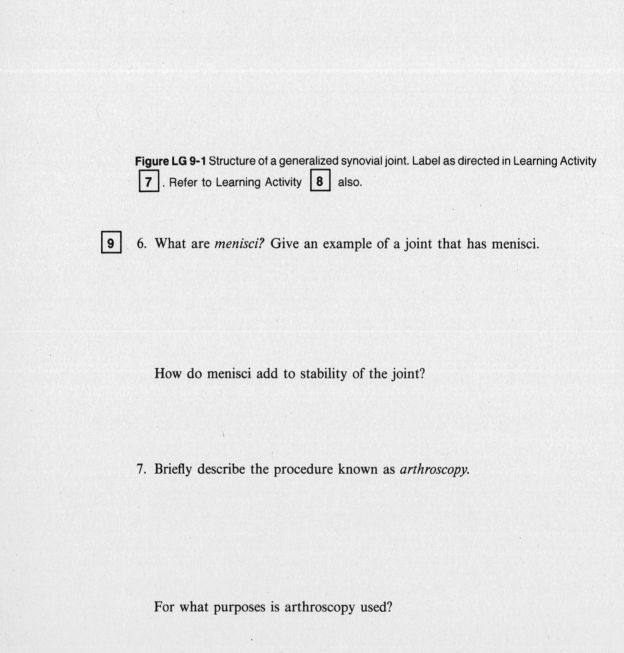

Figure LG 9-1 Structure of a generalized synovial joint. Label as directed in Learning Activity
⬜7. Refer to Learning Activity ⬜8 also.

⬜9 6. What are *menisci?* Give an example of a joint that has menisci.

How do menisci add to stability of the joint?

7. Briefly describe the procedure known as *arthroscopy.*

For what purposes is arthroscopy used?

8. What are *bursae?* Where are they located? What is their function? **163**

9. Define *bursitis.* What factors do you think might cause bursitis?

D. Synovial joints: movements (pages 178–181)

1. The design of synovial joints permits free movement of bones. However, if bones moved too freely, they could move right out of their joint cavities (dislocation). Describe how each of these factors accounts for limitation of movement at synovial joints.
   a. Apposition of soft parts

   b. Tension of ligaments

   c. Muscle tension

2. Choose the description that fits the type of movement in each case. Not all answers will be used. ☐10

164

_____ a. Decrease in angle between anterior surfaces of bones (or between posterior surfaces at knee and toe joints)

_____ b. Simplest kind of movement that can occur at a joint; no angular or rotary motion involved; example: ribs moving against vertebrae

_____ c. State of entire body when it is in anatomical position

_____ d. Movement away from the midline of the body

_____ e. Movement of a bone around its own axis

_____ f. Position of foot when heel is on the floor and rest of foot is raised

Abd. Abduction
Add. Adduction
C. Circumduction
D. Dorsiflexion
E. Extension
F. Flexion
G. Gliding
I. Inversion
P. Plantar flexion
R. Rotation

3. Perform the action described. Then write in the name of the type of movement.

a. Describe a cone with your arm, as if you are winding up to pitch a ball.

The movement is called _____.

b. Stand in anatomical position (palms forward). Turn your palms backward. This action is called _____.

c. Move your fingers from "fingers together" to "fingers apart" position.

This action is _____ of fingers.

d. Raise your shoulders, as if to shrug them. This movement is called

_____ of the shoulders.

e. Stand on your toes. This action at the ankle joint is called

_____.

f. Grasp a ball in your hand. Your fingers are performing the type of

movement called _____.

4. Identify the kinds of movements shown in Figure LG 9-2. Write the name of the movement below each figure. Use the following terms: *abduction, adduction, extension, flexion,* and *hyperextension.*

(pages 181–183) **E. Synovial joints: types**

1. After you have studied the types of joints, (Figure 9-5, pages 184–185, and Exhibit 9-2, page 186 in your text), check your understanding by choosing the type of synovial joint that fits the description. (Answers may be used more than once.)

_____ a. Monaxial joint; only rotation possible

_____ b. Joint between carpal and metacarpal of the thumb joint

_____ c. Shoulder and hip joints

_____ d. Spool-like surface articulated with concave surface

_____ e. Monaxial joint; only flexion and extension possible

_____ f. Biaxial joints (two answers)

B. Ball-and-socket
E. Ellipsoidal
G. Gliding
H. Hinge
P. Pivot
S. Saddle

165

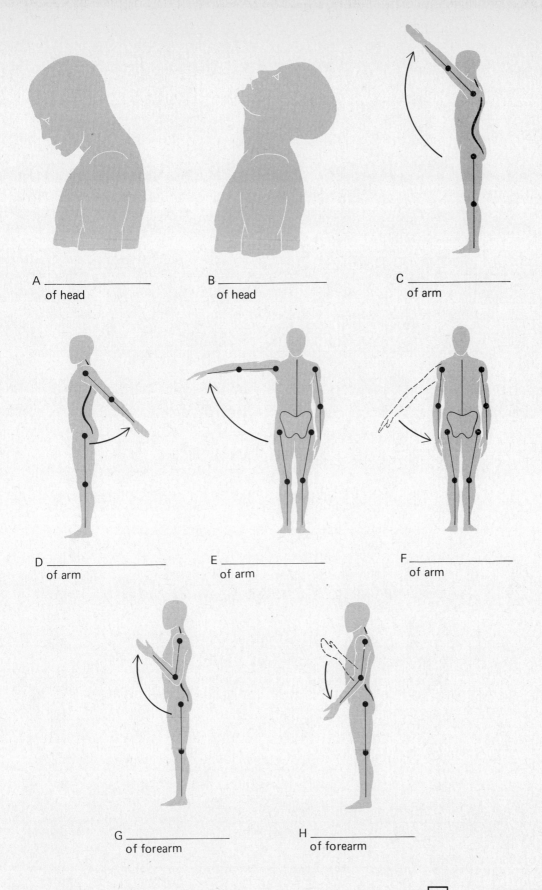

A _____ of head

B _____ of head

C _____ of arm

D _____ of arm

E _____ of arm

F _____ of arm

G _____ of forearm

H _____ of forearm

Figure LG 9-2 Movements at synovial joints. Letters refer to Learning Activity 2 and to questions in Chapter 11.

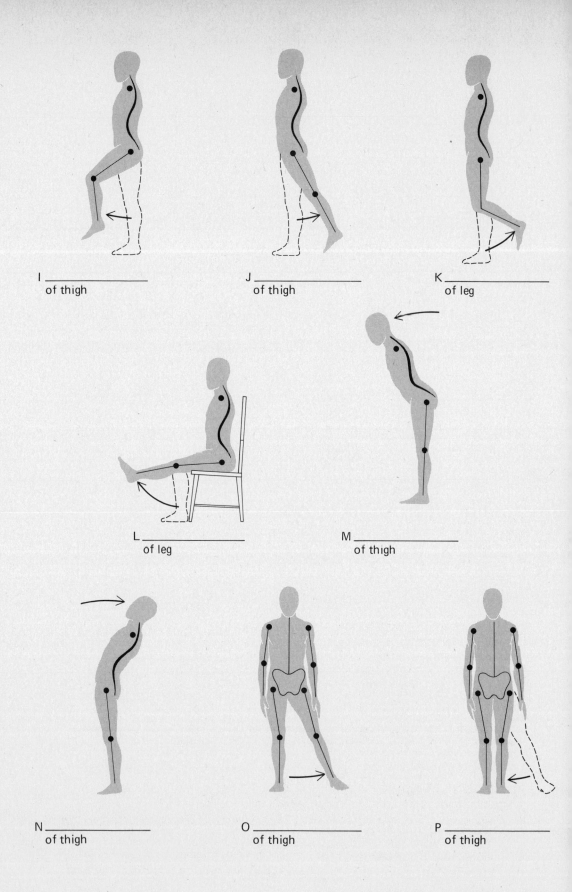

I _____
of thigh

J _____
of thigh

K _____
of leg

L _____
of leg

M _____
of thigh

N _____
of thigh

O _____
of thigh

P _____
of thigh

Figure LG 9-2 (Continued)

2. All synovial joints are classified functionally as *(diarthrotic? synarthrotic? amphiarthrotic?)*. 14

167

F. Selected articulations of the body   (pages 183–191)

1. Complete this exercise describing the shoulder, hip, and knee joints. Consult 15 Figures 9-6 to 9-8 on pages 187–190 of your text.

   a. The articular capsule is strengthened by ligaments, such as the

   _____-humeral ligament between the coracoid process of the scapula and the *(greater? lesser?)* tubercle of the humerus. Name one other ligament that forms part of this joint.

   b. The shoulder joint is further strengthened and stabilized by two other structures. One is the glenoid _____ which is a rim (or liplike structure) of fibrocartilage; it increases the concavity of the glenoid cavity, providing the head of the humerus with a more stable socket. The other consists of the set of deep muscles and their tendons known as the _____ cuff. These cross over this joint and strengthen it, but are also a common site of injury.

   c. List three bursae associated with the humeroscapular joint.

   d. The articular capsule of the coxal, or hip, joint includes circular fibers,

   called the _____ which form a protective collar around the head of the femur. Name the three ligaments that reinforce the articular capsule.

   e. The stability of the hip joint is further enhanced by the ligament of the head of the femur and the transverse ligament of the acetabulum. Both of these are *(intra? extra?)*-capsular ligaments. The hip joint *(does? does not?)* have a labrum.

   f. The knee (tibiofemoral) joint *(does? does not?)* include a complete capsule uniting the two bones. This factor contributes to the relative *(strength? weakness?)* of this joint.

g. A number of ligaments do provide some strength and support. Name two extra capsular ligaments: _____ and _____

_____.

h. Name two intracapsular ligaments of the knee joint. _____

_____ and _____. Which one is more often involved in serious knee injuries?

i. The knee joint *(does? does not?)* include a labrum. Actually it has two liplike cartilage discs called _____ which add stability to the joint.

j. There are *(no? three? over a dozen?)* bursae associated with the knee joint.

2. *For extra review.* Indicate whether structures listed below are associated with the shoulder, hip, or knee joints.

a. Medial and lateral menisci: _____

b. Anterior and posterior cruciate ligaments: _____

c. Pubofemoral ligaments: _____

d. Subdeltoid bursa: _____

e. Fibular collateral ligament: _____

f. Glenohumeral ligament: _____

3. Make a final comparison of the shoulder, hip, and knee joints by completing this exercise.

a. Which of these joints has the widest range of motion? _____

_____

b. Which has the most limited range of motion? _____

_____

c. Which is most stable and so is rarely dislocated? _____

_____

d. Which is least stable? _____

4. *Clinical challenge.* Which three knee structures beginning with C's are examined closely when a knee is injured?

G. **Disorders, medical terminology, drugs**

1. How are *arthritis* and *rheumatism* related? Choose the correct answer. 19
   A. Arthritis is a form of rheumatism.
   B. Rheumatism is a form of arthritis.

2. List several forms of arthritis. Name one symptom common to all forms of this ailment.

3. Contrast *rheumatoid arthritis* with *osteoarthritis.* Write *More, Less, Yes,* or *No* for answers. 20

	Rheumatoid Arthritis	Osteoarthritis
a. Which is more common?		
b. Which is usually more damaging?		
c. Is this an inflammatory condition?		
d. Is synovial membrane affected?		
e. Does articular cartilage degenerate?		
f. Does fibrous tissue join bone ends?		
g. Is movement limited?		

4. Briefly describe each of these disorders involving articulations by doing this exercise. 21

   a. Gouty arthritis is a condition due to an excess of _____

   _____ in the blood leading to deposit of

   _____ in joints. This condition is more common

   among *(males? females?).*

b. An acute chronic inflammation of a bursa is called _____

_____. Inflammation of certain bursae associated

with the knee joint is known as _____ knee.

c. *Luxation,* or _____, is displacement of a bone from

its joint with tearing of ligaments. Three most common sites of disloca-

tions are the _____, _____, and

_____ joints.

d. Pain in a joint is known as _____.

5. Contrast:

a. Sprain/strain

b. Rheumatology/arthrology

Answers to Numbered Questions in Learning Activities

1	(a) Immovable. (b) Amphiarthrotic. (c) Diarthrotic.
2	Fibrous, cartilage, synovial.
3	(a) Lack, fibrous. (b) Suture, frontal, synostosis. (c) More, syndesmoses. (d) Articulations of teeth in alveolar processes of maxilla and mandible bones.
4	(a) Hyaline, epiphyseal, is replaced by bone during adult life, immovable. (b) Symphysis, some, amphi-, discs between vertebrae, symphysis pubis between coxal bones.
5	The space between the articular bones and the absence of tissue between articulating surfaces, which might restrict movement, make the joints more freely movable.

6   (a) -, space between articulating bones, permits free movement of synovial joint. (b) Articular cartilage, -, provides flexible surface between ends of bones. (c) -, dense connective (collagenous) tissue, provides strength and flexibility to articular capsule. (d) Synovial membrane, loose connective tissue with elastic fibers and some adipose tissue, -. (e) -, hyaluronic acid and interstitial fluid with consistency similar to egg white, lubricates joint and provides nourishment for avascular articular cartilage. (f) -, elastic and collagenous fibers, provide strength and flexibility to joint.

7   Fibrous capsule (A), synovial membrane (B), synovial (joint) cavity (C), periosteum (D), articulating bone (E), articular cartilage (F).

8   Articular capsule.

9   Wedge-shaped pieces of fibrocartilage, knee joint.

10   (a) F. (b) G. (c) E. (d) Abd. (e) R. (f) D.

11   (a) Circumduction. (b) Pronation. (c) Abduction. (d) Elevation. (e) Plantar flexion. (f) Flexion.

12   Flexion (A); hyperextension (B); flexion (C); hyperextension (D); abduction (E); adduction (F); flexion (G); extension (H); flexion, while leg also slightly flexed (I); hyperextension, while leg extended (J); flexion, with hip extended (K); extension, with hip flexed (L); flexion (M); hyperextension (N); abduction (O); adduction (P).

13   (a) P. (b) S. (c) B. (d) H. (e) H. (f) E and S.

14   Diarthrotic.

15   (a) Coraco-, greater, glenohumeral or transverse humeral ligament. (b) Labrum, rotator. (c) Subscapular, subdeltoid, subacromial or subcoracoid. (d) Zona orbicularis; iliofemoral, ischiofemoral, pubofemoral. (e) Intra, does. (f) Does not, weakness. (g) Tibial collateral, fibular collateral, patellar, oblique popliteal, arcuate popliteal. (h) Anterior and posterior cruciates; anterior cruciate. (i) Does not, menisci. (j) Over a dozen.

16   (a) Knee. (b) Knee. (c) Hip. (d) Shoulder. (e) Knee. (f) Shoulder.

17   (a) Shoulder. (b) Knee. (c) Hip. (d) Knee.

18   Tibial collateral ligament, anterior cruciate ligament, and medial meniscus (cartilage).

19   A.

20   (a) Less, More. (b) More, Less. (c) Yes, No. (d) Yes, No. (e) Yes, Yes. (f) Yes, No. (g) Yes, Yes.

21   (a) Uric acid, sodium urate, males. (b) Bursitis, housemaid's or carpet layer's. (c) Dislocation; finger, thumb, shoulder. (d) Arthralgia.

MASTERY TEST: Chapter 9

Questions 1–14: Circle T (true) or F (false). If the statement is false, change the underlined word or phrase so that the statement is correct.

T   F   1. A fibrous joint is one in which there is <u>no joint cavity and bones are held together by fibrous connective tissue.</u>

172

T F 2. <u>Sutures, syndesmoses, and symphyses</u> are kinds of fibrous joints.

T F 3. Osteoarthritis is <u>less common and usually more damaging than</u> rheumatoid arthritis.

T F 4. <u>Ball-and-socket, gliding, pivot, and ellipsoidal joints</u> are all diarthrotic joints.

T F 5. All fibrous joints <u>are synarthrotic and all cartilaginous joints are amphiarthrotic.</u>

T F 6. In synovial joints synovial membranes <u>cover the surfaces of articular cartilages.</u>

T F 7. Bursae are <u>saclike structures that reduce friction</u> at joints.

T F 8. Synovial fluid becomes <u>more</u> viscous when there is friction at a joint.

T F 9. When your arm is in the supine position, your radius and ulna are <u>parallel (not crossed).</u>

T F 10. When you touch your toes, the major action you perform at your hip joint is called <u>hyperextension.</u>

T F 11. The <u>only type of joint that is triaxial</u> is the ball-and-socket.

T F 12. Abduction is movement <u>away from</u> the midline of the body.

T F 13. The <u>elbow, knee, and ankle</u> joints are all hinge joints.

T F 14. Joints that are very stable tend to have <u>more</u> mobility than joints that are unstable.

Questions 15–16: Arrange the answers in correct sequence.

_____ _____ _____ 15. From most mobile to least mobile:
- A. Amphiarthrotic
- B. Diarthrotic
- C. Synarthrotic

_____ _____ _____ 16. Stages in rheumatoid arthritis, in chronological order:
- A. Articular cartilage is destroyed and fibrous tissue joins exposed bone.
- B. The synovial membrane produces pannus which adheres to articular cartilage.
- C. Synovial membrane becomes inflamed and thickened, and synovial fluid accumulates.

_____ 17. Which structure is extracapsular in location?

      A. Acetabular labrum   B. Posterior cruciate ligament
      C. Ligament of the head of the femur
      D. Tibial collateral ligament

_____ 18. All of these structures are associated with the knee joint EXCEPT:

      A. Glenoid labrum   B. Patellar ligament
      C. Infrapatellar bursa   D. Medial meniscus
      E. Tibial collateral ligament

_____ 19. "Housemaid's knee" is:

      A. Sprained knee   B. Tendinitis
      C. Inflammation of a bursa of the knee
      D. Torn cartilage

_____ 20. A suture is found between:

      A. The two pubic bones   B. The two parietal bones
      C. Radius and ulna   D. Diaphysis and epiphysis
      E. Tibia and fibula (distal ends)

Questions 21–25: fill-ins. Write the word or phrase that best completes the statement.

_____ 21. _____ are saclike structures which reduce friction at joints.

_____ 22. _____ is the forcible wrenching or twisting of a joint with partial rupture of it, but without dislocation.

_____ 23. The action of pulling the jaw back from a thrust-out position so that it becomes in line with the upper jaw is the movement called _____.

_____ 24. Another name for a freely movable joint is _____.

_____ 25. The type of joint between the atlas and axis and also between proximal ends of the radius and ulna is a _____ joint.

10

Muscle Tissue

In the past four chapters you have learned about the structural framework of the body. The skeleton provides support, and its jointed arrangement makes movement possible. However, movement can occur only when muscles pull on bones. In this chapter you will look closely at the structure and function of muscle tissue (Objectives 1–4), learn how muscle contraction occurs (5–10), and contrast skeletal muscle with smooth and cardiac muscle (11). You will see the role of muscles in maintaining homeostasis (12–14). You will also study disorders and medical terminology associated with the muscular system (15–17).

Topics Summary

A. Characteristics, functions, types of muscle tissue
B. Skeletal muscle tissue
C. Contraction
D. Kinds of muscle contraction
E. Types of skeletal muscle fibers
F. Cardiac muscle tissue, smooth muscle tissue
G. Homeostasis
H. Aging and development of muscles
I. Disorders, medical terminology, drugs

Objectives

1. List the characteristics and functions of muscle tissue.
2. Compare the location, microscopic appearance; nervous control, and functions of the three kinds of muscle tissue.
3. Define fascia, epimysium, perimysium, endomysium, tendons, and aponeuroses and list their modes of attachment to muscles.
4. Explain the relationship of blood vessels and nerves to skeletal muscles.
5. Describe the principal events associated with the sliding-filament theory.
6. Describe the structure and importance of a neuromuscular junction (motor end plate) and a motor unit.
7. Identify the source of energy for muscular contraction.
8. Define the all-or-none principle of muscular contraction.
9. Describe different types of normal contractions performed by skeletal muscles.
10. Describe the phases of contraction in a typical myogram of a twitch contraction.
11. Compare the structure and function of the three types of skeletal muscle fibers.
12. Compare oxygen debt, fatigue, and heat production as examples of muscle homeostasis.
13. Explain the effects of aging on muscle tissue.
14. Describe the development of the muscular system.
15. Define such common muscular disorders as fibrosis, fibrositis, "charleyhorse," muscular dystrophy, and myasthenia gravis.
16. Compare spasms, cramps, convulsions, fibrillation, and tics as abnormal muscular contractions.
17. Define medical terminology associated with the muscular system.

Learning Activities

(page 195) **A. Characteristics, functions, types of muscle tissue**

☐1 1. List four characteristics of muscle tissue.

a. _____

b. _____

c. _____

d. _____

Now write a brief description of each characteristic in the space next to each line.

2. State three functions of muscle tissue that are important for maintenance of homeostasis.

☐2 3. Match the muscle types listed at right with descriptions below.

_____ a. Involuntary muscle found in blood vessels and intestine   C. Cardiac
Sk. Skeletal
_____ b. Involuntary striated muscle   Sm. Smooth
_____ c. Striated voluntary muscle attached to bones

(pages 195–199) **B. Skeletal muscle tissue**

1. Define *fascia*.

☐3 2. Contrast two kinds of fascia by indicating which of the following are characteristics of superficial fascia (S) or deep fascia (D).

_____ a. Immediately under the skin

_____ b. Composed of dense connective tissue

_____ c. Contains much fat, so provides insulation and protection

_____ d. Also called subcutaneous layer

☐4 3. Describe the extensive network of deep fascia associated with muscles in this matching exercise.

_____ a. Invagination of deep fascia that surrounds muscle cells and bundles and holds muscles into functional groups   A. Aponeurosis
EPE. Endomysium, perimysium, epimysium
T. Tendon
_____ b. Cord of dense connective tissue that attaches muscle into periosteum   TS. Tendon sheath

_____ c. Similar in function to tendon, but consists of a broad, flat layer of dense connective tissue

_____ d. Tube of fibrous connective tissue lined with synovial membrane that permits tendon to slide easily, as in wrist and ankle

4. Explain why good blood and nerve supplies are required for normal muscle function.

5. Carefully study a diagram of skeletal muscle tissue (Figure 10-2a, page 197, in your text). Then check your understanding of the sarcomere and its components by doing this exercise.

| 5 |

a. In muscle tissue fibers are *(cells? intercellular material?),* whereas in connective tissue fibers are *(cells? intercellular material?).*

b. Each muscle fiber (also known as a _____) is surrounded by a membrane called the _____ and contains cytoplasm called _____.

c. Nuclei and mitochondria within muscle cells are located close to the _____.

d. What is *sarcoplasmic reticulum?*

e. T tubules run *(parallel? perpendicular?)* to sarcoplasmic reticulum.

f. A T tubule, along with sarcoplasmic reticulum on either side, is called a _____.

g. The hundreds or thousands of myofibrils in a skeletal muscle cell each consist of bundles of thick and thin _____ arranged in compartments called _____.

h. An entire sarcomere of relaxed skeletal muscle is labeled in Figure LG 10-1 with letter _____. One end of the sarcomere is labeled with letter *E;* this point is known as the _____.

i. The central dark band (labeled *B*) is known as the *(A? I?)* band. Locate the area labeled *F.* This light band is known as the *(A? I?)* band. These alternating *A* and *I* bands give skeletal muscle its *(striated? nonstriated?)* appearance.

j. Locate the portion of the sarcomere which is labeled with letter *H.* It contains only *(thick? thin?)* filaments. It is called the _____ zone. In the center of this zone (labeled *G*) is the *(M? Z?)* line which consists of threads that appear to connect adjacent thick filaments.

Figure LG 10-1 Detail of sarcomere. Relaxed muscle. Complete as directed. Letters refer to Learning Activity 5 .

k. Which part of the sarcomere contains only thin filaments?

l. Thin myofilaments are composed mostly of molecules of _____

_____. Each action moecule contains a binding site for *(myosin? ATP?)*. Two other proteins in thin myofilaments are

_____ arranged in helical strands, and

_____ located at regular intervals along thin myofila-ments.

m.*(Thick? Thin?)* myofilaments are composed mainly of myosin. Myosin molecules are shaped much like *(basketballs? golf clubs?)*. The "head" of the club is called a cross bridge and contains binding sites for

_____ and _____.

 6 6. *For extra review.* Write labels next to all structures *(A–H)* in Figure LG 10-1.

7. Discuss several advantages and disadvantages of uses of anabolic steroids.

(pages 199–203) **C. Contraction**

1. Refer to Figure 10-4 (page 199) in your text and Figure LG 10-1, and do this exercise summarizing one theory of how muscle contraction occurs.

 7   (Exercise 11 describes more details of the theory.)

a. In order to effect muscle shortening (or contraction), heads (cross bridges) of _____ myofilaments act like oars

pulling on _____ molecules of thin filaments.

b. As a result, *(thick? thin?)* myofilaments move inward toward or across the *(H zone and M line? Z line?).*

c. This theory of muscle contraction is called the _____

_____ theory.

2. To describe the sarcomere during contraction, complete each statement with one of the following: lengthens, shortens, or stays the same length.

8

a. The sarcomere _____.

b. Each thick myofilament (A band) _____.

c. Each thin myofilament _____.

d. The I band _____.

e. The H zone _____.

3. Refer to Figure LG 10-2 and identify structures involved in nerve stimulation of muscle in this exercise.

9

a. A nerve impulse travels along *(an axon? a dendrite?)* of a nerve cell (at letter *A*) toward a muscle (letter ____). The axon is enlarged at its end to form a synaptic _____ (at letter *B*).

b. Inside the bulb are synaptic vesicles (letter ____) which release chemical transmitter (letter ____) into the synaptic cleft (letter ____).

c. The effectiveness of the transmitter is enhanced by the fact that the sarcolemma (letter ____) exhibits an extensive surface area providing the infolded _____ (letter *G*).

d. The axon terminal along with the closely approximated muscle fiber sarcolemma is known as a _____ (letter I).

4. Name the parts of a *motor unit.*

5. Muscles that control precise movements have *(more? fewer?)* muscle fibers per motor unit than muscles controlling gross movements.

10

6. Explain how recruitment of motor units helps the body to meet homeostatic requirements of the body at a given time.

7. Complete this exercise describing the principal events that occur during muscle contraction and relaxation.

11

a. A nerve impulse causes release of a chemical transmitter named _____ at a neuromuscular junction. This chemical combines with _____ on the muscle fiber membrane and initiates an impulse across the muscle membrane.

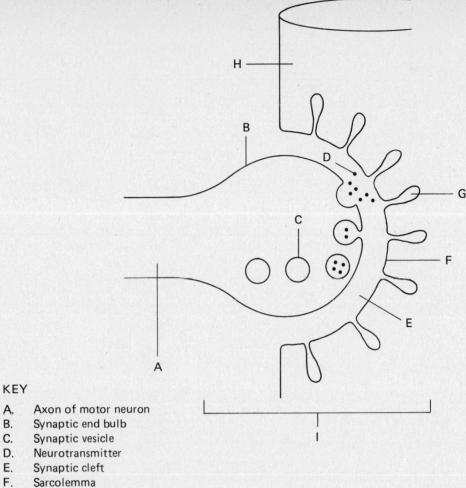

KEY

A. Axon of motor neuron
B. Synaptic end bulb
C. Synaptic vesicle
D. Neurotransmitter
E. Synaptic cleft
F. Sarcolemma
G. Subneural cleft
H. Skeletal muscle myofibril
I. Neuromuscular junction (motor end plate)

Figure LG 10-2 Diagram of a neuromuscular junction (motor end plate). Refer to Learning Activity 9 .

b. In a relaxed muscle the concentration of calcium ions (Ca^{2+}) in sarcoplasm is *(high? low?)*. The effect of the nerve impulse and transmitter is to cause Ca^{2+} to pass from storage areas in _____

_____ into the sarcoplasm surrounding myofilaments.

c. In relaxed muscle myosin cross bridges are not attached to the actin in thin filaments because _____ is bound to myosin cross bridges, while the tropomyosin-_____ complex blocks binding sites on actin.

d. The released calcium ions attach to *(myosin? troponin?)*, causing a structural change which leads to exposure of binding sites on *(myosin? actin?)*.

e. The nerve impulse also causes *(formation? breakdown?)* of ATP located on myosin cross bridges. This results in release of energy which activates myosin cross-bridges to bind to and move _____. The oarlike action of myosin cross bridges (heads of golf clubs) upon actin is called a _____ stroke.

f. Repeated power strokes slide actin filaments *(toward or even across? away from?)* the H zone and M line, and so shorten the sarcomere (and entire muscle).

g. Relaxation of a muscle occurs when an enzyme named _____

_____ destroys ACh. This terminates impulse conduction over the muscle. Calcium ions then move from sarcoplasm back into _____.

h. With a low level of Ca^{2+} now in the sarcoplasm surrounding myofilaments, _____ reforms from ADP on myosin cross bridges, while tropomyosin-troponin complex once again blocks binding sites on _____. As a result, thick and thin filaments detach, slip back into normal position, and the muscle is said to

_____.

i. The movement of Ca^{2+} back into sarcoplasmic reticulum ($\boxed{11}$g is *(an active? a passive?)* process. After death a supply of ATP *(is? is not?)* available, so an active transport process cannot occur. Explain why the condition of rigor mortis results.

$\boxed{12}$

8. Refer to Figure LG 10-3 and complete this exercise about energy sources for muscle contraction.

a. Breakdown of *(ADP? ATP?)* provides the energy muscles use for contraction. (*Hint*: Look at the right side of the figure.) Recall from Activity $\boxed{11}$e that ATP is attached to *(actin? myosin?)* cross bridges and so is available to energize the power stroke. Complete the chemical reaction showing ATP breakdown.

$ATP \rightarrow$

b. ATP must be regenerated constantly. One method involves use of ADP and energy from food sources. Complete that chemical reaction.

$ADP +$

In essence ADP is serving as a transport vehicle which can pick up and drop off energy stored in an extra high energy phosphate bond ($\sim P$).

c. But ATP is used for other cell activities such as _____

_____. To assure adequate energy for muscle work, muscle cells contain an additional molecule for transporting high energy phosphate; this is _____. Complete the reaction showing how the phosphocreatine (PC) and ADP transport "vehicles"

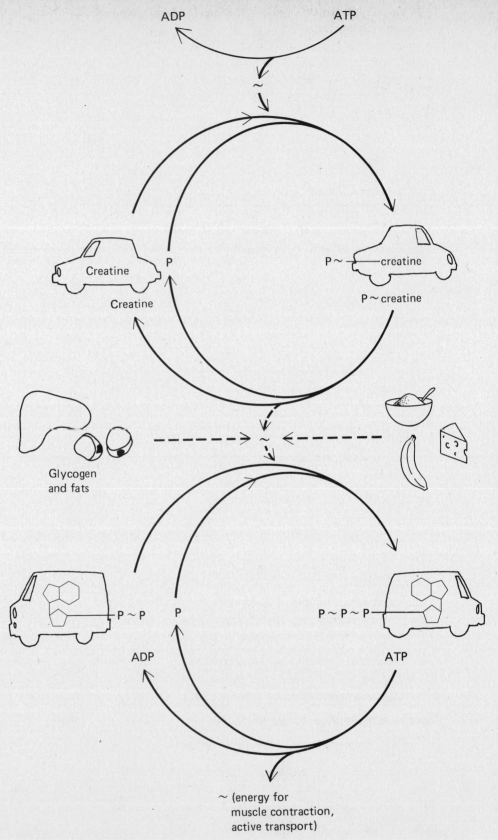

Figure LG 10-3 Diagram of energy sources for muscles. Refer to Learning Activity 12 . P = phosphate, ~ = energy.

can meet and transfer the high energy phosphate "trailer" so that more ATP is formed for muscle work.

$$PC + ADP \rightarrow$$

d. How is CP regenerated during times when muscles are at rest? *Hint*: Look at the far left of the figure.

e. ATP and PC available in a muscle provide only enough energy to power muscle activity for about *(an hour? 10 minutes? 15 seconds?)*. After that time, muscles rely on _____ and later _____ as energy sources.

9. Explain why a muscle must stretch adequately but not excessively in order to produce an optimal contraction.

10. State the all-or-none principle.

This principle applies to *(individual motor units? an entire muscle such as the biceps?)* [13]

11. List three factors that decrease the strength of a muscle contraction. [14]

12. Define *threshold (liminal) stimulus.*

184 (pages 203–205)

D. Kinds of muscle contraction

15 1. Match the terms at right with the definitions below.

_____ a. Rapid, jerky response to a single stimulus   L. Latent period

_____ b. Recording of a muscle contraction   M. Myogram

_____ c. Period between application of a stimulus and start of a contraction   R. Refractory period

_____ d. Period when a muscle is not responsive to a stimulus   T. Twitch

The listing of the terms in the right column:

L. Latent period
M. Myogram
R. Refractory period
T. Twitch

16 2. Choose the type of contraction that fits each descriptive phrase. Answers should each be used twice.

a. Sustained partial contraction of some portion of skeletal muscle; some fibers contracted, other not: _____

b. Sustained contraction due to stimulation at a rate of 30 stimuli per second: _____

c. More forceful contraction of skeletal muscle in response to same strength stimuli after muscle has contracted several times: _____

d. Phenomenon that is the principle behind athletic warmups _____

e. Necessary for maintaining posture _____

f. Voluntary contraction of muscles, such as biceps _____

Muscle tone
Treppe
Tetanus

3. Define *wave summation.* Explain how summation is involved in tetanic contractions.

17 4. Contrast isometric and isotonic contractions by doing this exercise.

a. A contraction in which a muscle shortens while tension (tone) of the muscle remains constant is known as an *(isometric? isotonic?)* contraction.

b. In an isometric contraction the muscle length *(shortens? stays about the same?),* and tension of the muscle *(increases? stays the same?).*

c. When you attempt to lift a piano, but cannot because of its weight, your arm muscles are undergoing iso-_____ contraction.

5. Contrast isometric with isotonic exercises according to advantages and disadvantages of each.

6. *A clinical challenge.* Complete this exercise about clinical aspects of muscle change.
   a. *(Spastic? Flaccid?)* is a term which refers to muscles with less than normal muscle tone. State a cause of flaccid muscles.

   b. _____ is a term which means decrease in size or wasting away of muscles. Explain how transcutaneous muscle stimulations (TMS) may prevent atrophy.

   c. Hypertrophy means an increase in *(number? size?)* of muscle fibers, as in development of muscles through exercise.

E. Types of skeletal muscle fibers (pages 205–206)

1. The substance in muscle that stores oxygen until oxygen is needed by mito- $\boxed{18}$ chondria is _____. This protein is structurally somewhat like the _____-globin molecule in blood which also binds to and stores oxygen. Both of these molecules have a _____ color which accounts for the color of blood and also of red muscle. White muscle fibers have a relatively *(high? low?)* amount of myoglobin.

19

2. Refer to the table below and contrast the three types of skeletal muscle by doing this Learning Activity.

Rate of Contraction	Color	Myoglobin Concentration	Mitochondria and Blood Vessels	Source of ATP	Fatigue Easily?
a. Slow	Red	High	Many	Aerobic	No
b. Fast	Red	Very high	Very high	Aerobic	Moderate
c. Fast	White	Low	Few	Anaerobic	Yes

a. Postural muscles (as in neck and back) are used *(constantly? mostly for short bursts of energy?)* Therefore it is appropriate that they be able to contract at a *(fast? slow?)* rate. Such muscle tissue consists mostly of slow red fibers which *(do? do not?)* fatigue easily. These appear red since they have *(much? little?)* myoglobin and *(many? few?)* blood vessels. They have *(many? few?)* mitochondria, and therefore can depend on *(aerobic? anaerobic?)* metabolism for energy.

b. Muscles of the arms are used *(constantly? mostly for short bursts of energy?)* as in lifting and throwing. Therefore they must contract at a *(fast? slow?)* rate. The arms consist mainly of fast twitch white fibers. These respond rapidly to nerve impulses and contract *(fast? slowly?).* They *(do? do not?)* fatigue easily since they are designed for the relatively inefficient processes of *(aerobic? anaerobic?)* metabolism.

c. Endurance exercises tend to transform fast twitch white muscle fibers into the more efficient (aerobic) _____ fibers, while weight lifting which requires short bursts of energy tends to develop _____ fibers.

(pages 206-209) **F. Cardiac muscle tissue, smooth muscle tissue**

1. Carefully examine the cardiac muscle tissue shown in Figure 10-11 (page 207) in your text. Point out structural differences between cardiac and skeletal muscle tissue by completing this table. Then state the significance

20

of differences for characteristics which have an asterisk (*).

Characteristic	Skeletal Muscle	Cardiac Muscle
a. Shape of muscle fibers		Quadrangular
b. Number of nuclei per fiber	Several (multinucleate)	
c. Location of nuclei in fiber	Peripheral (just under sarcolemma)	
d. Number of mitochondria per fiber (More or Fewer)*		
e. Striated appearance due to alternated actin and myosin myofilaments (Yes or No)	Yes	
f. Arrangement of muscle fibers (Parallel or Branching)		
g. Nerve stimulation required for contraction (Yes or No)*		
h. Length of refractory period (Long or Short)*		

2. Define and state the function of each of these cardiac muscle structures:
   a. Gap junction

b. Intercalated disc

21 3. Contrast smooth muscle with the muscle tissue you have already studied by circling the correct answers in this paragraph.

Smooth muscle fibers are *(cylinder? spindle?)*-shaped with *(several nuclei? one nucleus?)* per cell. They *(do? do not?)* contain actin and myosin. However, due to the irregular arrangement of these filaments, smooth muscle tissue appears *(striated? nonstriated or "smooth"?)*. In general, smooth muscle contracts and relaxes more *(rapidly? slowly?)* than skeletal muscle does, and smooth muscle holds the contraction for a *(shorter? longer?)* period of time than skeletal muscle does.

4. State the functions of *intermediate filaments* and *dense bodies* in smooth muscle tissue.

5. Compare the two types of smooth muscle.

Muscle Type	Structure	Spread of Stimulus	Locations
a. Visceral		Impulse spreads and causes contraction of adjacent fibers	
b.			Blood vessels, iris of eye

6. Contraction of *(skeletal? smooth?)* muscle is usually more prolonged. Explain why this is so.

7. Smooth muscle *(is? is not?)* normally under voluntary control. List three chemicals released in the body which can also lead to smooth muscle contraction. 22

8. Define *stress/relaxation* and explain how it relates to your own stomach as you are consuming a large meal. 23

G. Homeostasis

(page 201)

1. Complete this exercise about muscle contraction with or without oxygen. 24

   a. _____ is the main food source that is broken down by human cells to release energy.

   b. Glucose is readily converted to _____ acid with release of a small amount of energy. For complete breakdown of glucose (to _____ and _____ and much energy), oxygen *(is? is not?)* required.

   c. During strenuous exercise an adequate supply of oxygen may not be available to muscles. They must convert pyruvic acid to _____ _____ acid by an *(aerobic? anaerobic?)* process. Excessive amounts of lactic acid in muscle tissue may account for muscle _____.

   d. After exercising, a person breathes rapidly for a while. During this period sufficient oxygen enters the body to "pay back the _____ _____ debt," that is, to provide oxygen for the complete catabolism of stored lactic acid to _____, _____, and much energy in the form of _____.

2. Indicate what percent of energy used during muscular contraction may go toward each of these purposes: 25

   a. Mechanical work (contraction): _____ percent

   b. Heat production (to maintain body temperature): _____ percent

3. Contrast two types of heat production by muscles: initial heat and recovery heat.

(pages 210–211) **H. Aging and development of muscles**

1. Describe three changes in skeletal muscles that occur with normal aging.

☐26 2. Complete this exercise about the development of muscles.
   a. Which of the three types of muscle tissue develop from mesoderm? *(Skeletal? Cardiac? Smooth?).*
   b. Part of the mesoderm forms columns on either side of the developing nervous system. This tissue segments into blocks of tissue called _____. The first pair of somites forms on day *(10? 20? 30?)* of gestation. By day 30 a total of _____ pairs of somites are present.
   c. Which part of a somite develops into vertebrae? *(Myo-? Derma-? Sclero-?)* derm. The dermis of the skin and other connective tissues are formed from _____-tomes, while skeletal muscles develop from _____-tomes.
3. State three examples of different patterns by which myotomes in various locations develop into skeletal muscles.

I. Disorders, medical terminology, drugs

1. Match the name of the disorder at right with the correct description. 27

   a. "Charleyhorse": _____

   _____

   b. An inflammation of fibrous tissue, especially of fascia over muscle; called lumbago if occurs in lumbar region:

   _____

   c. Formation of fibrous tissue where it does not belong, for example, replacing skeletal muscle of cardiac tissue:

   _____

   d. Inherited, muscle-destroying disease causing atrophy of muscle tissue:

   _____

   Fibromyositis
   Fibrosis
   Fibrositis
   Muscular dystrophy

2. Describe myasthenia gravis in this exercise. 28
   a. In order for skeletal muscle to contract, a nerve must release the chemical

   _____ at the myoneural junction. Normally ACh

   binds to _____ on the muscle fiber membrane.
   b. It is believed that a person with myasthenia gravis · produces

   _____ that bind to these receptors, making them unavailable for ACh binding. Therefore, ACh *(can? cannot?)* stimulate the muscle, and it is weakened.
   c. One treatment for this condition employs a chemical called

   _____ which enhances muscle contraction by permitting the ACh molecules that *do* bind to act longer (and not be destroyed by AChE).

   d. Other treatments include _____-suppressants, which

   decrease the patient's antibody production, or _____

   _____, which segregates the patient's blood cells containing the harmful antibodies.
3. Define each of these types of abnormal muscle contractions.
   a. Spasm

   b. Cramp

   c. Convulsion

192

d. Fibrillation

e. Tic

29 4. Write the correct medical term after its description.

a. The study of muscles: _____

b. A muscle tumor: _____

c. The recording and study of electrical changes that occur in muscle tissue:

_____

d. Any disease of muscle: _____

e. Loss or impairment of motor or muscular function due to nerve or muscle

disorder: _____

Answers to Numbered Questions in Learning Activities

1 (a) Excitability. (b) Contractility. (c) Extensibility. (d) Elasticity.

2 (a) Sm. (b) C. (c) Sk.

3 (a) S. (b) D. (c) S. (d) S.

4 (a) EPE. (b) T. (c) A. (d) TS.

5 (a) Cells, intercellular material. (b) Myofiber, sarcolemma, sarcoplasm. (c) Sarcolemma. (d) Membrane-enclosed tubules, much like smooth endoplasmic reticulum. (e) Perpendicular. (f) Triad. (g) Myofilaments, sarcomeres. (h) A, Z line. (i) A. I, striated. (j) Thick, H, M. (k) I band which is labeled *F*. (l) Actin, myosin, tropomyosin, troponin. (m) Thick, golf clubs, actin and ATP.

6 Sarcomere *(A);* A band *(B);* thin filament *(C);* thick filament *(D);* Z line *(E);* I band *(F);* M line *(G);* H zone *(H).*

7 (a) Myosin, actin. (b) Thin, H zone and M line. (c) Sliding filament.

8 (a) Shortens. (b) Stays the same length. (c) Stays the same length. (d) Shortens. (e) Shortens or disappears.

9 (a) An axon, H, bulb. (b) *C, D. E.* (c) *F,* subneural clefts. (d) Neuromuscular junction or motor end plate.

10 Fewer.

11 (a) Acetylcholine (ACh), receptors. (b) Low, sarcoplasmic reticulum. (c) ATP, troponin. (d) Troponin, actin. (e) Breakdown, actin, power. (f) Toward or even across. (g) Acetylcholinesterase (AChE), sarcoplasmic reticulum. (h) ATP, actin (or thin filament), relax (or lengthen). (i) An active, is not; myosin cross bridges stay attached to actin and the muscles remain in a state of partial contraction (rigor mortis).

12 (a) ATP, myosin, ATP → ADP + P + energy. (b) ADP + P + energy (from foods) → ATP. (c) Active transport, creatine which combines with phosphate to form phosphocreatine (PC), PC + ADP → ATP + creatine (C). (d) ATP + creatine (C) → ADP + phosphocreatine (PC). (e) 15 seconds, glycogen, fats.

13 Individual motor units.

14 Fatigue, lack of nutrients, lack of oxygen.

15 (a) T. (b) M. (c) L. (d) R.

16 (a) Muscle tone. (b) Tetanus. (c) Treppe. (d) Treppe. (e) Muscle tone. (f) Tetanus.

17 (a) Isotonic. (b) Stays about the same, increases. (c) -Metric.

18 Myoglobin, hemo-, red, low.

19 (a) Constantly, slow, do not, much, many, many, aerobic. (b) Mostly for short bursts of energy, fast, fast, do, anaerobic. (c) Fast twitch red, fast twitch white.

20 (a) Cylindrical. (b) One. (c) Central. (d) Fewer, More since heart muscle requires constant generation of energy. (e) Yes. (f) Parallel, Branching. (g) Yes, No, so heart can contract without nerve stimulation, but nerves can increase or decrease heart rate. (h) Short, Long, so allows heart time to relax between beats.

21 Spindle, one nucleus, do, nonstriated or "smooth," slowly, longer.

22 Is not; hormones, pH changes, O_2 and CO_2 levels, and certain ions.

23 Smooth muscle fibers can stretch greatly without developing tension. So your stomach can accommodate a large meal without building up such tension in muscle fibers as to interfere with effective contractions.

24 (a) Glucose. (b) Pyruvic, carbon dioxide, water, is. (c) Lactic, anaerobic, fatigue. (d) Oxygen, carbon dioxide, water, ATP.

25 (a) 15. (b) 85.

26 (a) All three: skeletal, cardiac, and smooth. (b) Somites, 20, 44. (c) Sclero-, derma-, myo-.

27 (a) Fibromyositis. (b) Fibrositis. (c) Fibrosis. (d) Muscular dystrophy.

28 (a) Acetylcholine (ACh), receptors. (b) Antibodies, cannot. (c) Anticholinesterase (anti-AChE). (d) Immuno-, plasmapheresis.

29 (a) Myology. (b) Myoma. (c) Electromyography (EMG). (d) Myopathy. (e) Paralysis.

MASTERY TEST: Chapter 10

Questions 1–10: Circle T (true) or F (false). If the statement is false, change the underlined word or phrase so that the statement is correct.

T   F   1. Tendons are <u>cords of connective tissue, whereas aponeuroses are broad, flat bands</u> of connective tissue.

T   F   2. During contraction of muscle <u>both A bands and I bands</u> get shorter.

T   F   3. Fascia is <u>one type of skeletal muscle tissue.</u>

T   F   4. Gap junctions <u>inhibit</u> nerve conduction from one muscle fiber to another.

T   F   5. Myasthenia gravis is caused by <u>an excess of acetylcholine production at the myoneural junction.</u>

T   F   6. Muscle fibers remain relaxed if there are <u>no calcium ions in the sarcoplasm.</u>

T   F   7. <u>Atrophy</u> means decrease in muscle mass.

T   F   8. Muscles that are used mostly for quick bursts of energy (such as those in the arms) contain large numbers of <u>fast twitch white fibers.</u>

T   F   9. Most of the energy released during muscle contraction is used for <u>heat production.</u>

T   F   10. In a relaxed muscle fiber <u>thin and thick myofilaments overlap to form the A band.</u>

T   F   11. As a result of aging, muscle tissue is largely <u>replaced by fat,</u> and muscle strength <u>decreases.</u>

Questions 12–13: Arrange the answers in correct sequence.

_____ _____ _____ 12. According to the amount of muscle tissue they surround, from most to least:

        A. Perimysium
        B. Endomysium
        C. Epimysium

_____ _____ _____ 13. From largest to smallest:

        A. Myofibril
        B. Myofilament
        C. Muscle fiber

Questions 14–20: Choose the one best answer to each question.

_____ 14. All of the following molecules are parts of thin filaments EXCEPT:

        A. Actin   B. Myosin   C. Tropomyosin
        D. Troponin

_____ 15. Choose the one statement that is FALSE.

        A. *A band* refers to the anisotropic band.
        B. The A band is darker than the I band.
        C. Thick myofilaments reach the Z line in relaxed muscle.
        D. The H zone contains thick myofilaments, but not thin ones.
        E. Thick myofilaments are made of myosin.

_____ 16. Which of the following statements about the role of calcium in muscle contraction is FALSE?

    A. Calcium ions are not necessary for muscle contraction.

    B. Calcium ions are released from sarcoplasmic reticulum to sarcoplasm as a result of nerve stimulation.

    C. Calcium ions are actively transported into the sarcoplasmic reticulum by calsequestrin and calcium ATPase.

    D. Calcium ions permit the tropomyosin-troponin complex to split from thin filaments so myosin can bind to actin.

_____ 17. Which statement about muscle physiology in the relaxed state is FALSE?

    A. Myosin cross bridges are bound to ATP.

    B. Calcium ions are stored in sarcoplasmic reticulum.

    C. Myosin cross bridges are bound to actin.

    D. Tropomyosin-troponin complex is bound to actin.

_____ 18. The staircase phenomenon refers to _____ contractions.

    A. Tetanic   B. Treppe   C. Tonic   D. Isotonic

_____ 19. Choose the FALSE statement about cardiac muscle.

    A. Cardiac muscle has a long refractory period.

    B. Cardiac muscle has one centrally located nucleus per fiber.

    C. Cardiac muscle cells are called cardiac muscle fibers.

    D. Cardiac fibers are separated by intercalated discs.

    E. Cardiac fibers are spindle-shaped with no striations.

_____ 20. Most voluntary movements of the body are results of _____ contractions.

    A. Isometric   B. Fibrillation   C. Twitch

    D. Tetanic   E. Spasm

Questions 21–25, fill-ins. Complete each sentence with the word or phrase that best fits.

_____ 21. Intercalated discs are specially modified _____ of cardiac muscle.

_____ 22. Cardiac muscle remains contracted longer than skeletal muscle does because _____ is slower in cardiac than in skeletal muscle.

_____ 23. It is during the _____ period of a muscle contraction that calcium ions are released from sarcoplasmic reticulum and myosin cross bridge activity begins to occur.

_____ 24. _____ is the transmitter released from synaptic vesicles of axons supplying skeletal muscle.

_____ 25. ADP + phosphocreatine → _____. (Write the products.)

The Muscular System

In this last chapter of the musculoskeletal unit, you will learn how skeletal muscles, by means of their attachments to bones, produce the wide array of movements of the body (Objective 1). You will consider the significance of structural patterns of muscles and their arrangement in functional groups (2–4). You will study criteria which will facilitate your learning names of muscles (5). Then you will identify the major skeletal muscles of the body with regard to origin, insertion, action, and innervation (6). You will also study drug administration by intramuscular injections (7).

Topics Summary

A. How skeletal muscles produce movement
B. Naming skeletal muscles
C. Principal skeletal muscles of the head and neck
D. Principal skeletal muscles that act on the abdominal wall, muscles used in breathing, muscles of the pelvic floor and perineum
E. Principal skeletal muscles that move the shoulder girdle, upper extremity and the vertebral column
F. Principal skeletal muscles that move the vertebral column
G. Principal skeletal muscles that move the lower extremity
H. Intramuscular injections

Objectives

1. Describe the relationship between bones and skeletal muscles in producing body movements.
2. Define a lever and fulcrum and compare the three classes of levers on the basis of placement of the fulcrum, effort, and resistance.
3. Identify the various arrangements of muscle fibers in a skeletal muscle and relate the arrangements to the strength of contractions and range of movement.
4. Discuss most body movements as activities of groups of muscles by explaining the roles of the prime mover, antagonist, and synergist and fixator.
5. Define the criteria employed in naming skeletal muscles.
6. Identify the principal skeletal muscles in different regions of the body by name, origin, insertion, action, and innervation.
7. Discuss the administration of drugs by intramuscular injection.

(pages 216–218)   **A. How skeletal muscles produce movement**

1. What structures constitute the *muscular system?*

2. Refer to Figure LG 11-1 and consider flexion of your own forearm as you do this learning activity.

☐ 1

   a. In flexion your forearm serves as a rigid rod, or _____

   _____, which moves about a fixed point, called a

   _____ (your elbow joint, in this case).

   b. Hold a weight in your hand as you flex your forearm. The weight plus your forearm serve as the *(effort? fulcrum? resistance?)* during this movement.

   c. The effort to move this resistance is provided by contraction of a

   _____. Note that if you held a heavy telephone book

   in your hand while your forearm is flexed, much more _____

   _____ by your arm muscles would be required.

   d. In Figure LG 11-1 identify the exact point at which the muscle causing flexion attaches to the forearm. It is the *(proximal? distal?)* end of the *(humerus? radius? ulna?)*. Write an E and an I on the two lines next to the arrow at that point in the figure. This indicates that this is the site where the muscle exerts its effort (E) in the lever system, and it is also the insertion (I) end of the muscle. (More about insertions in a minute.)

   e. Each skeletal muscle is attached to at least two bones. As the muscle shortens, one bone stays in place and so is called the *(origin? insertion?)* end of the muscle. What bone in the figure appears to serve as the origin

   bone? _____ Write O on the line at that point in the figure.

   f. The attachment of the muscle to the bone that moves is called the

   _____ end of the muscle. As you already noted, this

   bone is the _____.

Figure LG 11-1 The lever-fulcrum principle is illustrated by flexion of the forearm. Complete the figure as directed in Learning Activity ☐ 1 .

3. Describe the roles of lever systems in the body by completing this exercise. $\boxed{2}$
   a. There are *(2? 3? 5? 7?)* classes of lever systems. Levers are categorized according to relative positions of effort (E), fulcrum (F), and resistance (R).
   b. Study Figure 11-2 (page 217) in your text, and identify a lever system which is arranged F E R from proximal to distal in the body. This is the arrangement of a *(first-? second-? third-?)* class lever. This is the *(most? least?)* common type of lever in the body. State one example.

   c. Now hyperextend your head as if to look at the sky (and refer to Figure 11-2a, page 217 of your text). The weight of your face and jaw serves as *(E? F? R?)*, while your neck muscles provide *(E? F? R?)*. The fulcrum is

   the joint between the _____ and the _____

   bones. This is an example of a _____-class lever.
   d. State one example of a movement involving a second-class lever.

4. Briefly describe the four basic patterns of muscle fiber arrangement within fasciculi (bundles). Give one example of each. As you study the muscles throughout this chapter, add other examples.
   a. Parallel

   b. Convergent

   c. Pennate

   d. Circular

5. Correlate fascicular arrangement with muscle power and range of motion of muscles. $\boxed{3}$
   a. A muscle with *(many? long?)* fibers will tend to have great strength. An example is the *(parallel? pennate?)* arrangement.
   b. A muscle with *(many? long?)* fibers will tend to have great range of motion. An example is the *(parallel? pennate?)* arrangement.
6. Refer again to Figure LG 11-1 and do this exercise about how muscles of the body work in groups. $\boxed{4}$

a. The muscle that contracts to cause flexion of the forearm is called a

_____. An example of a prime mover in this action

would be the _____ muscle.

b. The triceps brachii must relax as the biceps brachii flexes the forearm. The triceps is an extensor. Since its action is opposite to that of the biceps, the triceps is called *(a synergist? an agonist? an antagonist?)* of the biceps.

c. What would happen if the flexors of your forearm were functional, but not the antagonistic extensors?

d. What action would occur if both the flexors and extensors contracted simultaneously?

e. Muscles that assist or cooperate with the prime mover to cause a given

action are known as _____, while muscles that stabil-

izc a bone (such as the scapula) so that prime movers and synergists can

move another bone (such as the humerus) are called _____

_____.

(pages 218–219)   **B. Naming skeletal muscles**

1. If you study the names of muscles, you will find that most of them provide a good description of the muscle. For each of the following, indicate the type of clue that each part of the name gives. The first one is done for you.

| 5 |

_____ a. Rectus abdominis

_____ b. Gluteus maximus

_____ c. Biceps brachii

_____ d. Sternocleidomastoid

_____ e. Adductor longus

A. Action
D. Direction of fibers
L. Location
N. Number of heads or origins
P. Points of attachment of origin and insertion
S. Size or shape

| 6 |

2. Read the following descriptions of muscles, noting italicized words that give direct clues to muscle names. Guess the names of these muscles. As you proceed with your study of specific muscles, identify muscles whose names match these descriptions. Then return to this exercise and see if the names you gave initially were correct.

a. A *circular* muscle around the *mouth*

b. *Elevates* the *upper lip*

c. Attached to the *styloid* process of the temporal bone and also to the *tongue*

d. A muscle whose fibers run vertically *(straight)* up and down the *abdomen*

e. A muscle that *lifts* the *scapula*

f. The *broadest* muscle of the *back*

g. A *triangular* muscle like the Greek letter delta (Δ)

h. Located *above* the *spine* of the scapula

i. A *round* muscle that *pronates* the forearm

j. A muscle that *flexes* the *carpals* and lies along the *ulna*

k. A muscle that *tenses* a *wide* band of *fascia* on the thigh

l. A *four-headed* muscle lying over the *femur*

m A *large* muscle lying on the *lateral* side of the thigh

n. A *slender* muscle

o. A muscle that permits you to sit cross-legged, the position of *tailors*

p. A muscle that lies *anterior* to the *tibia*

q. A *long* muscle that *extends* the toes *(digits)*

C. Principal skeletal muscles of the head and neck (pages 219–231)

1. After studying Exhibit 11-2 (pages 222–223) in your text, check your understanding of the muscles of facial expression. Write the name of the muscle that answers each description. Locate muscles with asterisk (*) in Figure LG 11-2. Cover KEY and write name of each facial muscle next to its lettered lead line. [7]

   a. Elevates upper eyelid: _____

   b. Elevates upper lip: _____

   c. Produces a frowning expression: _____

   d. A flat muscle that causes pouting action; draws lower lip downward and backward: _____

   e. Major cheek muscle; allows you to blow air out of mouth and produce sucking action: _____

   f. Allows you to show surprise by raising your eyebrows and forming horizontal forehead wrinkles*: _____

   g. Muscle surrounding opening of your mouth; allows you to use your lips in kissing and in speech*: _____

   h. Muscle for smiling and laughing since it draws the outer portion of the mouth upward and outward*: _____

   i. Circular muscle around eye; closes eye*: _____

2. Essentially all of the muscles controlling facial expression receive nerve impulses via the _____ nerve, which is cranial nerve *(III? V? VII?)*. [8]

202

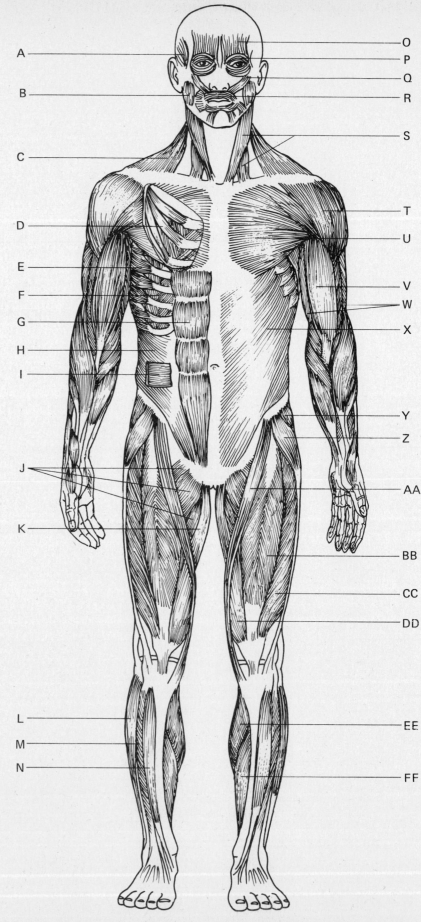

A

B

C

D

E

F

G

H

I

J

K

L

M

N

O
P
Q
R

S

T
U

V
W
X

Y
Z

AA

BB

CC
DD

EE

FF

KEY

A. Temporalis
B. Masseter
C. Trapezius
D. Pectoralis minor
E. Serratus anterior
F. Intercostals
G. Rectus abdominis
H. Internal oblique
I. Transversus abdominis
J. Adductor group
K. Gracilis
L. Peroneus longus
M. Extensor digitorum longus
N. Tibialis anterior
O. Frontalis
P. Orbicularis oculis
Q. Zygomaticus major
R. Orbicularis oris
S. Sternocleidomastoid
T. Deltoid
U. Pectoralis major
V. Biceps brachii
W. Brachialis
X. External oblique
Y. Gluteus medius
Z. Tensor fasciae latae
AA. Sartorius
BB. Rectus femoris
CC. Vastus lateralis
DD. Vastus medialis
EE. Gastrocnemius
FF. Soleus

Figure LG 11-2 Major muscles of the body, anterior view. Some deep muscles are shown on the left side of the figure. All muscles on the right side are superficial. Label as directed.

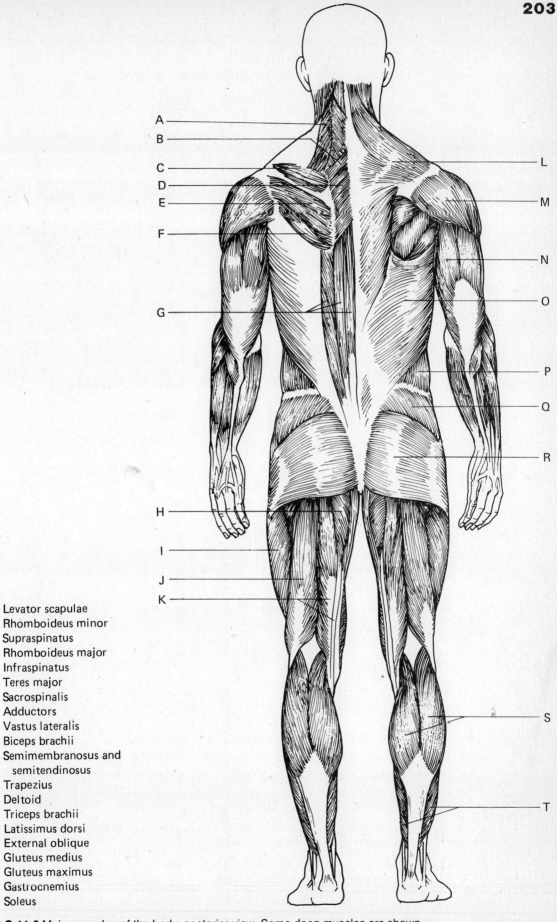

KEY

- A. Levator scapulae
- B. Rhomboideus minor
- C. Supraspinatus
- D. Rhomboideus major
- E. Infraspinatus
- F. Teres major
- G. Sacrospinalis
- H. Adductors
- I. Vastus lateralis
- J. Biceps brachii
- K. Semimembranosus and
     semitendinosus
- L. Trapezius
- M. Deltoid
- N. Triceps brachii
- O. Latissimus dorsi
- P. External oblique
- Q. Gluteus medius
- R. Gluteus maximus
- S. Gastrocnemius
- T. Soleus

Figure LG 11-3 Major muscles of the body, posterior view. Some deep muscles are shown on the left side of the figure. All muscles on the right side are superficial. Label as directed.

3. Place your index finger and thumb on the origin and insertion of each of the muscles that move your lower jaw. (Refer to Exhibit 11-3 and Figure 11-5, page 224 in the text for help.) Then do this learning activity.

   9

   a. Two large muscles help you to close your mouth forcefully, as in chewing. Both of these act by *(lowering the maxilla? elevating the mandible?)*. The

   _____ covers your temple and the _____

   _____ covers the ramus of the mandible.

   b. Most of the muscles involved in chewing are ones that help you to *(open? close?)* your mouth. Think about this the next time you go to the dentist and try to hold your mouth open for a long time, with only your

   _____ muscles to force your mouth wide open.

   c. The muscles that move the lower jaw, and so aid in chewing, are

   innervated by cranial nerve *(III? V? VII?)*, known as the _____

   _____ nerve.

   d. Refer to Figure LG 11-2. Muscles *A* and *B* are used primarily for *(facial expressions? chewing?)*, while muscles *O, P, Q,* and *R* are used mainly for *(facial expressions? chewing?)*.

4. *For extra review.* With the help of Figures 11-4 and 11-5 (pages 222–224) in your text and a mirror, locate each of the major facial muscles.

   10

5. Complete this exercise about muscles that move the eyeballs.

   a. Why are these muscles called *extrinsic* eyeball muscles?

   b. All of the *rectus* muscles move the eyeball in the direction *(opposite to? the same as?)* that given in the muscle name. For example, the superior rectus moves the eyeball *(superiorly? inferiorly?)*.

   c. The superior and inferior oblique muscles both move the eyeballs *(medially? laterally?)* as well as toward the direction *(opposite to? same as?)* that given in the name. For example, the inferior oblique movs the eyeball

   _____ .

   d. Only one muscle moves the eyeball at all in the medial direction. This is

   the _____ muscle.

6. *For extra review* of actions of eye muscles, work with a study partner. One person moves the eyes in a particular direction; the partner then names the eye muscles used for that action. Note: will both eyes use muscles of the same name? For example, as you look to your right, will both eyes contract the lateral rectus?

   11

7. Actions such as opening your mouth for eating or speaking and also swallowing require integrated action of a number of muscles. You have already studied some that allow you to open your mouth (such as masseter and temporalis). We will now consider muscles of the tongue, floor of the oral cavity, pharynx, and larynx.

   Identify locations of three bony structures relative to your tongue. Tell whether each is superior or inferior and anterior or posterior to your tongue. (For help refer to Figure 11-7, page 227 of the text.)

a. Styloid process of temporal bone

b. Hyoid bone

c. Chin (anterior portion)

8. Remembering the locations of these bony points and that muscles move a structure (tongue) by pulling on it, determine the direction that the tongue is pulled by each of these muscles. [12]
   a. Styloglossus
   b. Hyoglossus
   c. Genioglossus

9. The three muscles listed in Learning Activity [12] are are innervated by [13] cranial nerve _____; the name _____ nerve indicates that this pair of cranial nerves supplies the region "under the tongue."

10. Look at Figure 11-9, page 230 in your text and a human skull (or Figure [14] 7-9, page 145 in your text). Feel your own mandible. It *(is U-shaped? forms a solid, bony floor to your oral cavity?)*. Name three muscles that do form the floor of the oral cavity, just inferior to your tongue.

11. Match groups of muscles in each answer with descriptions below. [15]
    A. Styloglossus, hyoglossus, and genioglossus
    B. Superior, middle, and inferior constrictors
    C. Stylopharyngeus, salpingopharyngeus, and palatopharyngeus
    D. Stylohyoid, mylohyoid, and geniohyoid
    E. Thyrohyoid, sternohyoid, and omohyoid

    _____ a. Located superior to the hyoid (suprahyoid), these form the floor of the oral cavity.
    _____ b. "Strap muscles" that are infrahyoid, these muscles cover the anterior of the larynx and trachea. These permit you to elevate your thyroid cartilage ("Adam's apple") during swallowing to prevent food from entering your larynx. (Try it.)

_____ c. Forming part of the walls of the pharynx, these muscles also elevate the larynx during swallowing. And they close off the nasopharynx (so that food does not back up into the nasal cavity) during swallowing. Finally one of these muscles opens your auditory tube (leading to the middle ear) during swallowing, an action that helps to "pop" your ears while you are descending from a high altitude.

_____ d. Three muscles that squeeze the pharynx to propel a bolus of food into the esophagus during swallowing.

_____ e. Muscles permitting tongue movements.

12. Using a mirror, find the origin and insertion of your left sternocleidomastoid muscle. (See Figure 11-4, pages 222–223, in your text.) The muscle contracts when you pull your chin down and to the right; this diagonal muscle of your neck will then be readily located. Note that the left sternocleidomastoid pulls your face toward the *(same? opposite?)* side. It also *(flexes? extends?)* the head.

| 16 |

| 17 |

13. Alternately flex and extend vertebrae of your neck by looking at your toes and then toward the sky. As you look at the sky, you are *(flexing? extending and then hyperextending?)* your head and cervical vertebrae. On which surface of the neck would you expect to find extensors of the head and neck? *(Anterior? Posterior?)* Note that these muscles are the most superior muscles of the columns of extensors of the vertebrae. Find them on Figure 11-19 (page 247) in your text. Now name three of these muscles.

| 18 |

14. Turn to Figure LG 9-2 (page LG 165). Write in the names of muscles that produce the actions *A* and *B*.

15. *For extra review,* cover the KEY to Figure LG 11-2 and write labels of muscles of the head and neck.

(pages 232–237) **D. Principal skeletal muscles that act on the abdominal wall, muscles used in breathing, muscles of the pelvic floor and perineum**

| 19 |

1. Each half of the abdominal wall is composed of *(two? three? four?)* muscles. Describe these in the exercise below.

a. Just lateral to the midline is the rectus abdominis muscle. Its fibers are

(vertical? horizontal?), attached inferiorly to the _____

_____ and superiorly to the _____

_____. Contraction of this muscle permits you to bend *(forward? backward?);* this action is *(flexion? extension?)* of the vertebral column.

b. Name the remaining abdominal muscles that form the sides of the abdominal wall. List them from most superficial to deepest.

_____

_____

_____

c. Do all three of these muscles have fibers running in the same direction? *(Yes? No?)* Of what advantage is this?

2. List three sets of muscles used during breathing.

3. Answer these questions about muscles used for breathing. |20|

a. The diaphragm is _____-shaped. Its oval origin is located _____. Its insertion is not into bone, but rather into dense connective tissue forming the roof of the diaphragm; this tissue is called the _____

_____.

b. Contraction of the diaphragm flattens the dome, causing the size of the thorax to *(increase? decrease?)*, as occurs during *(inspiration? expiration?)*.

c. The name *intercostals* indicates that these muscles are located

_____. Which set are used during expiration? *(Internal? External?)*

4. *For extra review,* cover KEY to Figure LG 10-2 and write labels for muscles *F, G, H, I,* and *X.*

5. Look at the inferior of the human pelvic bones on a skeleton (or refer to Figure 8-7, page 165 in your text). Note that a gaping hole (outlet) is present. Pelvic floor muscles attach to the bony pelvic outlet. Name these muscles and state functions of the pelvic floor.

208

6. In the space below draw a diamond about the size and shape of the perineum. For help, again look at a skeleton and refer to Figures 28-22 (page
728) and 11-13 and 11-14 (pages 235 and 237) in your text.

21
a. Label the four points that demarcate the perineum.
b. Label the *urogenital triangle* and the *anal triangle*.
c. What openings are located in these triangles in males? In females? Draw
22
and label these in contrasting colors on your diagram.
d. Draw in and label the following muscles as they appear in females: *bulbo-
cavernosus, ischiocavernosus, levator ani,* and *sphincter ani externus.* Note
which of these form parts of the penis, and so will be considerably larger
in males than in females.

23
7. Fill in the blanks in this paragraph. *Diaphragm* means literally
_____ *(dia-)* _____

_____ *(-phragm).* You are familiar with the diaphragm that separates the _____ from the abdomen.
The *pelvic diaphragm* consists of all of the muscles of the _____

_____ floor plus their fasciae. It is a "wall" or barrier
between the inside and outside of the body.

(pages 238–245)
**E. Principal skeletal muscles that move the shoulder girdle and upper ex-
tremity**

1. On Figures LG 11-4 and LG 11-5, draw in pencil and label the muscles listed
in the table below. Draw superficial muscles on the right side and deep
muscles on the left side of the figures. Be sure to locate precise origins and

insertions. Mark *O* and *I* on the origin and insertion ends of each muscle. Think about how origins and insertions determine what movements occur. Then use the spaces provided here for writing principal actions of each muscle. (Note: Muscles of particular importance are marked with an asterisk.) For help, refer to Exhibit 11-14 and Figure 11-15, pages 000–000 in your text.

Muscle	Action
a. Subclavius	
b. Pectoralis minor	
c. Serratus anterior*	
d. Trapezius*	
e. Levator scapulae	
f. Rhomboideus major	
g. Rhomboideus minor	

2. All of the muscles listed in the table are involved with movement of the _____. All are *(superficial? deep?)* except the trapezius and parts of serratus anterior.
3. Using a red pencil or pen (for contrast) add to these figures drawings of the muscles listed below. Again, draw superficial muscles on the left and deep muscles on the right. Label *O* and *I*. List actions in the table. (*Note:* Muscles of particular importance are marked with an asterisk.)

Muscle	Action
a. Pectoralis major*	
b. Deltoid*	

Muscle	Action
c. Supraspinatus	
d. Infraspinatus	
e. Latissimus dorsi*	
f. Teres major	
g. Teres minor	

4. All of the muscles listed in the above table are directly involved with movement of the *(shoulder girdle? humerus? radius/ulna?)*. All of these muscles are *(superficial? deep?)* with the exception of the supraspinatus and infraspinatus.

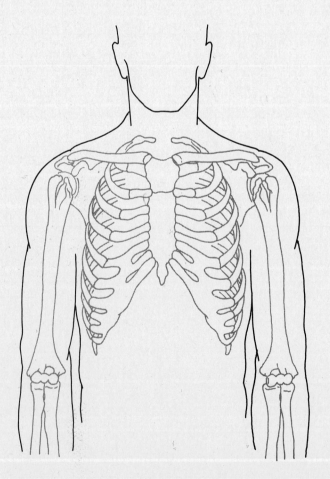

Figure LG 11-4 Anterior view of part of the skeleton. Draw and label muscles as directed.

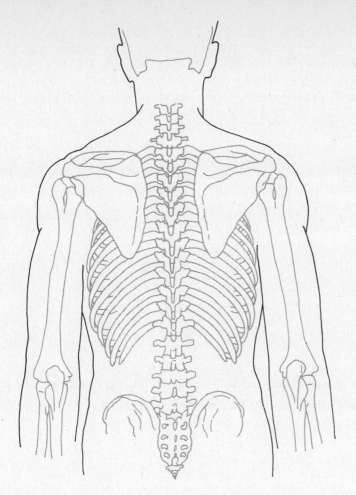

Figure LG 11-5 Posterior view of part of the skeleton. Draw and label muscles as directed.

5. Using green pencil or pen, draw and label three muscles that move the forearm, and mark *O* and *I,* use the right sides of Figures LG 11-4 and LG 11-5. Then list actions here.

6. *For extra review,* try to locate each of the muscles named in the above three lists:
   a. On a skeleton, find origins and insertions and visualize actions.
   b. On yourself or on a partner, observe surface anatomy and actions.
   c. On Figures LG 11-2 and 11-3, cover the KEY and write names of muscles next to each lettered lead line. Be sure that you can identify and label all muscles on those figures except omit those moving the vertebrae and the lower extremities.
7. Name one or two muscles that produce each of the actions *C-H* shown in Figure LG 9-2 (page LG 165). Write the names next to each figure. 24

25   8. Complete this exercise about muscles that move the wrist and fingers.

    a. Examine your own forearm, palm, and fingers. There is more muscle mass on the *(anterior? posterior?)* surface. You therefore have more muscles that can *(flex? extend?)* your wrist and fingers.

    b. Locate the flexor carpi ulnaris muscle on Figure 11-18, page 245 in your text. What action does it have other than flexion of the wrist?_____

    _____ What muscles would you expect to abduct the wrist?

    c. What is the difference in location between flexor digitorum superficialis and flexor digitorum profundus?

    d. What muscle helps you to point (extend) your index finger?

   9. *For extra review* of muscles that move the upper extremities, write the name

26   of one or more muscles that fit these descriptions.

    a. Covers most of the posterior of the humerus

    b. Turns your hand from palm down to palm up position

    c. Originates from upper eight or nine ribs; inserts on scapula; moves scapula laterally

    d. Two muscles other than the biceps brachii that flex the forearm

    e. Used when a baseball is grasped

    f. Antagonist to levator scapulae

    g. Largest muscle of the chest region; used to throw a ball in the air (flex humerus) and to adduct arm

    h. Raises or lowers scapula, depending on which portion of the muscle contracts

i. Controls action at the elbow for a movement such as the downstroke in hammering a nail

j. Hyperextends the humerus, as in doing the "crawl" stroke in swimming or exerting a downward blow; also adducts the humerus

F. Principal skeletal muscles that move the vertebral column   (pages 246–247)

1. Describe the muscles that comprise the sacrospinalis. Locate and label on Figure LG 11-3.

   a. The sacrospinalis muscle is also called the _____.

   b. The muscle consists of three groups: _____ (lateral),

   _____ (intermediate), and _____

   _____ (medial).

   c. In general, these muscles have attachments between _____

   _____.

   d. They are *(flexors? extensors?)* of the vertebral column, and so are *(synergists? antagonists?)* of the rectus abdominis muscles.

2. Explain why it is common for women in their final weeks of pregnancy to experience frequent back pains.   `27`

G. Principal skeletal muscles that move the lower extremity   (pages 248–254)

1. Complete the table about muscles that move the thigh. Locate origins and insertions on a skeleton (if available), and try to determine locations of these muscles on yourself.

Muscle	Origin	Insertion	Actions
a. Psoas major			
b.	Iliac fossa (anterior of ilium)	Tendon of psoas major	
c. Gluteus maximus			

Muscle	Origin	Insertion	Actions
d.	Ilium (posterior)	Femur (greater trochanter)	
e.	Ilium (posterior)	Femur (greater trochanter)	
f.	Iliac crest (lateral portion)		Flexes and abducts thigh
g. Adductor group			

2. Two large groups of muscles are involved with movements at the knee joint. Describe them in this exercise.

28

a. Quadriceps femoris is the main muscle mass on the *(anterior? posterior?)* surface of the thigh. The name of this muscle mass indicates it has

_____ heads of origin. All four converge to insert on

the _____ bone by means of the patellar ligament. All four therefore cause *(flexion? extension?)* of the leg. Feel their contraction as you extend your own leg.

b. Three of the heads of the quadriceps originate on the _____

_____. Their names are _____

_____, and _____. Since they cross only one joint, their only action is extension of the *(thigh? leg?)*.

c. The fourth head, named the _____, originates on the

_____ _____ spine of the

_____ bone. Since it crosses the hip joint, it has the

additional action of _____ the thigh.

d. The hamstrings are *(synergists? antagonists?)* of the quadriceps. Hamstrings are located on the *(anterior? posterior?)* surface of the thigh. Why are the hamstrings so named?

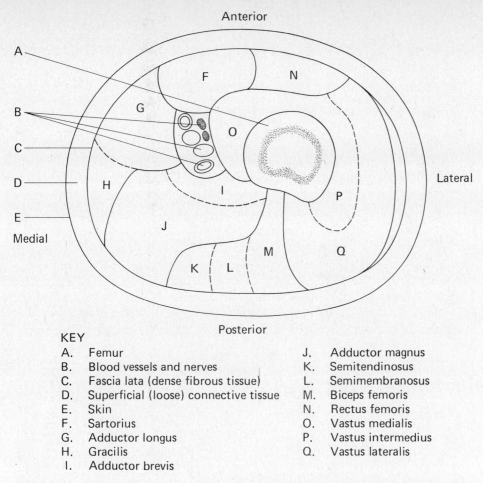

Anterior

A

B

C

D

E

Medial

Lateral

Posterior

KEY

A. Femur
B. Blood vessels and nerves
C. Fascia lata (dense fibrous tissue)
D. Superficial (loose) connective tissue
E. Skin
F. Sartorius
G. Adductor longus
H. Gracilis
I. Adductor brevis

J. Adductor magnus
K. Semitendinosus
L. Semimembranosus
M. Biceps femoris
N. Rectus femoris
O. Vastus medialis
P. Vastus intermedius
Q. Vastus lateralis

Figure LG 11-6 Cross section of right thigh midway between hip and knee joints. Structures *A* to *E* are nonmuscle; *F* to *Q* are skeletal muscles of the thigh. Color as directed.

e. The three muscles in the hamstring group are _____

_____.. All three of

these muscles originate on the _____
and insert just below the knee. The actions they cause are the *(flexion? extension?)* of the thigh and the *(flexion? extension?)* of the leg.

3. Examine the muscles of the right hip and thigh in Figures LG 11-2 and LG 11-3. Now refer to Figure LG 11-6.

a. Locate the major muscle groups of the right thigh in this cross section. Color the groups: adductors, *blue;* hamstrings, *red;* quadriceps femoris, *green.* ⬚29

b. Cover the KEY to Figure LG 11-6 and identify each muscle. Relate position of muscles in this figure to views of muscles in Figures LG 11-2 and LG 11-3.

4. Sit in the "tailor's position": place the outer ankle of your right leg on top of your left knee. The muscle that crosses the thigh obliquely and causes this

action of the right leg is _____. ⬚30

5. What muscles cause the actions *I-P* shown in Figure LG 9-2 (page LG 166)? Write the muscle names next to each diagram. ⬚31

216

32

6. Perform these actions of your foot. Feel which muscles are contracting. Name two or more muscles that cause each action.
   a. Stand on your toes.

   b. Stand on your heels with toes turned up.

   c. Evert your foot.

   d. Invert your foot.

33

7. Summarize muscles that move the lower extremity. Choose the muscles that fit the descriptions. Write one muscle name on each line provided.

a. Allow you to touch your toes by causing flexion of your thigh:

_____

_____

_____

b. Allow you to kick your leg backwards (hyperextension of thigh):

_____

_____

_____

c. Become large in dancers because they stand on their toes often:

_____

_____

d. Allow you to place your heel on your buttocks (flex leg):

_____

_____

e. Laterally rotate thigh, an important action in normal walking:

_____

_____

f. Antagonists of the hamstrings (with regard to action at knee joint):

_____

_____

Biceps femoris
Gastrocnemius
Gluteus maximus
Gluteus medius
Iliacus
Obturator internus
Psoas major
Rectus femoris
Semitendinosus
Soleus
Tensor fasciae latae
Vastus lateralis

g. Form the "calf" of the leg:

_____

_____

h. Form much of the mass of the buttocks:

_____

_____

H. Intramuscular injections

(page 255)

1. State three reasons why intramuscular (IM) injections may be the method of choice for administration of drugs.

2. *A clinical challenge.* Why is the gluteus medius considered a safer site for intramuscular injections than the gluteus maximus? |34|

Two other muscles commonly used for intramuscular injections are the

_____ in the thigh and the _____ in the upper extremity. |35|

Answers to Numbered Questions in Learning Activities

|1| (a) Lever, fulcrum. (b) Resistance. (c) Muscle, effort. (d) Proximal, radius. (e) Origin, scapula. (f) Insertion, radius.

|2| (a) 3. (b) Third-, most; flexion of the forearm. (c) *R, E,* atlas, occipital, first-. (d) Raising the body on the toes using the gastrocnemius muscle.

|3| (a) Many, pennate. (b) Long, parallel.

|4| (a) Prime mover (agonist), biceps brachii muscle. (b) An antagonist. (c) Your forearm would stay in the flexed position. (d) None; each opposing muscle would negate the action of the other. (e) Synergists, fixators.

|5| (b) L, S. (c) N, L. (d) P. (e) A, S.

|6| (a) Orbicularis oris. (b) Levator labii superioris. (c) Styloglossus. (d) Rectus abdominis. (e) Levator scapulae. (f) Latissimus dorsi. (g) Deltoid. (h) Supraspinatus. (i) Pronator teres. (j) Flexor carpi ulnaris. (k) Tensor fasciae latae. (l) Quadriceps femoris. (m) Vastus lateralis. (n) Gracilis. (o) Sartorius. (p) Tibialis anterior. (q) Extensor digitorum longus.

7   (a) Levator palpebrae superioris. (b) Levator labii superioris. (c) Corrugator supercilii. (d) Platysma. (e) Buccinator. (f) Frontalis portion of epicranius. (g) Orbicularis oris. (h) Zygomaticus major, (i) Orbicularis oculi.

8   Facial, VII.

9   (a) Elevating the mandible, temporalis, masseter. (b) Close, lateral pterygoid. (c) V, trigeminal. (d) Chewing, facial expression.

10   (a) They are outside of the eyeballs, not intrinsic like the iris. (b) The same as, superiorly. (c) Laterally, opposite to, superiolaterally. (d) Medial rectus.

11   No. The left eye contracts its medial rectus, while the right eye uses its lateral rectus and exerts some tension upon both oblique muscles.

12   (a) Up and back. (b) Down. (c) Down and forward.

13   XII, hypoglossal

14   Is U-shaped; digastric, stylohyoid, mylohyoid, geniohyoid

15   (a) D. (b) E. (c) C. (d) B. (e) A.

16   Opposite, flex.

17   Extending and then hyperextending; posterior; longissimus capitis, semispinalis capitis and splenius capitis.

18   Sternocleidomastoid *(A)*, three capitis muscles *(B)*.

19   Four. (a) Vertical, pubic crest and symphysis pubis, lower eight ribs, forward; flexion. (b) External oblique, internal oblique, transversus abdominis. (c) No; strength is provided by the three different directions.

20   (a) Dome, around the bottom of the rib cage and on lumbar vertebrae, central tendon. (b) Increase, inspiration. (c) Between ribs (costa), internal.

21   Symphysis pubis, ischial tuberosities, and coccyx.

22   Urogenital triangle: urethral orifice in both sexes and vaginal opening in females; anal triangle: anus.

23   Across, wall, thorax, pelvic.

24   Pectoralis major and coracobrachialis, also anterior portion of deltoid *(C);* latissimus dorsi and teres major *(D);* Deltoid and supraspinatus *(E);* pectoralis major, latissimus dorsi, and teres major *(F);* biceps brachii, brachialis, and brachioradialis *(G);* triceps brachii *(H)*.

25   (a) Anterior, flex. (b) Adducts wrist, flexor and extensor carpi radialis. (c) Superficialis is more superficial, and profundus lies deep. (d) Extensor indicis.

26   (a) Triceps brachii. (b) Supinator and biceps brachii. (c) Serratus anterior. (d) Brachialis and brachioradialis. (e) Flexor digitorum superficialis and profundus. (f) Pectoralis minor and lower fibers of trapezius. (g) Pectoralis major. (h) Trapezius. (i) Triceps brachii. (j) Latissimus dorsi.

27   Extra weight of the abdomen demands extra support (contraction) by sacrospinalis muscles.

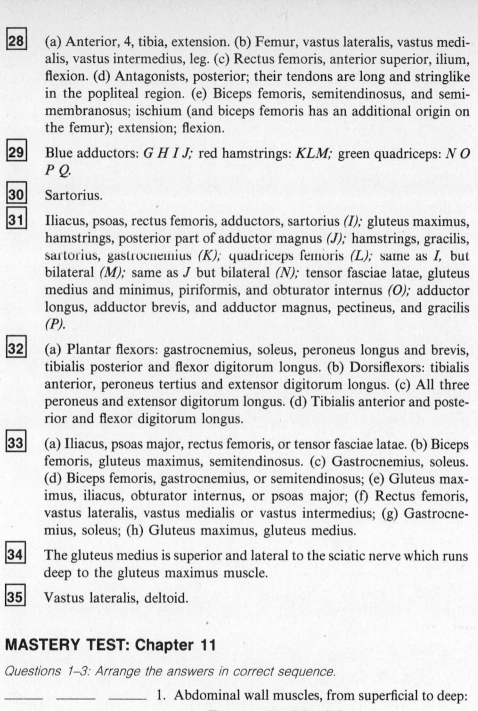

28 (a) Anterior, 4, tibia, extension. (b) Femur, vastus lateralis, vastus medialis, vastus intermedius, leg. (c) Rectus femoris, anterior superior, ilium, flexion. (d) Antagonists, posterior; their tendons are long and stringlike in the popliteal region. (e) Biceps femoris, semitendinosus, and semimembranosus; ischium (and biceps femoris has an additional origin on the femur); extension; flexion.

29 Blue adductors: *G H I J;* red hamstrings: *KLM;* green quadriceps: *N O P Q.*

30 Sartorius.

31 Iliacus, psoas, rectus femoris, adductors, sartorius *(I);* gluteus maximus, hamstrings, posterior part of adductor magnus *(J);* hamstrings, gracilis, sartorius, gastrocnemius *(K);* quadriceps femoris *(L);* same as *I*, but bilateral *(M);* same as *J* but bilateral *(N);* tensor fasciae latae, gluteus medius and minimus, piriformis, and obturator internus *(O);* adductor longus, adductor brevis, and adductor magnus, pectineus, and gracilis *(P).*

32 (a) Plantar flexors: gastrocnemius, soleus, peroneus longus and brevis, tibialis posterior and flexor digitorum longus. (b) Dorsiflexors: tibialis anterior, peroneus tertius and extensor digitorum longus. (c) All three peroneus and extensor digitorum longus. (d) Tibialis anterior and posterior and flexor digitorum longus.

33 (a) Iliacus, psoas major, rectus femoris, or tensor fasciae latae. (b) Biceps femoris, gluteus maximus, semitendinosus. (c) Gastrocnemius, soleus. (d) Biceps femoris, gastrocnemius, or semitendinosus; (e) Gluteus maximus, iliacus, obturator internus, or psoas major; (f) Rectus femoris, vastus lateralis, vastus medialis or vastus intermedius; (g) Gastrocnemius, soleus; (h) Gluteus maximus, gluteus medius.

34 The gluteus medius is superior and lateral to the sciatic nerve which runs deep to the gluteus maximus muscle.

35 Vastus lateralis, deltoid.

MASTERY TEST: Chapter 11

Questions 1–3: Arrange the answers in correct sequence.

_____ _____ _____ 1. Abdominal wall muscles, from superficial to deep:

      A. Transversus abdominis
      B. External oblique
      C. Internal oblique

_____ _____ _____ 2. From superior to inferior in location:

      A. Sternocleidomastoid
      B. Pelvic diaphragm
      C. Diaphragm and intercostal muscles

Questions 3-12: Circle T (true) or F (false). If the statement is false, change the underlined word or phrase so that the statement is correct.

T   F   3. In extension of the thigh the hip joint serves as the <u>fulcrum (F),</u> while the <u>hamstrings and gluteus maximus</u> serve as the effort (E).

T   F   4. The wheelbarrow and gastrocnemius are both examples of the action of <u>second</u>-class levers.

T   F   5. The hamstrings are <u>synergists</u> to the quadriceps femoris.

T   F   6. The name trapezius is based on the <u>action</u> of that muscle.

T   F   7. The most important muscle used for normal breathing is the <u>diaphragm.</u>

T   F   8. The insertion end of a muscle is the attachment to the bone that <u>does move.</u>

T   F   9. In general, adductors (of the arm and thigh) are located more on the <u>medial</u> than on the <u>lateral</u> surface of the body.

T   F   10. Both the <u>pectoralis major and latissimus dorsi muscles extend</u> the humerus.

T   F   11. The capitis muscles (such as splenius capitis) are <u>extensors</u> of the head and neck.

T   F   12. The biceps brachii and biceps femoris are both muscles with <u>two heads of origin located in the arm.</u>

Questions 13–20: Choose the one best answer to each question.

_____ 13. All of these muscles are located in the lower extremity EXCEPT:

      A. Hamstrings   B. Gracilis   C. Tensor fasciae latae
      D. Deltoid   E. Peroneus longus

_____ 14. All of these muscles are located on the anterior of the body EX-CEPT:

      A. Tibialis anterior   B. Rectus femoris
      C. Sacrospinalis   D. Pectoralis major
      E. Rectus abdominis

_____ 15. All of these structures are parts of the perineum EXCEPT:

      A. External anal sphincter   B. Bulbocavernosus
      C. Ischiocavernosus   D. Iliacus
      E. Urethral sphincter

_____ 16. All of these muscles are attached to ribs EXCEPT:

      A. Serratus anterior   B. Intercostals   C. Trapezius
      D. Iliocostalis   E. External oblique

_____ 17. All of these muscles are directly involved with movement of the scapulae EXCEPT:

      A. Levator scapulae   B. Pectoralis major
      C. Pectoralis minor   D. Rhomboideus major
      E. Serratus anterior

_____ 18. All of these muscles have attachments to the coxal bones EXCEPT:

      A. Adductor muscles (longus, magnus, brevis)
      B. Biceps femoris   C. Rectus femoris
      D. Vastus medialis   E. Latissimus dorsi

_____ 19. The masseter and temporalis muscles are used for:

    A. Chewing   B. Pouting   C. Frowning
    D. Depressing tongue   E. Elevating tongue

_____ 20. All of these muscles are used for facial expression EXCEPT:

    A. Zygomaticus major   B. Orbicularis oculi
    C. Platysma   D. Rectus abdominis
    E. Mentalis

Questions 21–25: fill-ins. Refer to Figures LG 11-2 and 11-3. Write the word or phrase or KEY letters of muscles that best complete the statement or answers the questions.

_____ 21. Muscles *G, S, U, V,* and *W* on Figure LG 11-2 all have in common the fact that they carry out the action of _____.

_____ 22. Muscles *G, N, O,* and *R* on Figure LG 11-3 all have in common the fact that they carry out the action of _____.

_____ 23. If you colored all flexors red and all extensors blue on these two figures, the view of the _____ surface of the body would appear more blue.

_____ 24. Write the letters of all of the muscles listed below which would contract as you raise your left arm straight in front of you, as if to point toward a distant mountain: Figure LG 11-2: *D T U V;* Figure LG 11-3: *N O P*

_____ 25. Write the letters of all of the muscles listed below that would contract as you raise your knee and extend your leg straight out in front of you, as if you are starting to march off to the distant mountain: Figure LG 11-2: *AA BB CC DD;* Figure LG 11-3: *H I J K R*

221

UNIT III

Control Systems of the Human Body

Nervous Tissue

The diversity and complexity of movements which you studied in Unit II point to the necessity for communication and control in the body. In Unit III you will learn about the mechanisms that provide such control over all body systems through nerve impulses and hormonal messages.

In the first chapter of this unit you will study functions and divisions of the nervous system (Objectives 1–2), and you will consider the cells that make up nervous tissue (3–4, 14). Then you will learn how nerve impulses take place, including the roles of synapses and chemical transmitters (5–15).

Topics Summary

A. Organization
B. Histology
C. Physiology: nerve impulse
D. Physiology: synapses
E. Regeneration of neurons
F. Organization of neurons

Objectives

1. Identify the three basic functions of the nervous system in maintaining homeostasis.
2. Classify the organs of the nervous system into central and peripheral divisions.
3. Contrast the histological characteristics and functions of neuroglia and neurons.
4. Classify neurons by structure and function.
5. Describe the conditions that contribute to the resting state of a neuron.
6. List the sequence of events involved in the generation and conduction of a nerve impulse.
7. Define the all-or-none principle of nerve impulse transmission.
8. Discuss the factors that determine the speed of impulse conduction.
9. Define a synapse and list the factors involved in the conduction of an impulse across a synapse.
10. Compare the functions of excitatory transmitter-receptor interactions and inhibitory transmitter-receptor interactions in helping to maintain homeostasis.
11. Describe the integration at synapses.
12. Define the roles of neurotransmitters in conducting an impulse across a synapse.
13. List the factors that may inhibit or block nerve impulses.
14. List the necessary conditions for the regeneration of nervous tissue.
15. Explain the organization of neurons in the nervous system.

Learning Activities

(pages 262–263) **A. Organization**

1. List three principal functions of the nervous system.

2. Contrast the roles of the nervous and endocrine systems in maintenance of homeostasis.

3. Write an outline summarizing the divisions of the nervous system.

4. Check your understanding of nervous system organization by doing this exercise.

   a. The central nervous system (abbreviated _CNS_) consists of two structures, the _brain_ and the _spinal cord_.

   b. Another name for *afferent* is (sensory? motor?).

   c. Efferent nerves pass from the central nervous system to _muscle_ and _glands_. Efferent nerves are often referred to as (sensory? motor?)

   d. Efferent nerves to skeletal (voluntary) muscles are called _somatic_ efferent nerves.

e. The ___*autonomic*___ nervous system (abbreviated ANS) consists of efferent nerves that innervate three types of tissue. These are ___*smooth*___, ___*cardiac*___, and ___*glands*___.

f. The two divisions of the autonomic nervous system (ANS) are ___*sympath*___ and ___*parasympath*___.

B. Histology (pages 263–268)

1. Write names of two kinds of cells that make up the nervous system.

2. Write *neurons* or *neuroglia* after descriptions of these cells.   [2]
   a. Conduct impulses from one part of the nervous system to another:
   ___*neurons*___

   b. Provide support and protection for the nervous system: ___*neuroglia*___
   _____

   c. Bind nervous tissue to blood vessels, form myelin, and serve phagocytic functions: ___*neuroglia*___

   d. Small in size, but more abundant in number: ___*neuroglia*___
   _____

3. Complete the table describing neuroglia cells.

Type	Description	Functions
a.	Star-shaped cells	
b. Oligodendroglia		
c.		Phagocytic
d.		Line ventricles of brain and central canal of spinal cord

3

4. Match the parts of a neuron listed at right with the following descriptions.

_____ a. Contains nucleus; cannot regenerate since lacks mitotic apparatus

_____ b. Yellowish pigment that increases with age; appears to be byproduct of lysosomes

_____ c. Provide energy for neurons

_____ d. Long, thin fibrils composed of microtubules; may function in transport

_____ e. Orderly arrangement of rough ER; site of protein synthesis

_____ f. Conducts impulses toward cell body

_____ g. Conducts impulses away from cell body; has synaptic knobs that secrete chemical transmitter

_____ h. Fine filaments which are branching ends of axon collaterals; known as axon terminals

A. Axon
CB. Cell body
CS. Chromatophilic substance (Nissl bodies)
D. Dendrite
L. Lipofuscin
M. Mitochondria
NF. Neurofibrils
T. Telodendria

5. *For extra review.* On separate paper draw a diagram of a motor neuron. Label: *cell body, nucleus, axon, axon hillock, axon collateral, synaptic knobs,* and *dendrites.* Then compare your drawing to the structure of a neuron in Figure 12–3 (page 265) of your text.

4

6. Contrast two types of transport within neurons by writing AF for axoplasmic flow or AT for *axonal transport* before each related description.

_____ a. Flow in one direction only from cell body to end of axon

_____ b. Supplies new axoplasm for growing, regenerating, and mature axons

_____ c. Faster of the two types of flow; may move organelles

_____ d. Moves materials in both directions along neuron; chemicals may be recycled or degraded in nucleus; mechanism by which Herpes and rabies viruses and tetanus toxin reach cell bodies

7. Examine the enlargement of a cross section and longitudinal section of a myelinated nerve fiber in Figure LG 12–1. Label: *neurofibrils, myelin sheath, neurilemma* (composed of *neurilemma of neurolemmocyte (Schwann cell),* and *neurofibral node (of Ranvier).*

5

8. Briefly explain how myelination relates to:
a. The color of white matter

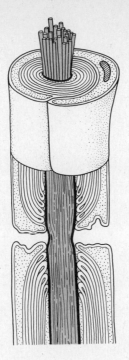

Figure LG 12-1 Diagram of a nerve fiber. Label as directed.

b. The color of gray matter

c. Speed of nerve impulse transmission

9. Describe how myelin is laid down on nerve fiber in the:
   a. PNS

   b. CNS

10. Name and give a brief description of three types of neurons based on structural characteristics and three types of neurons classified according to functional differences. State one example of each type of neuron.

6

Structural:

_____-polar

_____-polar

_____-polar

Functional:

_____-fferent

_____-fferent

_____

[7] 11. Identify phrases that describe a *neuron* and those that describe a *nerve*.
   a. A single nerve cell; consists of a cell body with axon and dendrites:

   _____

   b. Bundle of axons and dendrites of many neurons, both afferent and effer-
      ent, somatic and autonomic; contains no cell bodies: _____

   _____

   c. Located entirely outside of the CNS and in the PNS; macroscopic
      in diameter: _____

   d. Located partly in CNS and partly in PNS; microscopic in diameter:

   _____

[8] 12. Identify the type of nerve fiber that transmits each of the kinds of nerve
        impulses listed below.

   _____ a. All ANS fibers are of this type.
   _____ b. Pain from a thorn in your skin is
              sensed via fibers of this type.
   _____ c. Pain from a spasm of smooth mus-
              cle in the gallbladder is sensed by
              means of fibers of this type.
   _____ d. With your eyes closed, you can
              tell the position of your skeletal
              muscles and joints due to nerve
              impulses which pass along this
              type of fiber.
   _____ e. In order to increase your heart
              rate, impulses pass from your
              brain to your heart via this type of
              nerve fiber.
   _____ f. These nerve fibers carry impulses
              from the CNS to muscles in your
              fingers used in writing.

   GSA. General somatic affer-
        ent
   GSE. General somatic effer-
        ent
   GVA. General visceral affer-
        ent
   GVE. General visceral effer-
        ent

(pages 268–272) **C. Physiology: nerve impulse**

   1. Use Figure LG 12-2a to show the following.
      a. Differences in potassium ion (K^+) and sodium ion (Na^+) concentrations
         inside and outside of nerve cells
      b. How these concentrations are maintained by "pumps"

(a)

(b)

(c)

Figure LG 12-2 Diagrams of nerve fibers. Complete as directed.

2. Complete this exercise about membrane potential.

a. A neuron that is not conducting impulses is said to be a _____

_____ neuron.

b. Normally, the inside of a membrane of a resting neuron is more *(positive? negative?)* than the outside. Explain why and show this condition diagrammatically on Figure LG 12-2b.

c. Since there is a difference in electrical charge between the inside and the outside of the resting membrane, it is said to be *(polarized? depolarized?)*.

d. The difference in charge between the inside and the outside of the membrane is said to be the membrane _____

_____. In a resting membrane this is _____ millivolts (mV). This means that the voltage of the inside of the membrane is 70 mV *(more? less?)* than the voltage outside, reflecting the presence of relatively more negative ions than positive ions on the *(inside? outside?)* of the nerve membrane.

3. Define:

a. Excitability

b. Stimulus

10 4. Describe how a nerve impulse occurs by doing this exercise.

a. An adequate stimulus causes the nerve cell membrane to become *(more? less?)* permeable to Na^+. Consequently, Na^+ enters the cell as Na^+ _____ open.

b. At rest the membrane had a potential of _____ mV. As Na^+ enters the cell, the inside of the membrane becomes more *(positive? negative?)*. The potential will tend to go toward *(−80? −60?)* mV.

c. This level is said to be the _____ level, meaning that if the membrane potential is changed (becomes less negative) to this level, the process of *(polarization? depolarization?)* will proceed.

d. The membrane is completely depolarized at _____ mV. At this point there is _____ difference in electrical charge between inside and outside the cell. However, the stimulus usually causes sufficient Na^+ to enter the cell so that the membrane potential continues to rise to _____ mV.

e. These events of depolarization (and reversed polarization to $+30$ mV) occur at the point of stimulation and initiate similar electrical changes along the entire length of the nerve fiber. This is called the nerve impulse or _____ .

f. The nerve impulse is essentially a wave of negativity along the *(inside? outside?)* of the nerve cell membrane, with a related wave of positivity along the *(inside? outside?)* of the nerve cell membrane.

g. As the impulse (depolarization) spreads to successive points along the membrane, the original points become _____ . Repolarization is due to an increase in membrane permeability to *(Na^+? K^+?)*.

h. Since some of the K^+ ions which were concentrated inside the nerve cell now move outward, the outside of the membrane again becomes *(positive? negative?)*. The inside becomes *(positive? negative?)*.

i. So the membrane returns to a *(positive? negative?)* mV value, restoring the resting potential of _____ mV. Note, however, that the negativity inside the membrane is now due largely to the temporary exit of the cation _____ .

j. During the repolarization period a neuron is said to be recovering from depolarization. This time is also called the _____ period since the nerve is refractory to nerve impulses; this means that the nerve *(can? cannot?)* initiate another nerve impulse during this time.

k. Depolarization and repolarization can occur *(only one time? thousands of times?)* before sufficient Na^+ and K^+ have shifted across the neuron cell membrane so as to interfere with continued nerve impulse formation.

l. When that does finally occur, then Na^+ and K^+ are returned to their original sites. This occurs by an *(active? passive?)* process known as the

_____.

5. Write the correct letter label of Figure LG 12-3a next to each description. |11|

_____ a. The stimulus is applied at this point.

_____ b. Resting membrane potential is at this level.

_____ c. Membrane becomes so permeable to K^+ that K^+ diffuses rapidly out of the cell.

_____ d. The membrane is becoming more positive inside; its potential is −30 mV. The process of depolarization is occurring.

_____ e. The membrane is completely depolarized at this point (equally positive and negative on either side of membrane).

_____ f. The membrane is repolarizing at this point.

_____ g. Reversed polarization occurs; enough Na^+ has entered so that this part of the cell is more positive inside than outside.

_____ h. No potential difference (0 mV) exists across the membrane.

6. *For extra review.* Show the events that occur during initiation and transmission of a nerve impulse on Figure LG 12-2c. Compare your diagram to Figure 12-8e (page 270) in the text.

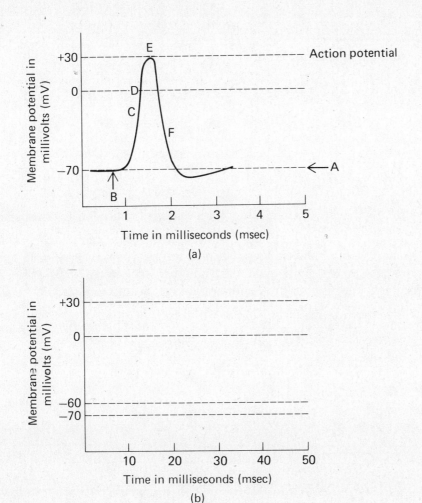

Figure LG 12-3 Diagrams for showing potentials associated with nerve transmission. (a) Identify letter labels in Learning Activity |11|. (b) Complete as directed.

7. Define these terms.
a. Threshold stimulus

b. All-or-none principle

c. Subthreshold stimulus

d. Summation

12 8. The absolute refractory period of a large nerve fiber is about
_____ msec; for a small nerve fiber this period is about
_____ msec. A nerve with a short refractory period can
transmit *(more? fewer?)* impulses per second than a nerve with a long
refractory period. So you can conclude that thick nerve fibers which have
relatively *(long? short?)* refractory periods transmit more *(rapidly? slowly?)*
than thin ones.

9. Saltatory transmission occurs along *(myelinated? unmyelinated?)* nerve
fibers. Explain how saltatory transmission occurs.

13 10. State two advantages of saltatory conduction.

11. Complete the table contrasting the three classes of nerve fibers that have different speeds of impulse transmission.

	A Fiber	B Fiber	C Fiber
a. Size (diameter)	Largest		
b. Myelination			No
c. Length of absolute refractory period	Shortest		
d. Speed of conduction	(Fastest)	10 m/sec	
e. Structures innervated			

D. Physiology: synapses

(pages 272–276)

1. Refer to Figure LG 13–3a (page LG 252). Examine the point within the spinal cord at which one neuron (sensory) comes in close proximity to another neuron (motor). This location is known as a _____

    _____. The neuron that transmits a nerve impulse toward the synapse is known as a *(pre-? post-?)* synaptic neuron, in this case, the sensory neuron. The motor neuron would serve as the _____

    _____-synaptic neuron.

2. Does a nerve impulse actually "jump" across a synapse? Explain briefly. |14|

3. Now label structures *A-G* on Figure LG 12-4. Use the following terms as |15| labels: *neurotransmitter, neurotransmitter receptor, postsynaptic neuron, presynaptic end bulb, presynaptic neuron, synaptic cleft, synaptic vesicle.*

4. Note the similarity of the synapse to the neuromuscular junction (Figure LG |16| 10-2, page LG 180).

    State one example of a *neuroglandular junction.*

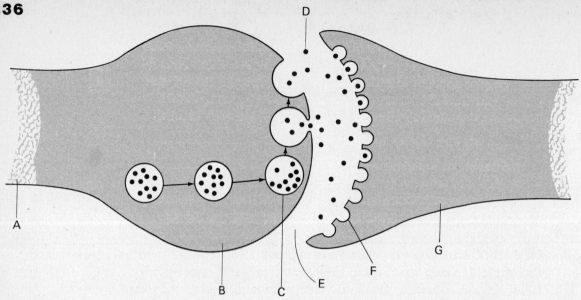

Figure LG 12-4 Diagram of a synapse. Identify letter labels in Learning Activity 15

5. Contrast the terms in the following pairs.
   a. Synapse/neuromuscular junction

   b. Axodendritic synapse/axosomatic synapse

   c. Divergence at synapse/convergence at synapse (See also Learning Activity 25, page LG 240.)

6. Define and state the significance of *one-way impulse conduction*.

7. Discuss briefly two theories proposed to explain how neurotransmitters are released into synaptic clefts.

8. Complete the following exercise.

   a. A transmitter-receptor interaction refers to the effect that a certain chemical has upon the receptors of a _____*post*_____-synaptic neuron.

   b. An excitatory interaction *(raises? lowers?)* the −70 mV resting potential of the postsynaptic neuron, for example, to −65 mV. Such a potential is called an EPSP which stands for _____

   _____

   _____ .

   c. Since the excitation is not sufficient to cause a threshold potential (_____ mV), for example, if an EPSP of −65 were produced, then _____ is said to have occurred. Diagram a facilitation EPSP in green on Figure LG 12–3b.

   d. If transmitter substance is released from several (or hundreds of) presynaptic knobs at one time, their combined effect may produce threshold EPSP. This phenomenon is known as *(spatial? temporal?)* summation. Temporal summation is that due to accumulation of transmitter from *(one? many?)* synaptic knob(s) over a period of time.

9. Describe two theories explaining how neurotransmitters may cause sodium ion channels of postsynaptic neurons to become more permeable to sodium ions and therefore initiate a nerve impulse.

10. Now that you have studied EPSP, define IPSP, and diagram an IPSP in red on Figure LG 12-3b. (Compare your drawing to Figure 12-10b, page 274 in the text.)

[18] 11. Explain why an IPSP is referred to as *hyperpolarization*.

12. The postsynaptic neuron is called an *integrator*. Briefly explain.

[19] 13. Seven statements about the action of the neurotransmitter acetylcholine (ACh) at a neuromuscular junction are listed below. Order these in correct sequence by writing numbers 1–7 on lines next to statements.

3 _2_ a. Quanta of ACh enter the synaptic cleft.

5 _6_ b. Alteration of the structure of receptors permits Na^+ to pass through channels into muscle cell. This depolarization leads to muscle contraction.

_1_ c. Calcium ions enter a presynaptic axon end bulb.

6 _5_ d. Acetylcholinesterase (ACHase) which is present in subneural clefts of muscle cell membranes inactivates ACh.

2 _3_ e. As a result, ACh is released from synaptic vesicles or cytoplasm of the axon.

_4_ f. ACh binds to protein receptors in a sarcolemma (muscle cell membrane).

_7_ g. In the absence of ACh, the muscle fiber is repolarized so that it stops contracting until it receives another nerve impulse.

[20] 14. What is the ultimate fate of the products of acetylcholine that result from the action of ACHase?

15. Name two transmitters that cause IPSP. 21 **239**

16. Alkalosis tends to *(stimulate? depress?)* CNS activity, leading to convul-
    sions, while acidosis tends to *(stimulate? depress?)* neuronal activity, lead-
    ing to _____. 22
17. Explain what may cause your right leg to "go to sleep" after you have
    crossed your right leg over your left one for some time.

18. Check your understanding of chemicals that affect transmission at synapses
    and neuromuscular or neuroglandular junctions by completing this activ-
    ity. Write E if the effect is excitatory or I if the effect is inhibitory. (Lines
    following descriptions are for the *extra review* activity below.) 23

    ___I___ a. A chemical that <u>inhibits</u> release of ACh _3_
    ___E___ b. A chemical that <u>competes</u> for the ACh receptor sites on post-
              synaptic neuron membrane _____
    ___I___ c. A chemical that <u>inactivates acetylcholinesterase</u> _____
    ___E___ d. A chemical that increases threshold (for example, from −60 to
              −40) of a neuron _____   *lower threshold letter*
    ___I___ e. A chemical that decreases threshold _____

    A clinical challenge. Match the following chemicals with related mech-
    anisms of actions. Write the number of the chemicals on lines to the right
    of the above descriptions. One is done for you.
    1. Hypnotics, tranquilizers, anesthetics
    2. Caffeine, benzedrine
    3. Botulism toxin, inhibiting muscle contraction
    4. Curare, a muscle relaxant
    5. Nerve gas such as diiospropyl fluorophosphate

E. Regeneration of neurons (page 276)

1. Which of the following can regenerate if destroyed? Briefly explain why in
   each case. 24
   a. Neuron cell body

b. CNS nerve fiber

c. PNS nerve fiber

(pages 276–278) **F. Organization of neurons**

1. Define *neuronal pool, discharge zone,* and *facilitated zone.*

2. Match the types of circuits at right with related descriptions. Answers may be used more than once.

25

_____ᴰ a. Impulse from a single presynaptic neuron causes stimulation of increasing numbers of cells along the circuit.

_____ᴰ b. An example is a single motor neuron in the brain stimulates many motor neurons in the spinal cord, therefore activating many muscle fibers

_____ᴿ c. Branches from a second and third neuron in a pathway may send impulses back to the first, so the signal may last for hours, as in coordinated muscle activities

_____ᶜ d. One postsynaptic neuron receives impulses from several nerve fibers

C. Converging
D. Diverging
P. Parallel afterdischarge
R. Reverberating

___P___ e. A single presynaptic neuron stimulates intermediate neurons which synapse with a common postsynaptic neuron, allowing this neuron to send out a stream of impulses, as in precise mathematical calculations

Answers to Numbered Questions in Learning Activities

1 (a) CNS, brain, spinal cord. (b) Sensory. (c) Muscles, glands, motor. (*Note:* Remember S A M E. *S*ensory = *A*fferent; *M*otor = *E*fferent.) (d) Somatic. (e) Autonomic, smooth muscle, cardiac muscle, glands. (f) Sympathetic, parasympathetic.

2 (a) Neurons. (b) Neuroglia. (c) Neuroglia. (d) Neuroglia.

3 (a) CB. (b) L. (c) M. (d) NF. (e) CS. (f) D. (g) A. (h) T.

4 (a) AF. (b) AF. (c) AT. (d) AT.

5 (a) Consists of myelinated fibers; myelin is white. (b) Consists of unmyelinated fibers, cell bodies, and neuroglia. (c) Speeds conduction by saltatory effect.

6 (a) By neurolemmocytes wrapping their cell membranes around nerve fibers. (b) By a similar process involving oligodendroglial cells.

7 (a) Neuron. (b) Nerve. (c) Nerve. (d) Neuron.

8 (a) GVE. (b) GSA. (c) GVA. (d) GSA. (e) GVE. (f) GSE.

9 (a) Resting. (b) Negative. Na^+ does not readily diffuse through the resting nerve cell membrane. If it does, it is pumped out, leaving the outside positive. The high concentration of negative ions (mostly proteins) inside more than compensates for the high K^+ level inside. See Figure 12-7 page 269 in your text. (c) Polarized. (d) Potential, −70, less, inside.

10 (a) More, channels. (b) −70, positive −60. (c) Threshold, depolarization. (d) 0, no, +30. (e) Action potential. (f) Outside, inside. (g) Repolarized, K^+. (h) Positive, negative. (i) Negative, −70, K^+. (j) Refractory, cannot. (k) Thousands of times. (l) Active, sodium-potassium pump.

11 (a) *B.* (b) *A.* (c) *F.* (d) *C.* (e) *D.* (f) *F.* (g) *E.* (h) *D.*

12 0.4, 4.0, more, short, rapidly.

13 Speed of saltatory impulse conduction is much greater than continuous (point-to-point) conduction. Saltatory conduction is more efficient: less energy is expended since fewer Na^+ and K^+ ions are displaced (that is, only those at neurofibral nodes of Ranvier), so less use of Na^+–K^+ pump is required.

14 No. The nerve impulse travels along the presynaptic axon. At the end of the axon it stimulates synaptic vesicles there to release chemicals (transmitters) which enter the synaptic cleft and may then affect the postsynaptic neuron (either excite it or inhibit it).

15 Presynaptic neuron *(A)*, presynaptic end bulb *(B)*, synaptic vesicle *(C)*, neurotransmitter *(D)*, synaptic cleft *(E)*, neurotransmitter receptor *(F)*, postsynaptic neuron *(G)*.

16 An axon transmitting a nerve impulse to a sweat gland or to a salivary gland.

17 (a) Post-, (b) Raises, excitatory postsynaptic potential. (c) −60, facilitation. (d) Spatial, one.

18 The inside of the postsynaptic membrane is even more negative (such as −90 mV) than the usual resting membrane due to its decreased Na^+ permeability and increased K^+ permeability.

19 3 5 1 6 2 4 7

20 Acetate and choline are recycled: they reenter the presynaptic axon and reform ACh as a result of the action of the enzyme choline acetyltransferase.

21 Gamma aminobutyric acid (GABA) and glycine.

22 Stimulate, depress, decreased level of consciousness with possible coma and death

23 (a) I (3). (b) I (4). (c) E (5). (d) I (1). (e) E (2).

24 (a) No. About the time of birth, the mitotic apparatus is lost. (b) No. CNS nerve fibers lack the neurilemma necessary for regeneration, and scar tissue builds up by proliferation of astroglia cells. (c) Yes. PNS fibers do have a neurilemma. (See Chapter 13 for more about PNS nerve fiber regeneration.)

25 (a) D. (b) D. (c) R. (d) C. (e) P.

MASTERY TEST: Chapter 12

Questions 1–2: Arrange the answers in correct sequence.

___A___ ___C___ ___B___ 1. In order of transmission across synapse, from first structure to last:
        A. Presynaptic end bulb
        B. Postsynaptic neuron
        C. Synaptic cleft

___A___ ___C___ ___B___ 2. Membrane potential values, from most negative to zero:
        A. Resting membrane potential
        B. Depolarized membrane potential
        C. Threshold potential

_B_ 3. Choose the one FALSE statement.

    A. Neurons in a discharge zone are more likely to receive enough transmitters to fire (start a nerve impulse) than neurons in a facilitated zone of a neuronal pool.

    B. In a resting membrane, permeability to K^+ ions is about 100 times less than permeability to Na^+ ions.

    C. C fibers are thin, unmyelinated fibers with relatively slow rate of nerve transmission.

    D. C fibers are more likely to innervate the heart and bladder than structures (such as skeletal muscles) that must make instantaneous responses.

_A_ 4. Choose the one FALSE statement about the membrane of a resting
_C_ neuron.

    A. It has a membrane potential of −70 mV.

    B. It is more negative inside than outside.

    C. It is depolarized.

    D. Its membrane potential is called resting potential.

_A_ 5. Presynaptic end bulbs are located:

    A. At ends of telodendria of axons   B. On axon hillocks

    C. On neuron cell bodies   D. At ends of dendrites

    E. At ends of both axons and dendrites

_B_ 6. Which of these is equivalent to a nerve fiber?
_C_

    A. A neuron   B. A neurofibril

    C. An axon or dendrite

    D. A nerve, such as sciatic nerve

_C_ 7. *AChE* is an abbreviation for:

    A. Acetylcholine   B. Norepinephrine

    C. Acetylcholinesterase   D. Serotonin

    E. Inhibitory postsynaptic potential

_D_ 8. A term that means the same thing as *afferent* is:
_E_

    A. Autonomic   B. Somatic   C. Peripheral   D. Motor

    E. Sensory

_X_ 9. An excitatory transmitter substance that changes the membrane
_B_ potential from −70 to −65 mV causes:

    A. Impulse conduction   B. Facilitation

    C. Inhibition   D. Hyperpolarization

Questions 10–20: Circle T (true) or F (false). If the statement is false, change the underlined word or phrase so that the statement is correct.

T   F   10. The concentration of potassium ions (K^+) is about 28 times <u>greater</u> inside a resting cell than outside of it.

T   F   11. Since CNS fibers contain no neurilemma, and the neurilemma produces myelin, CNS <u>fibers are all unmyelinated.</u>

T   F   12. Transmitter substances are released at <u>synapses and also at neuromuscular junctions.</u>

244

T F 13. Generally, release of excitatory transmitter by <u>a single presynaptic knob</u> is sufficient to develop an action potential in the postsynaptic neuron.

T (F) 14. <u>Neuroglia</u> are a common source of tumors of the nervous system.

T (F) 15. In the <u>converging</u> circuit, a single presynaptic neuron influences several postsynaptic neurons (or muscle or gland cells) at the same time.

T (F) 16. Action potentials are measured in <u>milliseconds, which are thousandths of a second.</u>

T (F) 17. Nerve fibers with a short absolute refractory period can respond to <u>more rapid</u> stimuli than nerve fibers with a long absolute refractory period.

T (F) 18. Most neurotransmitters are chemicals that are classified as <u>lipids.</u>

(T) F 19. A stimulus that is adequate will temporarily <u>increase permeability of the nerve membrane to Na⁺.</u>

T (F) 20. The <u>brain and spinal nerves</u> are parts of the peripheral nervous system (PNS).

Questions 21–25: fill-ins. Complete each sentence with the word or phrase that best fits.

_____integrator_____ 21. A postsynaptic neuron is called a(n) _____ since it receives both excitatory and inhibitory signals and determines its response according to the relative magnitude of these two types of signals.

_____slows down_____ 22. Application of cold to a painful area can decrease
speed of trans   pain in that area because _____.

_____axon-end_____ 23. One-way nerve impulse transmission can be explained on the basis of release of transmitters only from the _____ of neurons.

_____periphery_____ 24. The neurilemma is found only around nerve fibers of the _____ nervous system, so that these fibers can regenerate.

_____cell body_____ 25. In an axosomatic synapse, an axon's synaptic
soma   end bulb transmits nerve impulses to the _____ of a postsynaptic neuron.

The Spinal Cord and the Spinal Nerves

In this chapter you will begin to see the organization of nervous tissue into functional units, exemplified by the spinal cord and nerves. You will define important terms related to the cord and nerves (Objective 1). You will describe principal features of the spinal cord (2, 8) as well as protective coverings (3–5). You will study reflexes (6, 7, 9–11) and examine the structure, distribution, and function of spinal nerves (12–16). You will also consider injuries and disorders involving the spinal cord and nerves (17–19).

Topics Summary

A. Grouping of neural tissue
B. Spinal cord: general features, protection and coverings
C. Spinal cord: conduction pathway (tracts)
D. Spinal cord: reflex center
E. Spinal nerves
F. Disorders

Objectives

1. Describe how neural tissue is grouped.
2. Describe the gross anatomical features of the spinal cord.
3. Explain how the spinal cord is protected.
4. Describe the structure and location of the spinal meninges.
5. Discuss the location, general technique, purpose, and significance of a spinal puncture.
6. Describe the structure of the spinal cord.
7. Explain the functions of the spinal cord as a conduction pathway and a reflex center.
8. List the location, origin, termination, and function of the principal ascending and descending tracts of the spinal cord.
9. Discuss the components of a reflex arc and its relationship to homeostasis.
10. Compare the functional anatomy of a stretch reflex, tendon reflex, flexor reflex, and crossed extensor reflex.
11. List and describe several clinically important reflexes.
12. Name the 31 pairs of spinal nerves.
13. Describe the composition and coverings of a spinal nerve.
14. Explain how a spinal nerve branches upon leaving the intervertebral foramen.
15. Define a plexus and an intercostal nerve.
16. Define a dermatome and its clinical importance.
17. Describe spinal cord injury and list the immediate and long-range effects.
18. Identify the effects of peripheral nerve damage and conditions necessary for its regeneration.
19. Explain the causes and symptoms of neuritis, sciatica, and shingles.

(page 281) **A. Grouping of neural tissue**

1. 1. Locations that contain concentrations of cell bodies appear *(gray? white?)*. Nerves that contain mostly unmyelinated nerve fibers look *(gray? white?)*, while those that contain myelinated fibers appear *(gray? white?)*.

2. 2. Complete the table about structures in the nervous system.

Structure	Color (Gray or White)	Composition (Cell Bodies or Nerve Fibers)	Location (CNS or PNS)
a. Nerve	White	*nerve fibers*	*PNS*
b. Tract	*white*	*nerve fibers*	CNS
c. Ganglia	*gray*	Cell bodies	*PNS*
d. Nucleus	Gray	*cell bodies*	*CNS*

(pages 281–284) **B. Spinal cord: general features, protection and coverings**

1. Describe the spinal cord according to:
   a. Shape

   b. Location, length (estimate its extent on a partner)

   c. Enlargements

3. 2. Match the names of the structures with descriptions given.

_____ a. Tapering inferior end of spinal cord

_____ b. Any region of spinal cord from which one pair of spinal nerves arises

_____ c. Nonnervous extension of pia mater; anchors cord in place

_____ d. "Horse's tail"; extension of spinal nerve roots in lumbar and sacral regions within subarachnoid space

Ca. Cauda equina
Co. Conus medullaris
F. Filum terminale
S. Spinal segment

3. In what ways is the spinal cord protected? List four structures or other factors that are protective. ☐4

4. Check your understanding of the protective coverings over the spinal cord by labeling A–F on Figure LG 13–1 using the following terms: *arachnoid, brain* or *spinal cord tissue, dura mater, pia mater, subarachnoid space,* and *subdural space.* ☐5

5. Now color the following structures on Figure LG 13-1: dura mater, blue; arachnoid, yellow; pia mater, green; blood vessels, red.

6. Complete this exercise about the spaces related to meninges. ☐6

    a. The epidural space is located *(outside of? deep to?)* the dura mater. Its

    contents include _____ _____.
    Explain why an "epidural" anesthetic can exert its effects on nerves relatively slowly and with prolonged effects.

    State one example of a procedure for which an epidural anesthetic may be used.

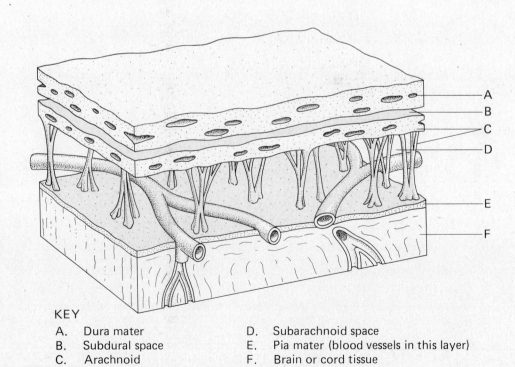

KEY

A.	Dura mater	D.	Subarachnoid space
B.	Subdural space	E.	Pia mater (blood vessels in this layer)
C.	Arachnoid	F.	Brain or cord tissue

Figure LG 13-1 Meninges. Color and label as directed.

b. What is the location of the subdural space?

It normally contains only a small amount of _____.

c. Cerebrospinal fluid is located in the _____ space
which is between the arachnoid and _____.

7. Name the structures that anchor the spinal cord in its position within the
vertebral canal. Superiorly, the _____ is attached to the
spinal cord; laterally, _____ ligaments extend from pia
to dura; and inferiorly the _____ anchors the cord to the
coccyx.

8. *A clinical challenge.*
a. Contrast *pachymeningitis* and *leptomeningitis.*

b. Describe a myelogram and tell why this procedure may be performed.

9. Do the following exercise about spinal (lumbar) puncture.
a. At what level of the vertebral column is the needle inserted?

b. Why is the puncture done here rather than at level L1 to L2?

c. The needle enters the *(subarachnoid? subdural? epidural?)* space where
cerebrospinal fluid is located.

(pages 284–285) **C. Spinal cord: conduction pathway (tracts)**

1. State two functions of the spinal cord.

2. A cross section of the spinal cord reveals an inner H-shaped area of *(gray?*
white?) matter surrounded by _____ matter.

3. Label the lettered structures in Figure LG 13-2a. Use these terms for labels:
*anterior median fissure, posterior median sulcus, gray commissure, central
canal, anterior gray horn, lateral gray horn, posterior gray horn, anterior
white column, lateral white column,* and *posterior white columns.*

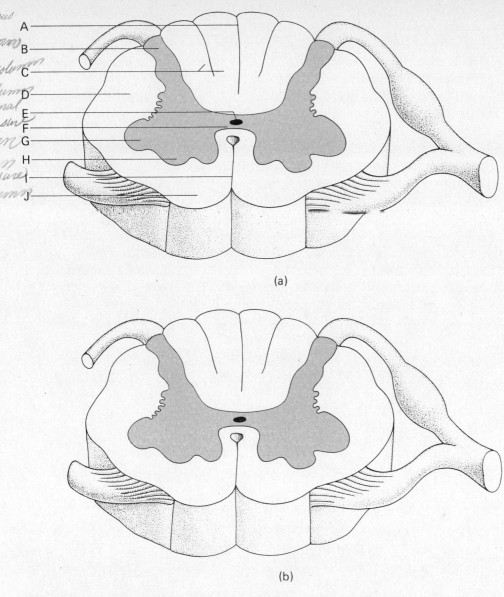

ulous
t horns
t column
dum
anal
ussure
horn
horn
issure
lumn

A
B
C
D
E
F
G
H
I
J

(a)

(b)

Figure LG 13-2 Outline of spinal cord, roots, and nerves. (a) Identify structures with letters in Learning Activity ⑨ . (b) Complete as directed.

4. Circle or fill in the correct answers about the tracts ⑩
   a. Tracts conduct nerve impulses in the *(central? peripheral?)* nervous system. Functionally they are comparable to _____ in the peripheral nervous system.
   b. Tracts are located in *(gray horns? white columns?)*. They appear white since they consist of bundles of *(myelinated? unmyelinated?)* nerve fibers.
   c. Ascending tracts are all *(sensory? motor?)*, conveying impulses between the spinal cord and the _____.
   d. All motor tracts in the cord are *(ascending? descending?)*.
   e. The name *lateral corticospinal* indicates that the tract is located in the _____ white column, that it originates in the *(cerebral cortex? thalamus? spinal cord?)*, and that it ends in the _____
   _____.
   f. The lateral corticospinal tract is *(ascending, sensory? descending, motor?)*.

250

5. Draw and label the following tracts on Figure LG 13-2b: *anterior corticospinal, anterior spinothalamic, fasciculus cuneatus, fasciculus gracilis, lateral corticospinal,* and *lateral spinothalamic.* Color ascending (sensory) tracts red, descending (motor) tracts blue. Compare your drawing to Figure 13-3 (page 286) in your text.

6. *For extra review.* Label these tracts on Figure LG 13-2b and color them as described above: *anterior spinocerebellar, posterior spinocerebellar, rubrospinal, tectospinal,* and *vestibulospinal.*

|11| 7. Match the names of the tracts with the descriptions given.

_____ a. Conveys impulses that tell you that you touched a hot stove _R_

_____ b. Allow you to be conscious of the position of your body parts; help you to assess weight and shape of objects _____

_____ c. Located in anterior white column; starts in spinal cord and ends in thalamus _____

_____ d. Sends impulses from your brain to enable you to move your skeletal muscles voluntarily _____

_____ e. Conveys impulses from one side of the medulla to the same side of your head to regulate head movements, for example, when you are trying to maintain your balance _____

_____ f. Controls head movements on opposite side, so the right tract causes you to turn to your left when you hear or see something toward your left _____

AS. Anterior spinothalamic
C. Anterior and lateral corticospinal
F. Fasciculus cuneatus and fasciculus gracilis
LS. Lateral spinothalamic
T. Tectospinal
V. Vestibulospinal

|12| 8. *A clinical challenge.* An injury which destroys one-half of the spinal cord (hemisection) would result in different losses of functions on each side of the body. A patient has a LEFT side hemisection. Write R (right) or L (left) next to each of the tract functions listed in Learning Activity |11| to indicate the side of the body that would be impaired in each of those functions. One is done for you. (*Hint:* It may help to refer to Figures 15-5 to 15-7 and 15-9, pages 346–350 in the text.)

(pages 285–292) **D. Spinal cord: reflex center**

|13| 1. Describe the two major functions of the spinal cord in this exercise.

a. The first main function of the spinal cord is to permit _____

_____ between the periphery, such as arms and legs, and the brain by means of the *(tracts? nerves? ganglia?)* located in the white columns of the cord.

b. The second main function of the cord is to serve as a _____

_____ center by means of spinal _____

_____. These are attached to the cord by ___ roots.

The _____ root contains sensory nerve fibers, and the

anterior (or ventral) root contains _____ fibers.

c. Cell bodies of *(sensory? motor?)* neurons are located in the dorsal root ganglia. Cell bodies of motor neurons leading to skeletal muscles are

located in the _____ gray horns of the cord, while

visceral efferent cell bodies lie in the _____ gray horns.

2. Write in order the basic components of a reflex arc. Note that the arrangement involves nerve impulses entering and leaving the spinal cord. (The first component is done for you.)

☐14

a. Receptor → b. _____

                c. _____

e. _____ ← d. _____

3. The receptor in a reflex arc is the distal end of a(n) *(axon? dendrite?)*. The center is located in the *(gray? white?)* matter of the spinal cord. The effector

is either a _____ or a gland.

☐15

4. Define each of these terms.
a. Spinal reflex

b. Somatic reflex

c. Visceral reflex

5. Label structures *A–H* on Figure LG 13–3, using the following terms: *effector, motor neuron axon, motor neuron cell body, receptor, sensory neuron axon, sensory neuron cell body, sensory neuron dendrite,* and *synapse.* Note that these structures are lettered in alphabetical order along the conduction pathway of a reflex arc.

☐16

6. *For extra review.* Draw and label a reflex arc on the right side of Figure LG 13-2a. Add arrows showing the direction of nerve transmission in the arc.

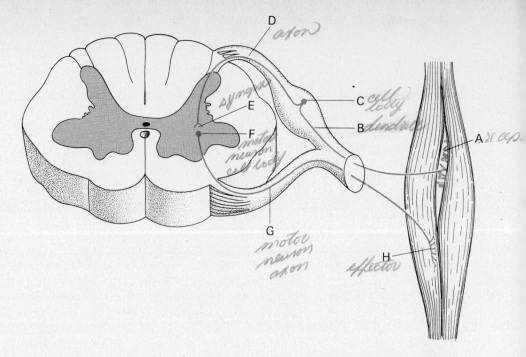

Figure LG 13-3 Reflex arc: stretch reflex. Letters refer to Learning Activity 16 .

Compare your drawing to Figure LG 13-4 and Figure 13-4 (page 288) in your text.

17 7. Answer these questions about the conduction pathway in Figure LG 13-3.

a. How many neurons does this reflex contain? ___2___ The neuron that conveys the impulse toward the spinal cord is a _____ neuron; the one that carries the impulses toward the effector is a _____ neuron.

b. This is a *(monosynaptic? polysynaptic?)* reflex arc. The synapse, like all somatic synapses, is located in the *(CNS? PNS?)*.

c. Receptors, in this case located in skeletal muscle, are called _____. They are sensitive to changes in _____. This type of reflex might therefore be called a _____ reflex.

d. Since sensory impulses enter the cord on the same side as motor impulses leave, the reflex is called *(ipsilateral? contralateral? intersegmental?)*.

e. What structure is the effector? _____

f. One example of a stretch reflex is the _____ in which stretching of the patellar tendon initiates the reflex.

8. Explain how the brain may get the message that a stretch reflex (or other type of reflex) has occurred. [18]

253

9. Contrast stretch, flexor, and crossed extensor reflexes in this learning activity. [19]
   a. A flexor reflex *(does? does not?)* involve association neurons, and so it is *(more? less?)* complex than a stretch reflex.
   b. A flexor reflex sends impulses to *(one? several?)* muscle(s), whereas a stretch reflex, such as the knee jerk, activates *(one? several?)* muscle(s), such as the quadriceps. A flexor reflex is also known as a

   _____ reflex.
   c. If you simultaneously contract the flexor and extensor (biceps and triceps) muscles of your forearm with equal effort, what action occurs? (Try it.)

   _____ In order for movement to occur, it is necessary for extensors (triceps) to be inhibited while flexors (biceps) are stimulated. The nervous system exhibits such control by a phenomenon known   as

   _____ innervation.
   d. Reciprocal innervation also occurs in the following instance. Suppose you step on a tack under your right foot. You quickly withdraw that foot by *(flexing? extending?)* your right leg using your hamstring muscles. (Stand up and try it.) What happens to your left leg? You *(flex? extend?)* it. This

   is an example of reciprocal innervation involving a _____

   _____ reflex.
   e. A crossed extensor reflex is *(ipsilateral? contralateral?)*. It *(may? may not?)* be intersegmental. Therefore many muscles may be contracted to extend your left thigh and leg to shift weight to your left side and provide balance during the tack episode.
10. Do this exercise describing how tendon reflexes protect tendons. [20]
    a. Receptors located in tendons are named *(muscle spindles? Golgi tendon organs?)*. They are sensitive to changes in muscle *(length? tension?)*, as when the hamstring muscles are contracted excessively, pulling on tendons.
    b. When this occurs, association neurons cause *(excitation? inhibition?)* of this same muscle, so that the hamstring fibers *(contract further? relax?)*.
    c. Simultaneously, other association neurons fire impulses which stimulate *(synergistic? antagonistic?)* muscles (such as the quadriceps in this case). These muscles then *(contract? relax?)*.
    d. The net effect of such a *(mono? poly?)*-synaptic tendon reflex is that the tendons are *(protected? injured?)*.

11. Why are deep tendon reflexes (that involve stretching a tendon such as that of the quadriceps femoris muscle in the patellar reflex) particularly helpful diagnostically?

21

12. Describe how reinforcement can be used to increase reflex activity.

22

13. Complete the table about reflexes that are of clinical significance.

	Patellar Reflex	Achilles Reflex	Babinski Sign
a. Procedure used to demonstrate reflex	Tap patellar ligament		
b. Nature of positive response		Plantar flexion	
c. Spinal nerves and muscles evaluated by procedure in reflex			Determines if corticospinal tract is myelinated yet
d. Cause of negative response			Normally negative after 1½ years old
e. Cause of exaggerated response		Damage to motor tracts in S1 or S2 segments of cord	

E. Spinal nerves

1. Complete this exercise about spinal nerves.

23

   a. There are _____ pairs of spinal nerves. Write the number of pairs in each region.

     _____ Cervical   _____ Sacral

     _____ Thoracic   _____ Coccygeal

     _____ Lumbar

   b. Which of these spinal nerves form the cauda equina?

   c. Spinal nerves are attached by two roots. The posterior root is *(sensory? motor? mixed?)*, the anterior root is _____, whereas the spinal nerve is _____.

   d. Individual nerve fibers are wrapped in a connective tissue covering known as *(endo-? epi-? peri-?)* neurium. Groups of nerve fibers are held in bundles (fascicles) by ____-neurium. The entire nerve is wrapped with ____-neurium.

   e. Spinal nerves branch when they leave the intervertebral foramen. These branches are called _____. Since they are extensions of spinal nerves, rami are *(sensory? motor? mixed?)*.

   f. Which ramus is larger? *(ventral? dorsal?)* What areas does it supply?

   g. What area does the dorsal ramus innervate?

   h. Name two other branches (rami) and state their functions.

2. Match the plexus names with descriptions. Refer to Figure 13-1a (page 282 in your text) for help.

24

_____ a. Provides the entire nerve supply for the arm

_____ b. Contains origin of phrenic nerve (nerve that supplies diaphragm)

_____ c. Forms median, radial, and axillary nerves

_____ d. Not a plexus at all, but rather segmentally arranged nerves

_____ e. Supplies nerves to scalp, neck, and part of shoulder and chest

_____ f. Supplies fibers to the femoral nerve which innervates the quadriceps, so injury to this plexus would interfere with actions such as touching the toes

_____ g. Forms the largest nerve in the body (the sciatic) which supplies posterior of thigh and the leg

B. Brachial
C. Cervical
I. Intercostal
L. Lumbar
S. Sacral

3. *A clinical challenge.* If the cord were completely transected (severed) just below the C7 spinal nerves, how would the functions listed below be affected? (Remember that nerves which originate below this point would not communicate with the brain, so would lose much of their function.) Explain your reasons in each case. (*Hint:* Refer to Figure 13-1a, page 282 in the text.)

a. Breathing via diaphragm

b. Movement and sensation of thigh and leg

c. Movement and sensation of the arm

d. Use of muscles of facial expression and muscles that move jaw, tongue, eyeballs

4 What is a *dermatome?*

5 Describe the general pattern of dermatomes:
   a. In the trunk

   b. In the extremities

F. Disorders

(pages 302–303)

1. List several causes of spinal cord injury.

2. Match the terms at right with descriptions below. | 26 |

_____ a. Paralysis of one extremity only
_____ b. Paralysis of both arms or both legs
_____ c. Paralysis of both arms and both
        legs
_____ d. Paralysis of the arm, leg, and
        trunk on one side of the body

H. Hemiplegia
M. Monoplegia
P. Paraplegia
Q. Quadriplegia

3. Contrast complete transaction with hemisection.

4. *For extra review* of spinal cord functions affected by hemisection, look
   again at Learning Activity | 12 |, page LG 250.

5. Briefly describe changes that occur following cord transaction and discuss surgical intervention. Include these terms: *spinal shock, areflexia, delayed nerve grafting.*

27 6. Answer these questions about nerve regeneration.
    a. In order for damaged neurons to be repaired, they must have an intact cell body and also a _____.
    b. Can axons in the CNS regenerate? Explain.

    c. When a nerve fiber (axon or dendrite) is injured, the changes that follow in the cell body are called *(axon reaction? Wallerian degeneration?)*. Those that occur in the portion of the fiber distal to the injury are known as _____.

7. Explain how peripheral nerves regenerate by describing each of these events.
    a. Axon reaction, chromatolysis

    b. Wallerian degeneration

    c. Neurolemmocytes "tunnel" formation

d. Accelerated protein synthesis

8. Shingles is an infection of the *(central? peripheral?) nervous system. The* [28] *causative virus is also the agent of (chickenpox? measles? cold sores?).* Following recovery from chickenpox, the virus remains in the body in the *(spinal cord? dorsal root ganglia?).* At times it is activated and travels along *(sensory? motor?) neurons, causing (pain? paralysis?).*
9. Match names of disorders at right with definitions below. [29]

_____ a. Inflammation of a single nerve
_____ b. Lack of reflex activity
_____ c. Acute inflammation of the nervous system by *Herpes zoster* virus
_____ d. Neuritis of a nerve in the posterior of hip and thigh; often due to a slipped disc in the lower lumbar region

A. Areflexia
N. Neuritis
Sc. Sciatica
Sh. Shingles

Answers to Numbered Questions in Learning Activities

[1] Gray, gray, white.

[2] (a) Nerve fibers, PNS. (b) White, nerve fibers. (c) Gray, PNS. (d) Cell bodies, CNS.

[3] (a) Co. (b) S. (c) F. (d) Ca.

[4] Vertebrae, meninges, cerebrospinal fluid, ligaments and fat in epidural space.

[5] Dura mater *(A),* subdural space *(B),* arachnoid *(C),* subarachnoid space *(D),* pia mater *(E),* brain or spinal cord tissue *(F).*

[6] (a) Outside of; fat, other connective tissues and blood vessels; anesthetic penetrates these other epidural tissues and all three meningeal layers before reaching nerve fibers; childbirth. (b) Deep to the dura mater, serous fluid. (c) Subarachnoid space, pia mater.

[7] Brain, denticulate, filum terminale.

[8] (a) Between L3 and L4 (or L4 and L5). (b) Spinal cord ends at about L1, so the cord cannot be injured at this lower level. (c) Subarachnoid.

[9] Posterior median sulcus *(A),* posterior gray horn *(B),* posterior (dorsal) white columns *(C),* lateral white column *(D),* central canal *(E),* gray commissure *(F),* lateral gray horn *(G),* anterior (ventral) gray horn *(H),* anterior median fissure *(I),* anterior white column *(J).*

10 (a) Central, nerves. (b) White columns, myelinated. (c) Sensory, brain. (d) Descending. (e) Lateral, cerebral cortex, spinal cord (in anterior gray horn). (f) Descending, motor.

11 (a) LS. (b) F. (c) AS. (d) C. (e) V. (f) T.

12 (b) L. (c) R. (d) L (since L lateral corticospinal is lost and some R since L anterior corticospinal is lost). (e) L. (f) R.

13 (a) Conduction, tracts. (b) Reflex, nerves, 2, posterior (dorsal), motor. (c) Sensory, anterior, lateral.

14 (b) Sensory neuron. (c) Center, usually in CNS. (d) Motor neuron. (e) Effector.

15 Dendrite, gray, muscle.

16 Receptor *(A)*, sensory neuron dendrite *(B)*, sensory neuron cell body *(C)*, sensory neuron axon *(D)*, synapse *(E)*, motor neuron cell body *(F)*, motor neuron axon *(G)*, effector *(H)*.

17 (a) 2, sensory, motor. (b) Monosynaptic, CNS. (c) Neuromuscular spindles, length (or stretch), stretch. (d) Ipsilateral. (e) Skeletal muscle. (f) Knee jerk (patellar reflex).

18 Branches from axons of sensory or association neurons can travel through tracts to the brain.

19 (a) Does, more. (b) Several, one, withdrawal. (c) No action, reciprocal. (d) Flexing, extend, crossed extensor. (e) Contralateral, may.

20 (a) Golgi tendon organ, tension. (b) Inhibition, relax. (c) Antagonistic, contract. (d) Poly, protected.

21 They can readily pinpoint a disorder of a specific spinal nerve or plexus, or portion of the cord, since they do not involve the brain.

22

	Patellar Reflex	Achilles Reflex	Babinski Sign
a. Procedure used to demonstrate reflex	Tap patellar ligament	Tap on calcaneal (achilles) tendon	Light stimulation of outer margin of sole of foot
b. Nature of positive response	Extension of leg	Plantar flexion	After age 1½, negative (toes curl under)
c. Spinal nerves and muscles evaluated by procedure in reflex	L2–4 (quadriceps muscles)	L4–5, S1–3 (gastrocnemius muscles)	Determines if corticospinal tract is myelinated yet
d. Cause of negative response	Chronic diabetes mellitus, neurosyphilis	Diabetes, neurosyphilis, alcoholism	Normally negative after 1½ years old and positive (great toe extends), before age 1½, showing myelination incomplete

	Patellar Reflex	Achilles Reflex	Babinski Sign
e. Cause of exaggerated response	Injury to corticospinal tracts	Damage to motor tracts in S1 or S2 segments of cord	Positive Babinski after age 1½ indicates interruption of corticospinal tracts

23 (a) 31, 8, 12, 5, 5, 1. (b) Lumbar, sacral and coccygeal. (c) Sensory, motor, mixed. (d) Endo-, peri-, epi-. (e) Rami, mixed. (f) Ventral: all of the extremities and the ventral and lateral portions of the trunk. (g) Muscles and skin of the back. (h) Meningeal branch supplies primarily vertebrae and meninges; rami communicantes have autonomic functions.

24 (a) B. (b) C. (c) B. (d) I. (e) C. (f) L. (g) S.

25 (a) Not affected since (phrenic) nerve to the diaphragm originates from the cervical plexus (at C3–C5), higher than the transection. So this nerve continues to receive nerve impulses from the brain. (b) Complete loss of sensation and paralysis since lumbar and sacral plexuses originate below the injury, and therefore no longer communicate with the brain. (c) Most arm functions are not affected. As shown on Figure 13-12, page 296 of the text, the brachial plexus originates from C-5 through T-1, so most nerves to the arm (those from C-5 through C-7) still communicate with the brain. (d) Not affected since all are supplied by cranial nerves which originate from the brain.

26 (a) M. (b) P. (c) Q. (d) H.

27 (a) Neurilemma. (b) No. They lack a neurilemma. (c) Axon reaction, Wallerian degeneration.

28 Peripheral, chickenpox, dorsal root ganglia, sensory, pain.

29 (a) N. (b) A. (c) Sh. (d) Sc.

MASTERY TEST: Chapter 13

Questions 1–4: Arrange the answers in correct sequence.

__C__ __B__ __A__ 1. From superficial to deep:
    A. Subarachnoid space
    B. Epidural space
    C. Dura mater

__B__ __C__ __A__ 2. From anterior to posterior in the spinal cord:
    A. Fasciculus gracilis and cuneatus
    B. Anterior spinothalamic tract
    C. Central canal of the spinal cord

__C__ __B__ __A__ __D__ 3. The plexuses, from superior to inferior:
    A. Lumbar
    B. Brachial
    C. Cervical
    D. Sacral

262

___D___ ___B___ ___C___ ___A___ ___E___ 4. Order of structures in a conduction pathway, from origin to termination:

        A. Motor neuron
        B. Sensory neuron
        C. Center
        D. Receptor
        E. Effector

Questions 5–10: Choose the one best answer to each question.

_____ 5. During the axon reaction phase of nerve regeneration, all of the following occur EXCEPT:

    A. Chromatophilic substance (Nissl bodies) breaks down into finely granular masses
    B. The cell body shrinks in size
    C. The number of free ribosomes increases as the number of them on endoplasmic reticulum decreases
    D. The nucleus is notably off-center in position

___D___ 6. All of these tracts are sensory EXCEPT:

    A. Anterior spinothalamic   B. Lateral spinothalamic
    C. Fasciculus cuneatus   D. Lateral corticospinal
    E. Posterior spinocerebellar

___C___ 7. Choose the FALSE statement about the spinal cord.

    A. It has enlargements in the cervical and lumbar areas.
    B. It lies in the vertebral foramen.
    C. It extends from the medulla to the sacrum.
    D. It is surrounded by meninges.
    E. In cross section an H-shaped area of gray matter can be found.

___A___ 8. All of these structures are composed of white matter EXCEPT:

    A. Posterior root (spinal) ganglia   B. Tracts
    C. Lumbar plexus   D. Sciatic nerve
    E. Ventral ramus of a spinal nerve

___D___ 9. Which is a FALSE statement about the patellar reflex?

    A. It is also called the knee jerk.
    B. It involves a two-neuron, monosynaptic reflex arc.
    C. It results in extension of the leg by contraction of the quadriceps femoris.
    D. It is contralateral.

___B___ 10. The cauda equina is:

    A. Another name for the cervical plexus
    B. The lumbar and sacral nerve roots extending below the end of the cord
    C. The inferior extension of the pia mater
    D. A denticulate ligament
    E. A canal running through the center of the spinal cord

Questions 11–20: Circle T (true) or F (false). If the statement is false, change the underlined word or phrase so that the statement is correct.

T (F) 11. In order for regeneration of a nerve to occur, the injured nerve must have an <u>intact cell body and a myelin sheath.</u>

T ⓕ 12. The two main functions of the spinal cord are that it serves as a <u>reflex center</u> and it is <u>the site where sensations are felt.</u>

ⓣ F 13. Dorsal roots of spinal nerves are <u>sensory,</u> ventral roots are motor, and spinal nerves are mixed.

ⓣ F 14. A tract is a bundle of nerve fibers <u>inside of the central nervous system (CNS).</u>

T ⓕ 15. Synapses <u>are</u> present in posterior (dorsal) root ganglia.

ⓣ F 16. After a person reaches 18 months, the Babinski sign should be <u>negative, as indicated by plantar flexion (curling under of toes and foot).</u>

T F 17. Visceral reflexes are used diagnostically <u>more often than somatic ones since it is easy to stimulate most visceral receptors.</u>

ⓣ F 18. Transection of the spinal cord at level C-6 will result in <u>greater</u> loss of function than transection at level T-6.

T F 19. The ventral root of a spinal nerve contains <u>axons and dendrites</u> of <u>both motor and sensory</u> neurons.

T ⓕ 20. A lumbar puncture (spinal tap) is usually performed at about the level of vertebrae <u>L1 to L2</u> since the cord ends between about <u>L3 and L4.</u>

Questions 21–25: fill-ins. Complete each sentence with the word or phrase that best fits.

_____ganglia_____ 21. Clusters of cell bodies located outside of the central nervous system (CNS) are known as _____.

_____meningitis_____ 22. An inflammation of the dura mater, arachnoid, and/or pia mater is known as _____.

_____poly ipsi_____ 23. Tendon reflexes are _____-synaptic and _____-lateral.

_____ 24. Lateral gray horns are found only in _____ regions of the spinal cord.

____ pia ____ 25. The filum terminale and denticulate ligaments are both composed of _____ mater.

The Brain and the Cranial Nerves

In the previous chapter you began a study of major nervous structures with a consideration of the spinal cord and its nerves. In this chapter you will learn about the brain, the structure involved in higher levels of integration and control. You will first identify principal parts of the brain (Objective 1) and the factors that are protective to the brain (2–4). Then you will consider in more detail the major parts of the brain (5–13) and the cranial nerves (14). Next you will study the effects of aging (15). And finally you will examine certain disorders and medical terminology associated with the central nervous system (15–17).

Topics Summary

A. Brain: principal parts, protection and coverings, cerebrospinal fluid, blood supply
B. Brain: brain stem, diencephalon
C. Brain: cerebrum
D. Cerebellum
E. Transmitter substances of the brain
F. Cranial nerves
G. Aging and developmental anatomy of the nervous system
H. Disorders, medical terminology, drugs

Objectives

1. Identify the principal parts of the brain.
2. Describe how the brain is protected.
3. Explain the formation and circulation of cerebrospinal fluid (CSF).
4. Describe the blood supply to the brain and the concept of the blood–brain barrier (BBB).
5. Compare the components of the brain stem with regard to structure and function.
6. Identify the structure and functions of the diencephalon.
7. Describe the surface features, lobes, tracts, and basal ganglia of the cerebrum.
8. Describe the structure and functions of the limbic system.
9. Compare the sensory, motor, and association areas of the cerebrum.
10. Describe the principal waves of an electroencephalogram (EEG) and explain its significance in the diagnosis of certain disorders.
11. Explain the concept of brain lateralization.
12. Describe the anatomical characteristics and functions of the cerebellum.
13. Discuss the various neurotransmitter substances found in the brain, as well as the different types of neuropeptides and their functions.
14. Define a cranial nerve and identify the 12 pairs of cranial nerves by name, number, type, location, and function.
15. Describe the effects of aging on the nervous system.
16. List the clinical symptoms of these disorders of the nervous system: brain tumors, poliomyelitis, cerebral palsy, Parkinson's disease, multiple sclerosis (MS), epilepsy, cerebrovascular accidents (CVAs), dyslexia, Tay-Sachs disease, headache, trigeminal neuralgia, Reye's syndrome (RS), and Alzheimer's disease (AD).
17. Define medical terminology associated with the central nervous system.

Learning Activities

(pages 307–310) **A. Brain: principal parts, protection and coverings, cerebrospinal fluid, blood supply**

| 1 | 1. The average adult brain weighs about _____ g (_____ lb).

| 2 | 2. List the four principal parts of the brain. _____

_____   _____

_____

| 3 | 3. Identify brain structures on Figure LG 14-1. Then complete this exercise.

a. Structures 1–3 are parts of the _____.

b. Structures 4 and 5 together form the _____.

c. Structure 6 is the largest part of the brain, the _____

_____.

d. The second largest part is structure 7, the _____.

| 4 | 4. Name three ways in which the brain is protected.

5. Review the layers of the meninges covering the brain and cord by listing them here.

For extra review. Study Learning Activity | 5 | in Chapter 13 (page LG 247).

| 5 | 6. Circle all correct answers about CSF.

a. How much CSF is found in the entire nervous system?
    A. 16 oz (1 pint)   B. 8 oz (1 cup)   C. 4 oz (½ cup)
    D. 1 oz (2 tablespoons)

b. The color of CSF is:
    A. Yellow   B. Clear, colorless   C. Red   D. Green

c. Choose the function(s) of CSF.
    A. Serves as a shock absorber for brain and cord
    B. Contains red blood cells
    C. Contains white blood cells called lymphocytes
    D. Contains nutrients

d. Which statement(s) describe its formation?
    A. It is formed by diffusion of substances from blood.
    B. It is formed by filtration and secretion.
    C. It is formed from blood in capillaries called choroid plexuses.
    D. It is formed in all four ventricles.

e. Which statement(s) describe its pathway?
    A. It circulates around the brain, but not the cord.
    B. It flows below the end of the spinal cord.

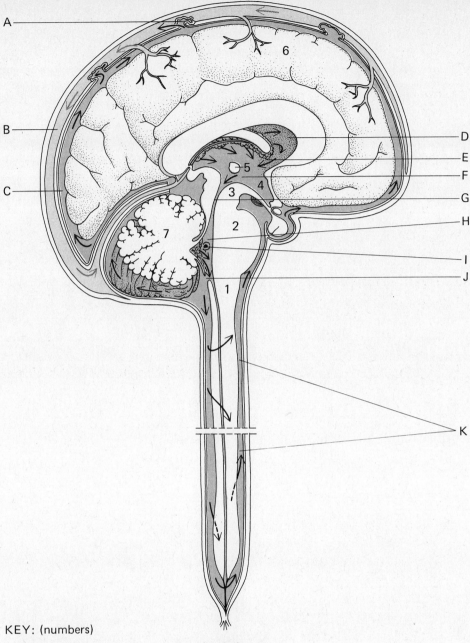

KEY: (numbers)

1. Medulla oblongata
2. Pons varolii
3. Midbrain
4. Hypothalamus
5. Thalamus
6. Cerebrum
7. Cerebellum

KEY; (letters)

A. Arachnoid villus
B. Cranial venous sinus
C. Subarachnoid space of brain
D. Lateral venticle
E. Interventricular foramen
F. Third ventricle
G. Cerebral aqueduct
H. Fourth ventricle
I. Lateral aperture
J. Median aperture
K. Subarachnoid space of spinal cord

Figure LG 14-1 Brain and meninges seen in sagittal section. Parts of the brain are numbered; refer to Learning Activity [3] . Letters indicate structures in pathway of cerebrospinal fluid; refer to Learning Activity [6] .

C. It bathes the brain by flowing through the epidural space.

D. It passes via projections (villi) of the arachnoid into blood vessels (venous sinuses) surrounding the brain.

E. It is formed initially from blood, and finally flows back to blood.

7. To check your understanding of the pathway of CSF, list in order the structures through which it passes. Use the key letters on Figure LG 14-1. Start at the site of formation.

6

8. Contrast *internal hydrocephalus* with *external hydrocephalus*.

9. What important substances are carried by blood to the brain?

Briefly state results of oxygen starvation of the brain for about 4 minutes.

Explain the role of lysosomes in this process.

Now list effects of glucose deprivation of the brain.

7 10. Increase in the carbon dioxide level or decrease in oxygen level of blood circulating through the brain has the effect of *(dilating? constricting?)* cerebral blood vessels, and therefore helping to supply the brain with a better blood flow. The amount of dilation of these vessels is directly related to the increase in the ____ ion in the cerebrum. Show the chemical reaction which explains how increased CO_2 level leads to increased H^+ level (or lower pH).

$$CO_2 + \boxed{} \rightarrow \boxed{} \rightarrow \boxed{} + H^+$$

11. Describe the blood–brain barrier (BBB) in this exercise. 8 **269**

   a. Blood capillaries supplying the brain have *(more? less?)* densely packed

   cells plus a layer of _____ cells and a continuous

   _____ membrane.

   b. What advantages are provided by this membrane?

   c. List two types of substances that do normally cross the blood–brain barrier.

   d. What problems may result from the fact that some chemicals cannot cross this barrier?

B. Brain: brain stem, diencephalon (pages 310–316)

1. Describe the principal functions of the medulla in this exercise. 9

   a. The medulla serves as a _____ pathway for all ascending and descending tracts. Its white matter therefore transmits *(sensory? motor? both sensory and motor?)* impulses.

   b. Included among these tracts are the triangular _____

   _____ tracts which are the principal *(sensory? motor?)* pathways. The main fibers that pass through the pyramids are the *(spinothalamic? corticospinal?)* tracts.

   c. Crossing (or _____) of fibers occurs in the medulla. This explains why movements of your right hand are initiated by motor neurons which originate in the *(right? left?)* side of your cerebrum. The *(axons? dendrites?)* of these motor neurons decussate in the medulla to proceed down the right _____-spinal tract.

   d. The medulla contains gray areas as well as white matter. Several important nuclei lie in the medulla. Two of these are synapse points for sensory axons that have ascended in the dorsal columns of the cord. These two

   nuclei, named _____ and _____,

   transmit impulses for sensations of _____,

   _____, and _____.

   e. Other medullary nuclei include an olive-shaped nucleus, named the

   _____, with connections to the cerebellum and the

   _____ nuclear complex associated with equilibrium.

   f. You maintain a conscious state as a result of all kinds of sensory input.

   Sensory impulses pass into the _____ formation, which consists of a dispersed area in the brain stem, diencephalon, and spinal cord.

   g. Some input to the medulla arrives by means of cranial nerves; these nerves may serve motor functions also. Which cranial nerves are attached to the

   medulla? _____

   (Functions of these nerves will be discussed later in this chapter.)

270

h. A hard blow to the base of the skull can be fatal since the medulla is the site of three vital centers: the _____ center, regulating the heart; the _____ center, adjusting the rhythm of breathing; and the _____ center, regulating blood pressure by altering diameter of blood vessels.
i. List five important, but nonvital, centers in the medulla.

10   2. Summarize important aspects of the pons in this learning activity.

a. The name *pons* means _____. It serves as a bridge in two ways. It contains longitudinally arranged fibers that connect the _____ and with the upper parts of the brain. It has transverse fibers that connect the two sides of the _____

_____.

b. Cell bodies associated with fibers in cranial nerves numbered _____ lie in nuclei in the pons.
c. *(Respiration? Heartbeat? Blood pressure?)* is controlled by the pneumotaxic and apneustic areas of the pons.

11   3. Match the name of the midbrain structure listed at right with its related description. Use each answer only once.

_____ a. Change of pupil size in response to light shone in the eye is a check on the health of this specific area of the brain where cranial nerves III and IV to eye muscles originate.

_____ b. This is the main connection for sensory and motor tracts through the midbrain portion of the brainstem.

_____ c. A tumor in the midbrain could block this passageway for CSF between the third and fourth ventricles, causing internal hydrocephalus.

_____ d. A continuation of fibers from cuneatus and gracilis nuclei, these convey impulses to the thalamus related to fine touch, proprioception, and vibrations.

_____ e. Contained within the tectum, these four bodies (also called colliculi) are the starting point of the tectospinal tracts; they control head movements in response to visual and auditory stimuli

CA. Cerebral aqueduct
CP. Cerebral peduncles
CQ. Corpora quadrigemina
ML. Medial lemniscus
OTN. Oculomotor and trochlear nuclei
RN. Red nucleus

_____ f. This is the origin of rubrospinal
      tracts controlling muscle tone and
      posture.

4. Describe the thalamus in this exercise.

   a. If you look for the thalamus in most figures of the brain, you will not find
      it. It is located *(deep within? on the surface of?)* the cerebrum.
   b. The thalamus is H-shaped. The cross bar of the H is known as the

      _____ mass. It passes through the center of the slit-

      like _____ ventricle. (See Figure LG 14-1.) The two
      side bars of the H form the lateral walls of the third ventricle.
   c. The thalamus is the principal relay station for *(motor? sensory?)* impulses.
      For example, spinothalamic and lemniscal tracts convey general sensa-

      tions such as pain, _____,

      _____, _____, and temperature
      to the thalamus where they are relayed to the cerebral cortex. Special
      sense impulses (for vision and hearing) are relayed through the *(genicu-
      late? reticular? ventral posterior?)* nuclei of the thalamus.
   d. The thalamus *(does? does not?)* contain nuclei controlling motor func-
      tions. However, its principal role is conveying *(motor? sensory?)* impulses.

5. The name hypothalamus indicates that this structure lies *(above? below?)* the
   thalamus, forming the floor and part of the lateral walls of the

   _____ ventricle.

6. Expanding on the key words listed below, write a sentence describing major
   hypothalamic functions.
   a. Regulator of visceral activities

   b. Psychosomatic

   c. Regulating factors to anterior pituitary

   d. ADH and oxytocin to posterior pituitary

   e. Thermostat

   f. Thirst

g. Feeding and satiety center

h. Reticular formation: arousal and consciousness

i. Other

7. *For extra review.* Check your understanding of these parts of the brain stem
and diencephalon by matching them with the descriptions given below.

|13|

_____ a. It is the main regulator of visceral activities since it acts as a liaison between cerebral cortex and autonomic nerves that control viscera.

_____ b. It is the site of the red nucleus, the origin of rubrospinal tracts concerned with muscle tone and posture.

_____ c. Cranial nerves III–IV attach to this brain part.

_____ d. Cranial nerves V–VIII attach to this brain part.

_____ e. Cranial nerves VIII–XII attach to this brain part.

_____ f. Feelings of hunger, fullness, and thirst stimulate centers here so that you can respond accordingly.

_____ g. All sensations except smell are relayed through here.

_____ h. Regulation of heart, blood pressure, and respiration occurs by centers located here.

_____ i. It constitutes four-fifths of the diencephalon.

_____ j. It lies under the third ventricle, forming its floor.

_____ k. It forms most of side walls of the third ventricle.

_____ l. Tumor in this region could compress cerebral aqueduct and cause internal hydrocephalus.

_____ m. It releases chemicals called regulating factors that control hormones.

H. Hypothalamus
Med. Medulla
Mid. Midbrain
P. Pons
T. Thalamus

(pages 316–323) **C. Brain: cerebrum**

|14|

1. Complete this exercise about cerebral structure.

a. The outer layer of the cerebrum is called _____. It is composed of *(white? gray?)* matter. This means that it contains mainly *(cell bodies? neurons?)*.

b. In the margin, draw a line the same length as the thickness of the cerebral cortex. Use a metric ruler. Note how thin the cortex is.

c. The surface of the cerebrum looks much like a view of tightly packed mountains or ridges, called _____. The parts where the cerebral cortex dips down into valleys are called _____

_____ (deep valleys) or _____ (shallow valleys).

d. The cerebrum is divided into halves called _____. Connecting the two hemispheres is a band of *(white? gray?)* matter called the _____. Notice this structure in Figure LG 14-1 and in Figure 14-6a, page 314 of your text.

e. The falx cerebri is composed of *(nerve fibers? dura mater?)*. Where is it located?

At its superior and inferior margins, the falx is dilated to form channels for venous blood flowing from the brain; these enclosures are called

_____.

2. Label the following cerebral structures on the lateral view in Figure LG 14-2: *frontal lobe, parietal lobe, occipital lobe, temporal lobe, central sulcus, lateral sulcus, precentral gyrus,* and *postcentral gyrus*.

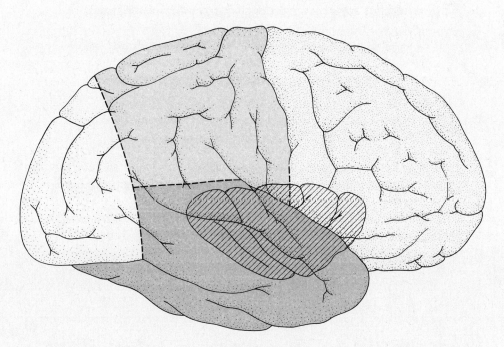

Figure LG 14-2 Right lateral view of lobes and fissures of the cerebrum. Color and label as directed.

15 3. Match the three types of white matter fibers with these descriptions.

       _____ a. The corpus callosum contains   A. Association
these fibers and connects the two   C. Commissural
cerebral hemispheres.   P. Projection

       _____ b. Sensory and motor fibers passing
between cerebrum and other parts
of the CNS are this type of fiber;
the internal capsule is an example.

       _____ c. These fibers transmit impulses
among different areas of the same
hemisphere.

4. *For extra review,* label the corpus callosum on Figure LG 14-1. (Verify the location by checking Figure 14-2a, page 309 in the text.)

5. Look at Figures 14-6 (page 314) and 14-9 (page 320) in the text and identify

16 parts of the *corpus striatum* by completing this exercise.

a. A structure with a head and a tail (cauda) is the _____

   _____ nucleus.

b. The two parts of the lentiform nucleus are the _____

   _____ and the _____ .

c. The caudate and lentiform nuclei, together with several other nuclei, are

   called the _____ . They are islands of *(gray? white?)* matter embedded deep within the cerebrum.

d. Separating the caudate from the lentiform nucleus, is a band of white

   fibers called the _____ . It consists of important sensory and motor tracts. All of these structures together present a striped appearance (gray caudate, white internal capsule, gray lentiform nucleus)

   called the _____ .

6. *For extra review,* list and describe four other structures that are considered parts of the basal ganglia.

7. In general, what are the functions of the basal ganglia?

What are the results of damage to it?

8. Describe the limbic system in this exercise.
   a. This system consists of portions of the cerebrum, thalamus, and hypothalamus. Name the component structures.

17

   b. The limbic system is shaped much like a _____ surrounding the brainstem.
   c. Explain why the limbic system is sometimes called the "visceral" or "emotional" brain.

   d. One other function of the limbic system is _____.
      Forgetfulness, such as inability to recall recent events, results partly from impairment of this system.
9. Write the terms *concussion, contusion,* or *laceration* after related definitions. 18

   a. Tearing of the brain with rupture of blood vessels: _____

   _____
   b. Abrupt, temporary loss of consciousness due to blow on head or sudden cessation of head movement; involves no visible brain bruising:

   _____
   c. Visible bruising of brain leading to extended loss of consciousness:

   _____
10. Use this list of functional areas of the cerebral cortex in two ways. First, write a sentence describing the functions of that area next to each name. Second, color (as indicated below) and label the areas on Figure LG 14-2.
    a. Primary sensory (somesthetic) area, blue

    b. Somesthetic association area, light blue

    c. Primary visual area, green

    d. Visual association area, light green

e. Primary auditory area, black. Write HI on the region which responds to high-pitched tones, and write LO on the specific area which receives impulses related to low pitches.

f. Auditory association area, gray

g. Primary gustatory area, purple

h. Primary motor area, red

i. Premotor area, pink

j. Frontal eyefield area, brown

k. Broca's area, yellow

11. Check your understanding of functional areas of the cerebral cortex by doing this matching exercise. Answers may be used more than once.

19

_____ a. In the occipital lobe
_____ b. In the postcentral gyrus
_____ c. Known as the somesthetic area; receives sensations of pain, touch, pressure, and temperature
_____ d. In the parietal lobe
_____ e. In the temporal lobe; interprets sounds
_____ f. Integrates general and special sensations to form a common thought about them
_____ g. In precentral gyrus of the frontal lobe; controls specific muscles or groups of muscles
_____ h. In the frontal lobe; translates thoughts into speech
_____ i. Controls smell

B. Broca's area
G. Gnostic area
PA. Primary auditory area
PM. Primary motor area
PO. Primary olfactory area
PS. Primary somesthetic area
PV. Primary visual area

12. *For extra review.* List the sequence of events that occurs when you listen to a person ask a question and then you respond to it orally. (Notice how many parts of the cerebrum are involved.) Use separate paper.

13. Contrast:
    a. Aphasia/agraphia

    b. Word blindness/word deafness

14. Discuss functions of association areas of the cerebrum.

15. Define learning disability and list several examples of types of learning disabilities.

16. Identify the type of brain waves associated with each of the following situations. [20]

    _____ a. Occur in persons experiencing stress and in certain brain disorders

    _____ b. Lowest frequency brain waves, normally occurring when adult is in deep sleep; presence in awake adult indicates brain damage

    _____ c. Highest frequency waves, noted during periods of mental activity

    A. Alpha
    B. Beta
    T. Theta
    D. Delta

_____ d. Present when awake but resting, these waves are intermediate in frequency between alpha and delta waves

17. List several criteria which may be used to determine brain death. Include the nature of the EEG.

18. Below is a list of cerebral functions. Indicate whether these are primarily left or right hemisphere functions by writing L or R in the correct column. The first two are done for you.

[21]

_____ _R_ a. Control of the left hand and the left foot

_L_ _____ b. Language, both spoken and written

_____ _____ c. Musical awareness

_____ _____ d. Insight and imagination

_____ _____ e. Numerical and scientific skills

_____ _____ f. Artistic awareness; space and pattern perception

[22] 19. *A clinical challenge.* You are helping to care for Laura, a hospital patient with a contusion of the right side (but not the left side) of the brain. Which of the following would you be most likely to observe in Laura? Circle letters of correct answers.

A. Her right leg is paralyzed.

B. Her left arm is paralyzed.

C. She cannot speak out loud to you nor write a note to you.

D. She used to sing well, but cannot seem to stay on tune now.

E. She has difficulty with concepts involving numbers.

(pages 323–325) **D. Cerebellum**

1. Describe the cerebellum.
   a. Where is it located?

b. Describe its structure. Identify the *vermis* and *hemispheres, arbor vitae,* and *peduncles.*

c. Describe its functions. Use these key terms: *coordinated movements, posture, equilibrium,* and *emotions.*

2. *For extra review.* Arrange the following sentences in correct order to form a paragraph describing how coordinated movement occurs. Write letters in order on these lines: ＿＿ ＿＿ ＿＿ ＿＿ ＿＿   23
    A. Cerebellum sends subconscious motor impulses along inferior cerebellar peduncles to medulla and spinal cord.
    B. Once movement has started, cerebral cortex receives proprioceptive information by means of dorsal columns, nuclei cuneatus and gracilis, medial lemniscal tracts, and thalamus.
    C. Cerebral cortex activates the brain stem (pons and midbrain) which relays impulses over middle and superior cerebellar peduncles to cerebellum.
    D. Impulses from medulla and spinal cord pass downward to motor neurons in anterior gray horn of cord; these finally stimulate skeletal muscle.
    E. Motor areas of cerebral cortex voluntarily initiate muscle contraction.
3. Define *ataxia.*

4. Assess a study partner for cerebellar damage. Note: symptoms of such   24
damage can be expected to be *(ipsi? contra?)*-lateral.

280 (page 325–327) **E. Transmitter substances of the brain**

1. Complete the table about neurotransmitter substances in the brain.

Transmitter Name	Abbreviation	Excitatory Inhibitory	Sites Where Released	Functions
	ACh	Usually		
				Induces sleep, controls mood, temperature
	GABA	Inhibitory (the most common one in brain)		
Dopamine				Controls gross, subconscious movements of skeletal muscles; if DA lacking, results in Parkinsonism
		Mostly inhibitory, but excitatory in some instances		Locus coeruleus which sends axons to hypothalamus, cerebellum, cerebral cortex, and spinal cord

2. Describe and contrast the following peptides. Explain their role in suppressing pain.
   a. Enkephalins

   b. Endorphins (Explain relationship to P substance.)

   c. Dynorphin

3. Match the following peptides with related descriptions.   [25]

   _____ a. Produced by the lining of the small intestine, it is known to stimulate the pancreas and liver, and may control feeding centers of the brain
   _____ b. Produced by the hypothalamus; regulate release of hormones by the anterior pituitary gland
   _____ c. Produced as a result of action of a kidney enzyme renin; may assist brain's regulation of its own blood pressure

   A. Angiotensin
   C. Cholecystokinin
   R. Regulating factors

F. Cranial nerves   (page 327)

1. Answer these questions about cranial nerves.   [26]

   a. There are _____ pairs of cranial nerves. They are all attached to the _____; they leave the _____ via foramina.
   b. They are numbered by Roman numerals in the order that they leave the cranium. Which is most anterior? *(I? XII?)* Which is most posterior *(I? XII?)*

c. All spinal nerves are *(purely sensory? purely motor? mixed?).* Are all cranial nerves mixed? *(Yes? No?)*

2. Pathways involving cranial nerves are quite similar to those of spinal nerves, but are shorter since they are closer to the brain. Describe the typical sensory and motor pathways of cranial nerves in this exercise.

   a. Cranial nerves with somatic motor fibers have relatively simple pathways, as shown in Figure LG 14-3a. Like spinal nerve pathways, they consist of just one neuron between the CNS (brain stem in this case) and effector. (Autonomic pathways involve two neurons; these will be discussed in Chapter 16.) Motor cell bodies that lie in *(ganglia? nuclei?)* in the brain stem send axons via cranial nerves to effectors. (These cell bodies could be stimulated by a variety of neurons in other parts of the brain.)

   b. The typical sensory pathway is shown in Figure LG 14-3b. It involves *(axons? dendrites?)* that extend from receptor organs to cell bodies in ganglia situated just *(inside? outside?)* the CNS. Axons from these cell bodies proceed into the CNS (brain stem) and the cranial nerves terminate

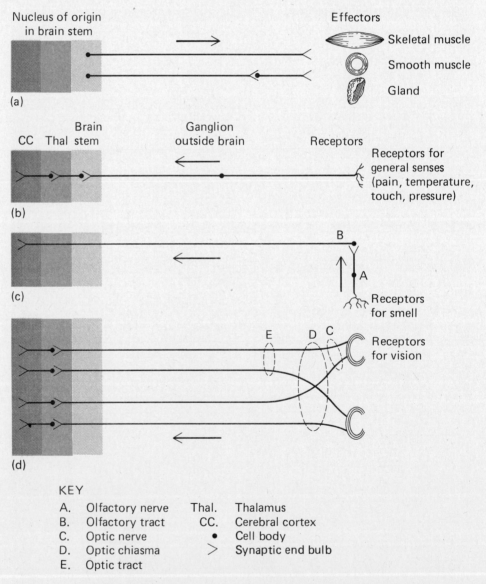

Figure LG 14-3 Conduction pathways of cranial nerves. (a) Typical motor pathway. (b) Typical sensory pathway. (c) Olfactory pathway. (d) Visual pathway.

in (*ganglia? nuclei?*) there. Then neurons relay impulses via tracts to

the _____, which passes messages to the cerebral

_____. The entire conduction pathway from receptor

to cerebral cortex involves at least _____ neurons.

3. Pathways for smell and sight are more complicated. The olfactory pathway is shown on Figure LG 14-3c. Describe it in this exercise.

☐28

a. Cell bodies of olfactory nerves are located in the _____

_____.

b. How many neurons are in the olfactory pathway from the nose to the

olfactory area of the cerebral cortex? _____

c. Which fibers are longer? (*Olfactory nerves? Olfactory tract?*)

4. Answer these questions about the visual pathway, shown on Figure LG 14-3d.

☐29

a. There are _____ layers of neurons in the retina of the eye.

b. Axons of the third layer, namely the _____ cells,

form the second cranial nerve, known as the _____ nerve.

c. The optic nerve exits from the eye and joins its partner from the other eye

at the optic _____. Here some fibers cross to the opposite side; all optic fibers proximal to this point are located in the optic

_____.

d. Most optic tract fibers terminate in the _____ where

impulses are relayed to the visual areas of the _____

_____.

e. Some optic tract fibers do not go to the thalamus, but rather lead to the

_____, where they cause reflex action that causes eye

_____ in response to visual stimuli.

5. Complete the table about cranial nerves. (For extra review on locations of

cranial nerves, refer back to Chapter 7, Learning Activity ☐7 , page LG 129.)

Number	Name	Functions
a.	Olfactory	
b.		Vision (not pain or temperature of the eye)
c. III		

Number	Name	Functions
d.	Trochlear	
e. V		
f.		Stimulates lateral rectus muscle to abduct eye; proprioception of the muscle
g.	Facial	
h.		Hearing; equilibrium
i. IX		
j.	Vagus	
k. XI		
l.		Supplies muscles of the tongue with motor and sensory fibers

6. Check your understanding of cranial nerves by completing this exercise. Write the name of the correct cranial nerve following the related description. $\boxed{30}$

   a. Differs from all other cranial nerves in that it originates from the brain stem and from the spinal cord: _____

   b. Eighth cranial nerve (VIII): _____

   c. Is widely distributed into neck, thorax, and abdomen: _____

   d. Senses toothache, pain under a contact lens, wind on the face: _____

   e. The largest cranial nerve; has three parts (ophthalmic, maxillary, and mandibular): _____

   f. Controls contraction of muscle of the iris, causing constriction of pupil: _____

   g. Innervates muscles of facial expression: _____

   h. Two nerves that contain taste fibers and autonomic fibers to salivary glands: _____ , _____

   i. Three purely sensory cranial nerves: _____ , _____ , _____

7. Match the ganglia at right with correct descriptions below. $\boxed{31}$

   _____ a. Nerve impulses for taste sensa-   G. Geniculate
   tions from the anterior two-thirds   Sp. Spiral
   of the tongue pass along dendrites   V. Vestibular
   to cell bodies here.

   _____ b. Cell bodies in equilibrium pathway are located here.

   _____ c. Located in a spiral arrangement
   within the cochlea, these cell bodies function in the pathway for
   hearing.

8. Write the number of the cranial nerve related to each of the following disorders. $\boxed{32}$

   _____ a. Bell's palsy

   _____ b. Inability to shrug shoulders or turn head

   _____ c. Anosmia

   _____ d. Strabismus and diplopia

   _____ e. Blindness

   _____ f. Trigeminal neuralgia (tic douloureux)

   _____ g. Paralysis of vocal cords; loss of sensation of many organs

   _____ h. Vertigo and nystagmus

G. Aging and developmental anatomy of the nervous system

☐33 1. Do this exercise describing effects of aging on the nervous system.

   a. The total number of nerve cells _____-creases with age. Since conduction velocity *(increases? slows down?)*, the time required for a typical reflex is _____-creased.

   b. Parkinson's disease is the most common *(motor? sensory?)* disorder that is likely to occur in the elderly.

   c. Problems with sense organs include presbyopia which is an inability to focus on objects that are *(near? far away?)* and presbycussis which is impaired *(hearing? taste?)*. The sense of touch is likely to be *(impaired? heightened?)* in old age, placing the older person at *(higher? lower?)* risk for burns or other skin injury.

☐34 2. Check your understanding of the early development of the nervous system by completing this exercise.

   a. The nervous system begins to develop during the third week of gestation when the _____-derm forms a thickening called the _____ plate. Soon a longitudinal depression is found in the plate; this is the neural _____. As neural folds on the sides of the groove grow and meet, they form the neural _____.

   b. Three types of cells form the walls of this tube. Two of these are the marginal layer which forms *(gray? white?)* matter and the mantle layer which becomes *(gray? white?)* matter.

   c. The neural crest forms *(peripheral? central?)* nervous system structures such as ganglia as well as spinal and cranial _____.

   d. The anterior portion of the neural plate and tube develops into three enlarged _____ during the fourth week. These fluid-filled cavities eventually become the _____ which are filled with _____ fluid.

   e. By week five the primary vesicles have flexed (bent) to form a total of *(five? ten?)* secondary vesicles. (More about these in the next exercise.)

☐35 3. Fill in the blanks in this table outlining brain development.

a. Primary brain vesicles (3)	b. Secondary brain vesicles (5)	c. Parts of brain formed (7)
Prosencephalon (_____-brain) →	1. Diecephalon 2. _____ →	{ 1A. _____ 1B. _____ 2A. Cerebrum
_____-encephalon → (Midbrain)	3. Mesencephalon →	3A. _____
_____-encephalon (_____) →	4. _____-encephalon 5. Met-_____ →	4A. _____ { 5A. Pons variolii 5B. _____

H. Disorders, medical terminology, drugs

1. Do this exercise about brain tumors.
   a. Tumors arising from the brain itself (such as the cerebrum) are known as *(intra? extra?)*-axial neoplasms. *(Neurons? Neuroglia?)* are the sources of such tumors. State one example of an extraaxial tumor.

   b. Write three signs or symptoms of brain tumors.

2. For each of the following disorders, give a brief description. Choose pertinent facts, such as cause(s) of the disorder, if they are known; age group or population most afflicted; parts of the nervous system affected; symptoms; and treatment. If the disorder occurs in a progressive manner, describe the stages. Also answer extra questions listed for some disorders.
   a. Poliomyelitis. How does the meaning of the term *poliomyelitis* indicate its effects?

   b. Cerebral palsy (CP)

c. Epilepsy. Contrast types of seizures: *grand mal, petit mal, psychomotor,* and *idiopathic.*

37 3. Complete this Learning Activity about Parkinson's disease.
   a. This condition involves a decrease in the neurotransmitter

   _____. It is normally produced by cell bodies located

   in the substantia nigra which is part of the _____.

   Axons lead from here to the _____ where dopamine (DA) is released. DA is an *(excitatory? inhibitory?)* transmitter.

   b. The most common sign of Parkinsonism is _____.
   Explain how tremor as well as muscle rigidity are related to low DA level.

   c. Voluntary movements may be slower than normal; this is the condition of *(brady? tachy?)*-kinesia.

38 4. Do this exercise describing some main points about multiple sclerosis (MS)
   a. MS affects *(cell bodies? myelin sheaths?)* within the *(central? peripheral?)* nervous system. At sites along the nerves where myelin is destroyed, many

   scars (or _____) form; this is the basis for the name MS.

   b. Initial symptoms are most likely to occur at ages *(10–15? 20–40? 60–70?)* MS affects *(motor? sensory? both motor and sensory?)* neurons. List signs or symptoms that may suggest MS.

   c. How do immunosuppressants such as ACTH or prednisone help MS?

5. CVA refers to _____.
   CVA's are *(rare? common?)* brain disorders. Describe three causes of
   CVA's: ruptured aneurysm (with hemorrhage), embolism, and atherosclerosis.

6. Contrast headaches of intracranial and extracranial origins.

7. Describe the following aspects of Alzheimer's disease (AD):
   a. Progressive changes in behavior

   b. Pathological findings of brain tissue

   c. One theory of causation

8. Arrange in order from highest to lowest level of consciousness. Patients 39
   who are experiencing: ____ ____ ____
   A. Stupor   B. Lethargy and torpor   C. Coma

40 9. *For extra review.* Match the name of the disorder with the related description.

Cerebral palsy	Multiple sclerosis
Cerebrovascular accident (CVA)	Neuralgia
Dyslexia	Parkinsonism
Epilepsy	Poliomyelitis
Encephalitis	Reye's syndrome
Migraine	Tay-Sachs disease

a. Degeneration of myelin sheath to form hard plaques in many regions causing loss of motor and sensory function: _____

b. Degeneration of dopamine-releasing neurons in basal ganglia; characterized by tremor or rigidity of muscles: _____

c. A type of throbbing headache: _____

d. Most common brain disorder; also called stroke: _____

e. Characterized by difficulty in handling words and symbols, for example, reversal of letters *(b* for *d);* cause unknown: _____

f. Inflammation of the brain: _____

g. A disease of children causing swelling of brain cells and fatty infiltration of liver; usually follows viral infection, especially if aspirin is taken: _____

h. Inherited disease especially among Jewish populations due to excessive lipids in brain cells; affects infants: _____

i. Characterized by seizures due to abnormal discharges of electricity from neurons of the brain: _____

j. Of viral origin; often affects only the respiratory system; when nervous system involved, affects movement, but not sensation: _____

k. Motor disorder caused by damage during fetal life, birth, or infancy; not progressive; apparent mental retardation often actually only speaking or hearing disability: _____

l. Attack of pain along an entire nerve, for example, tic douloureux: _____

Answers to Numbered Questions in Learning Activities

1 1,300, 3.

2 Diencephalon, cerebrum, brainstem, cerebellum.

3 (a) Brain stem. (b) Diencephalon. (c) Cerebrum. (d) Cerebellum.

4 Skull bones, meninges, and cerebrospinal fluid (CSF).

5 (a) C. (b) B. (c) A, C, D. (d) B, C, D. (e) B, D, E.

6 *D, E, F, G, H, I, J, K, C, A, B.*

7 Dilating, H^+, $CO_2 + H_2O \rightarrow H_2CO_3 \rightarrow HCO_3^- + H^+$.

8 (a) More, neuroglia, basement. (b) The brain is protected from many substances which could harm the brain, but are kept from the brain by this barrier. (c) Glucose, oxygen, and some ions. (d) Certain helpful medications, such as some antibiotics, cannot cross this barrier. (Fortunately, at the time they are most needed—as in a brain infection—the BBB is more permeable, and so may permit passage of these drugs.)

9 (a) Conduction, both sensory and motor. (b) Pyramid, motor, corticospinal. (c) Decussation, left, axons, cortico-. (d) Cuneatus, gracilis, fine touch (two-point discrimination), proprioception, vibrations. (See Exhibit 13-1, page 286, of text.) (e) Olivary, vestibular. (f) Reticular. (g) Part of VIII, as well as IX–XII. (h) Cardiac, medullary rhythmicity, vasomotor. (i) Coughing, hiccuping, sneezing, swallowing, and vomiting.

10 (a) Bridge, spinal cord, medulla, cerebellum. (b) V–VII and part of VIII. (c) Respiration.

11 (a) OTN. (b) CP. (c) CA. (d) ML. (e) CQ. (f) RN.

12 (a) Deep within. (b) Intermediate, third. (c) Sensory; touch, pressure, proprioception; geniculate. (d) Does, sensory.

13 (a) H. (b) Mid. (c) Mid. (d) P. (e) Med. (f) H. (g) T. (h) Med. (i) T. (j) H. (k) T. (l) Mid. (m) H.

14 (a) Cerebral cortex, gray, cell bodies. (b) 2 to 4 mm. (c) Gyri, fissures, sulci. (d) Hemispheres, white, corpus callosum. (e) Dura mater, between cerebral hemispheres, superior and inferior sagittal sinuses (See Figure 14-2, pages 309–310, in the text.)

15 (a) C. (b) P. (c) A.

16 (a) Caudate. (b) Putamen, globus pallidus. (c) Basal ganglia, gray. (d) Internal capsule, corpus striatum.

17 Wishbone.

18 (a) Laceration. (b) Concussion. (c) Contusion.

19 (a) PV. (b) PS. (c) PS. (d) PS. (e) PA. (f) G. (g) PM. (h) B. (i) PO.

20 (a) T. (b) D. (c) B. (d) A.

21 (c) R. (d) R. (e) L. (f) R.

22 B D

23 E B C A D

24 Ask partner to walk (check gait), touch finger to nose while eyes closed. Effects are ipsilateral.

25 (a) C. (b) R. (c) A.

26 (a) 12, brain, cranium. (b) I, XII. (c) Mixed; no (three are sensory).

27 (a) Nuclei. (b) Dendrites, outside, nuclei, thalamus, cortex, 3.

28 (a) Nose (nasal mucosa). (b) 2. (c) Olfactory tract.

29 (a) 3. (b) Ganglion, optic. (c) Chiasma, tracts. (d) Thalamus, cerebral cortex. (e) Midbrain, movement.

30 (a) Accessory. (b) Vestibulocochlear. (c) Vagus. (d) Trigeminal. (e) Trigeminal. (f) Oculomotor. (g) Facial. (h) Facial, glossopharyngeal. (i) Olfactory, optic, vestibulocochlear.

31 (a) G. (b) V. (c) Sp.

32 (a) VII. (b) XI. (c) I. (d) III. (e) II. (f) V. (g) X. (h) VIII (vestibular branch).

33 (a) De, slows down, in. (b) Motor. (c) Near, hearing, impaired, higher.

34 (a) Ecto, neural, groove, tube. (b) White, gray. (c) Peripheral, nerves. (d) Primary vesicles, ventricles of brain, cerebrospinal (CSF). (e) Five.

35 Read down columns in table. (a) Fore-, mes-, rhomb-, hindbrain. (b) Telencephalon (2), myel- (4), -encephalon (5). (c) Thalamus (1A), hypothalamus (1B), midbrain (3A), medulla (4A), cerebellum (5B).

36 (a) Intra; neuroglia; meninges or pituitary gland. (b) Headache, decreased level of consciousness, papilledema (optic disc edema), vomiting. See others on page 332 of the text.

37 (a) Dopamine, midbrain, basal ganglia, inhibitory. (b) Tremor; lack of inhibitory DA causes a double negative effect: extra unnecessary movements occur (tremor) which may then interfere with normal movements (rigidity), (c) Brady.

38 (a) Myelin sheaths, central, multiple scleroses. (b) 20– 40; both motor and sensory; lack of control in writing, walking, double vision, bladder infections. (c) Suppress autoimmune response in which the body is making antibodies which destroy the neuroglia necessary for CNS myelin formation.

39 B A C.

40 (a) Multiple sclerosis. (b) Parkinsonism. (c) Migraine. (d) CVA. (e) Dyslexia. (f) Encephalitis (g) Reye's syndrome. (h) Tay-Sachs disease. (i) Epilepsy. (j) Poliomyelitis. (k) Cerebral palsy. (l) Neuralgia.

MASTERY TEST: Chapter 14

Questions 1–8: Choose the one best answer to each question.

_____ 1. All of these are located in the medulla EXCEPT:

    A. Nucleus cuneatus and nucleus gracilis
    B. Cardiac center, which regulates heart
    C. Site of decussation (crossing) of pyramidal tracts
    D. Coughing, sneezing, and hiccuping centers
    E. Origin of cranial nerves V to VIII

_____ 2. Which of these is a function of the postcentral gyrus?

    A. Controls specific groups of muscles, causing their contraction
    B. Receives general sensations from skin, muscles, and viscera
    C. Receives olfactory impulses
    D. Primary visual reception area
    E. Somesthetic association area

_____ 3. All of these structures contain cerebrospinal fluid EXCEPT:

     A. Subdural space   B. Ventricles of the brain
     C. Central canal of the spinal cord   D. Subarachnoid space

_____ 4. Damage to the accessory nerve would result in:

     A. Inability to turn the head or shrug shoulders
     B. Loss of normal speech function
     C. Hearing loss
     D. Changes in heart rate
     E. Anosmia

_____ 5. Which statement about the trigeminal nerve is FALSE?

     A. It sends motor fibers to the muscles used for chewing.
     B. Pain in this nerve is called tic douloureux.
     C. It carries sensory fibers for pain, temperature, and touch from the face, including eyes, lips, and teeth area.
     D. It is the smallest cranial nerve.

_____ 6. All of these are functions of the hypothalamus EXCEPT:

     A. Control of body temperature
     B. Release of chemicals (regulating factors) that affect release or inhibition of hormones
     C. Principal relay station for sensory impulses
     D. Center for mind-over-body (psychosomatic) phenomena
     E. Involved in maintaining sleeping or waking state

_____ 7. Damage to the occipital lobe of the cerebrum would most likely cause:

     A. Loss of hearing
     B. Loss of vision
     C. Loss of ability to smell
     D. Paralysis
     E. Loss of feeling in muscles (proprioception)

_____ 8. Choose the one FALSE statement.

     A. Multiple sclerosis is a progressive disorder that worsens as time lapses.
     B. Cerebral palsy (CP) is a progressive disorder that gets worse with time.
     C. About 70 percent of CP victims appear mentally retarded, but, in fact, this may be a reflection of inability to speak or walk well.
     D. Poliomyelitis may affect motor nerves, but does not affect sensations.

Questions 9–12: Arrange the answers in correct sequence.

_____ _____ _____ 9. From anterior to posterior:

     A. Fourth ventricle
     B. Pons and medulla
     C. Cerebellum

_____ _____ _____ 10. From superior to inferior:

     A. Thalamus

     B. Hypothalamus

     C. Corpus callosum

_____ _____ _____ 11. Order in which impulses are relayed in conduction pathway for vision:

     A. Optic nerve

     B. Optic tract

     C. Optic chiasma

_____ _____ _____ _____ _____ 12. Pathway of cerebrospinal fluid, from formation to final destination:

     A. Choroid plexus in ventricle

     B. Subarachnoid space

     C. Cranial venous sinus

     D. Arachnoid villi

Questions 13–20: Circle T (true) or F (false). If the statement is false, change the underlined word or phrase so that the statement is correct.

T   F   13. GABA, serotonin, and acetylcholine all function as excitatory transmitter substances.

T   F   14. The thalamus, hypothalamus, and cerebrum are all developed from the forebrain (prosencephalon).

T   F   15. Endorphins, enkephalins, and dopamine are all chemicals which are considered the "body's own pain killers."

T   F   16. The language areas are located in the cerebellar cortex.

T   F   17. All of these coverings of the brain are composed of dura mater: falx cerebri, falx cerebelli, tentorium cerebelli.

T   F   18. The limbic system functions in control of behavior.

T   F   19. The reticular formation controls arousal and consciousness.

T   F   20. Delta brain waves have the slowest frequency of the four kinds of waves produced by normal individuals.

Questions 21–25: fill-ins. Complete each sentence with the word or phrase that best fits.

_____ 21. Portions of the cerebrum that connect sensory and motor areas, and are concerned with memory, personality, judgment, and intelligence are called ____ areas.

_____ 22. The superior and inferior colliculi, associated with movements of eyeballs and head in response to visual stimuli, are located in the

____.

_____ 23. If the brain is deprived of oxygen for more than 4 minutes, ____ of brain cells break open and release enzymes that can self-destruct these cells.

_____ 24. If blood flowing through the hypothalamus is warmer than normal, the hypothalamus directs heat loss activities such as ____.

_____ 25. The transverse fissure of the brain separates the ____ from the ____.

The Sensory, Motor, and Integrative Systems 15

In the previous chapter you learned about the structure and function of the many individual parts of the brain. In this chapter you will see how these parts work together with the spinal cord, nerves, receptors, and effectors to facilitate homeostasis. You will study how the body receives sensory information (Objectives 1–10). You will examine the motor pathways that control movement and secretion (11–14). Finally, you will consider the integrative functions of the body, such as memory, sleep, and wakefulness (15).

Topics Summary

A. Sensations
B. General senses
C. Sensory pathways
D. Motor pathways
E. Integrative functions

Objectives

1. Define a sensation and list the four prerequisites necessary for its transmission.
2. Describe the characteristics of sensations.
3. Classify receptors on the basis of location, stimulus detected, and simplicity or complexity.
4. List the location and function of the receptors for tactile sensations (touch, pressure, vibration), thermoreceptive sensations (heat and cold), and pain.
5. Distinguish somatic, visceral, referred, and phantom pain and describe the various methods used to relieve pain.
6. Identify the proprioceptive receptors and indicate their functions.
7. Describe the composition of the somatosensory cortex.
8. Discuss the origin, neuronal components, and destination of the posterior column and spinothalamic pathways.
9. Explain the neural pathways for pain and temperature; light touch and pressure; and discriminative touch, proprioception, and vibration.
10. Contrast the roles of the cerebellar tracts in conveying sensations.
11. Describe how sensory input and motor responses are linked in the central nervous system.
12. Describe the composition of the motor cortex.
13. Compare the course of the pyramidal and extrapyramidal motor pathways.
14. List the functions of the lateral corticospinal, anterior corticospinal, corticobulbar, rubrospinal, tectospinal, and vestibulospinal tracts.
15. Compare integrative functions such as memory, wakefulness, and sleep.

Learning Activities

A. Sensations

1. Think of the ways in which your senses help you make homeostatic adjustments during the day. Suppose that you suddenly lost your sight and your hearing; then your ability to sense smell and taste disappeared; then you lost perception of touch, hot and cold, and pain sensations. Consider how you would survive with such sensory deprivation.

2. Define:
   a. Sensation

   b. Perception

3. List and describe four prerequisites for perception of a sensation.

4. Contrast *generator (receptor) potential* with *action potential.*

5. Do the following exercise about receptors.
   a. Define *receptor.*

   1   b. What do all receptors have in common regardless of complexity?

6. Defend or dispute this statement: "You *seem* to see with your eyes, and hear with your ears, but this is not actually so."

7. Match the characteristics of sensations with the related descriptions. $\boxed{2}$

_____ a. The statement you just discussed (question 6) is explained by this characteristic.

_____ b. After a while in the presence of a strong-smelling aroma, you do not seem to notice it as much.

_____ c. With your eyes closed you can distinguish sensations: whether you are hearing a voice, feeling a breeze, experiencing pain.

_____ d. You look into a flash camera while your picture is taken, and even moments later you see a light spot in the field of view where the flash was. Since the sensation persists even after the stimulus is removed, this phenomenon is the reverse of adaptation.

Ad. Adaptation
Af. Afterimage
M. Modality
P. Projection

8. Identify the class of receptor that fits each description given. $\boxed{3}$

_____ a. Receptors in this class are located near the body surface.

_____ b. These receptors inform you that you are hungry or thirsty.

_____ c. Your ears, eyes, and receptors in your skin for pain, touch, hot, and cold are of this type.

_____ d. With your eyes closed you can tell your exact body position, thanks to these receptors.

_____ e. Receptors in your blood vessels inform your brain of adjustments in heart rate and blood pressure needed to maintain homeostasis.

E. Exteroceptor
P. Proprioceptor
V. Visceroceptor

9. Complete these sentences describing types of receptors based on nature of stimulus detected. $\boxed{4}$

a. Receptors which are stimulated by pain are called _____

_____ .

300

b. Light activates _____ receptors.

c. You can maintain balance by means of inner ear _____

_____-receptors sensitive to change in position.

d. Different tastes and smells are distinguished with the help of

_____-receptors in nose and mouth.

e. _____-receptors inform you of the temperature around you.

10. Choose the descriptions associated with general (G) or special (S) receptors.

⬛ 5

_____ a. These are numerous and widespread in the body.
_____ b. Relatively complex receptors are of this type.
_____ c. Sight, hearing, smell, and taste involve receptors of this type.
_____ d. Sensations of the skin (cutaneous) are of this type.

(pages 340–345) **B. General senses**

1. Name the cutaneous senses.

2. Explain how the two-point discrimination test determines the density of cutaneous receptors in different parts of the body.

⬛ 6

3. Do the following exercise about tactile sensations.
a. Name three categories of tactile sensations.

b. *(Discriminative? Light?)* touch refers to ability to determine that something has touched the skin, but its precise location, shape, size, or texture cannot be distinguished.

c. In general, pressure receptors are more *(superficial? deep?)* in location than touch receptors.

d. Sensations of _____ result from rapid repetition of sensory impulses from tactile receptors.

⬛ 7

4. Match receptors with descriptions.

_____ a. Sensitive to movement of hair shaft
_____ b. Egg-shaped masses located in dermal papillae, especially in fingertips, palms, and soles

FNE. Free nerve endings
MC. Meissner's corpuscles
MD. Merkel's discs
PC. Pacinian corpuscles
RHP. Root hair plexuses

_____ c. Onion-shaped structures sensitive to pressure and high frequency vibration

_____ d. Touch receptors that may be located in epidermis (2 answers)

5. *For extra review.* Refer to Figure LG 5-1 (page LG 85). Identify and label types of receptors on that figure. Add your own drawings of three other types of receptors in appropriate layers of skin. Label those receptors also.

6. One example of a painful stimulus is excessive stimulation of any sense organ. Write five more examples of stimuli which may lead to pain.

7. The body quickly adapts to certain stimuli, such as clothing on the body or odors in the environment. Does the body adapt rapidly to pain?

How is this advantageous?

8. State examples of each of these types of pain:
   a. Superficial somatic pain

   b. Deep somatic pain

   c. Visceral pain

9. *A clinical challenge.* Explain why patients experiencing visceral pain (as during a heart attack or gallbladder attack) may feel pain in locations quite distant from those two organs.   8

a. Pain impulses that originate in the heart, for example, during a "heart attack," enter the spinal cord at the same level as do sensory fibers from skin covering the _____

_____ .

This level of the cord is from about ____ to ____.

b. Refer to Figure 15-2 (page 343) in your text. Notice that some liver and gallbladder pain refers to a region quite distant from these organs, that is, to the _____ _____ region. These organs lie just inferior to the dome-shaped _____

_____ which is supplied by the phrenic nerve from the *(cervical? brachial? lumbar?)* plexus. A painful gallbladder can send impulses to the cord via the phrenic nerve which happens to enter the cord at the same level (about C____ or C____) as nerves from the neck and shoulder. So gallbladder pain is said to be _____

_____ to the neck and shoulder area.

10. What is *phantom pain?*

11. List four surgical methods for reducing pain. Give about a five-word definition of each.

12. A friend asks you how acupuncture is used to control pain. Write your answer. Briefly mention:
a. The procedure

b. How acupuncture may exert its effect: by enhancing the release of *(enkephalin? substance P?)* which then acts to inhibit release of transmitter _____ from sensory neurons in pain pathways.

c. Three purposes for which this procedure is used in the United States

13. Define *proprioception.*

14. Name the type of receptor used in each case. *For extra review,* write a [10] description of the structure of each receptor in spaces provided.
    a. Indicates changes in position of your hip and ankle joints as you do a sit-up: _____

    b. Senses tension applied to a tendon, as when you isometrically contract and relax your gluteus muscles: _____

    c. Senses the amount of stretching of your biceps and triceps brachii muscles while you flex your forearm: _____

15. Arrange these levels of sensation from simplest to most complex. [11] Write letters on these lines: _____ _____ _____
    A. Sensory impulse reaching the spinal cord
    B. Sensation reaching the cerebral cortex
    C. Sensory pathway reaching the brain stem or thalamus

C. Sensory pathways

(pages 345–348)

1. Do this learning activity about sensory reception in the cerebral cortex. [12]
   a. Picture the brain as roughly similar in shape to a small, rounded loaf of bread. The crust would be comparable to the cerebral _____ _____. The bread can be sliced to produce an uneven surface appearance (somewhat like _____ and sulci).

b. Suppose you pull out the slice of bread in midloaf which represents the postcentral gyrus. It looks somewhat like the "slice" or section of brain shown at the top of Figure LG 15-1. Locate the *Z* on that figure. This portion (next to the _____ fissure) receives sensations from the *(face? hand? hip? foot?)* (Refer to Figure 15-4, page 346 of your text for help.)

c. Suppose a person suffered a loss of blood supply (CVA or "stroke") to the area of the brain marked *X* in Figure LG 15-1. Loss of sensation in the *(face? hand? leg?)* area would result. Damage to the area marked *W* would be likely to lead to loss of the sense of *(vision? hearing? taste?)* since the W is located on the temporal lobe.

d. The letters *W, X, Y,* and *Z* are marked on the *(right? left?)* side of the brain. (Note that this is an anterior view of this brain section. If a friend stands and faces you, the *right* side of your friend's brain will be in your *left* field of view.) What effects would brain damage to area *Y* of the postcentral gyrus have? Loss of *(sensation? movement?)* to the *(face? hand? foot?)* on the *(right? left?)* side of the body.

c. Certain regions of the body are represented by large areas of somesthetic or _____ cortex, indicating that these areas *(are? are not?)* very sensitive. Name two of these areas.

13 2. Complete Table LG 15-1a–c, describing pathways controlling sensation of the left hand. Give the name of the pathway and the locations of cell bodies and axons in each case. Be sure to indicate whether cell bodies and axons are on the left or right side of the body. Refer to Figures 15-5 to 15-7 (pages 346–348) in the text for help.

3. Check your understanding of these pathways by drawing them according to the directions below. Try to draw them from memory; do not refer to the text figures or the table until you finish. Label the *first-, second-,* and *third-order neurons.*

a. Use Figure LG 15-1 to show the pathway for pain and temperature in the left hand. Draw the pathway in red.

b. Also draw on Figure LG 15-1 the pathway for fine touch, two-point discrimination, and proprioception in the left hand. Use blue for this pathway.

14 4. Look at the pathways you have just drawn and also at Table LG 15-1a–c. From these you can make the generalization that it is always the *(first? second? third?)* neuron in a sensory pathway that crosses to the opposite side of the nervous system.

15 5. Answer these questions about the spinocerebellar tracts.

a. They are concerned with *(conscious? subconscious?)* muscle sense.

b. The spinocerebellar tracts permit reflex adjustments for _____ and _____.

c. The left posterior spinocerebellar tract conveys impulses from muscles on *(the left side? the right side? both sides?)* of the body.

d. The left anterior spinocerebellar tract conveys impulses from muscles on *(the left side? the right side? both sides?)* of the body.

Table LG 15-1.
Principal Sensory and Motor Pathways Controlling Left Side of Body.

Function	Location of Cell Bodies (CB) and Axons		
	First Order	Second Order	Third Order
a. Sense of pain and temperature in left hand: _____ pathway	CB: Posterior root ganglion, left side. Axon enters posterior gray horn of cord, leftside.	CB: Posterior gray horn, left side. Axon crosses and then ascends through right lateral spinothalamic tract to thalamus.	CB: Thalamus, right side. Axon passes to postcentral gyrus of cerebral cortex
b. Light touch and pressure in left hand: _____ pathway			
c. Two-point discrimination and proprioception in left hand: posterior column pathway			
	Upper Motor Neuron		Lower Motor Neuron
d. Movement of left hand: lateral cortico-spinal pathway			
e. Movement of left hand: _____ pathway			

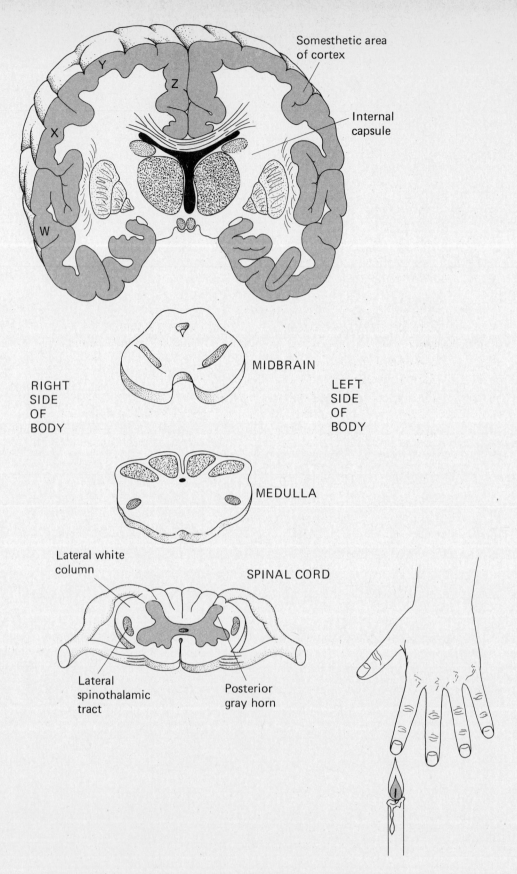

Figure LG 15-1 Diagram of central nervous system. Letters refer to Learning Activity 12 . Draw sensory pathways as directed.

D. Motor pathways

1. Match the structures with their roles in controlling movement. 〔16〕

_____ a. Controls precise, discrete movements
_____ b. Integrates semivoluntary movements, such as walking and swimming
_____ c. Not a control center, but assists in making movements smooth and coordinated

B. Basal ganglia
C. Cerebellum
CC. Cerebral cortex

2. The linking of sensory to motor response occurs at different levels in the 〔17〕 central nervous system. Indicate which is the highest level required for each of these responses.

_____ a. Quick withdrawal of a hand when it touches a hot object
_____ b. Unconscious responses to proprioceptive impulses
_____ c. Playing piano or writing a letter

B. Brainstem
C. Cerebellum
CMC. Cerebral motor cortex
S. Spinal cord

3. Refer to Figure LG 15-2 and do the following exercise about motor control. 〔18〕
   a. This "slice" or section of the brain containing the primary motor cortex is located in the *(frontal? parietal?)* lobe. It is known as the _____ gyrus.
   b. Area *P* in this figure controls *(sensation? movement?)* to the *(left? right?)* *(arm? leg? side of the face?)*. Motor neurons in area *P* would send impulses to the pons to activate cranial nerve ____ to facial muscles.
   c. Describe effects of damage to area *R*.

4. Show the route of impulses along the principal pyramidal pathway by listing in correct sequence the structures that comprise the pathway. Refer to Figure 15-9 (page 350) in the text if you have difficulty with this activity.

〔19〕

____ ____ ____ ____ ____ ____ ____ ____

A. Anterior gray horn (lower motor neuron)
B. Midbrain and pons
C. Effector (skeletal muscle)
D. Internal capsule

E. Lateral corticospinal tract
F. Medulla, decussation site
G. Precentral gyrus (upper motor neuron)
H Ventral root of spinal nerve

5. Now draw this pathway on Figure LG 15-2. Show the control of muscles in the left hand by means of neurons in the lateral corticospinal pathway. Draw in black pen or pencil. Label the *upper* and *lower motor neurons*.
6. Why are the lower motor neurons called the *final common pathway*?

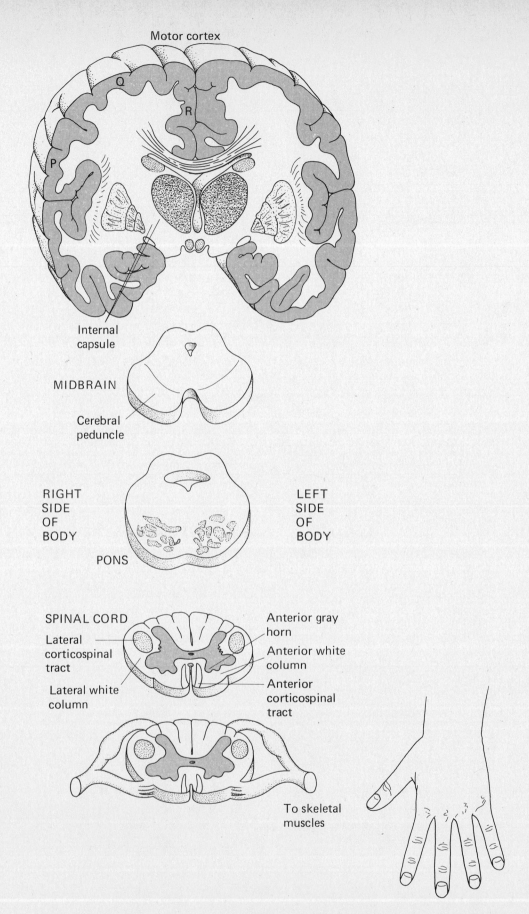

Q

R

P

Internal
capsule

MIDBRAIN

Cerebral
peduncle

RIGHT
SIDE
OF
BODY

LEFT
SIDE
OF
BODY

PONS

SPINAL CORD

Lateral
corticospinal
tract

Lateral white
column

Anterior gray
horn

Anterior white
column

Anterior
corticospinal
tract

To skeletal
muscles

Figure LG 15-2 Diagram of central nervous system. Letters refer to Learning Activities 18 and 20. Draw motor pathways as directed.

7 *For extra review.* What effect would destruction of the following structures [20] have on the *left* hand? Match the effect (A–F) with the related structure.
   A. Can perceive no sensations at all; movements unaffected
   B. Cannot perceive pain; other sensations and movements unaffected
   C. Cannot sense discriminating touch or proprioception; other sensations and movements unaffected
   D. Flaccid paralysis; sensations unaffected
   E. Spasticity; sensations unaffected
   F. Both sensations and movements affected little or none

   _____ a. Right side of thalamus
   _____ b. Upper motor neurons on right side of cerebral cortex, area Q in Figure LG 15-2
   _____ c. Right side of precentral gyrus, area P in Figure LG 15-2
   _____ d. Left side of cerebral cortex
   _____ e. Anterior gray horn on left side of spinal cord from C4 to T4, as from polio
   _____ f. Left lateral corticospinal (pyramidal) tract at level C4 of cord
   _____ g. Left lateral corticospinal tract at level T6 of cord
   _____ h. Entire right lateral corticospinal tract
   _____ i. Nucleus cuneatus on left side of cord
   _____ j. Posterior white columns on entire right side of spinal cord
   _____ k. Entire right lateral spinothalamic tract

8. Complete this exercise about the anterior corticospinal tract. [21]

   a. Only about _____ percent of the upper motor neurons pass through anterior corticospinal tracts; _____ percent pass through the large _____ corticospinal (pyramidal) tracts.

   b. Left anterior corticospinal tracts consist of axons that originated in upper motor neurons on the *(right? left?)* side of the motor cortex.

   c. Contrast the two different corticospinal (pyramidal) tracts by completing Table LG 15-1d–e, page LG 305.

9. What is the function of corticobulbar tracts?

Why are these tracts so named? [22]

10. In general, the extrapyramidal tracts begin in the _____

_____ and end in the _____.
Name three of these tracts and their functions.

11. Summarize how the basal ganglia and cerebellum, together with association neurons, help you to carry out coordinated, precise movements, such as those involved in a game of tennis.

(pages 351–354) **E. Integrative functions**

23 1. List three examples of activities that require complex integration processes by the brain.

2. Define *learning* and *memory*.

3. List several parts of the brain that are considered to assist with memory.

24 4. Do this exercise on memory.
   a. Looking up a phone number, dialing it, and then quickly forgetting it is an example of *(activated? long-term?)* memory. One theory of such memory is that the phone number is remembered only as long as a

_____ neuronal circuit is active.
   b. Repeated use of a telephone number can commit it to long-term memory.

Such reinforcement is known as memory _____.

(Most? Very little? of the information that comes to conscious attention goes into long-term memory, since the brain is selective about what it retains.

c. Some evidence indicates that *(activated? long-term?)* memory involves electrical and chemical events rather than anatomical changes. Supporting this idea is the fact that chemicals used for anesthesia and shock treatments interfere with *(recent? long-term?)* memory.

5. Discuss the following anatomical or biochemical changes that may be associated with enhanced long-term memory:

a. Presynaptic neurons and neurotransmitter

b. Postsynaptic neuron receptors

c. DNA and RNA

6. Complete this exercise about sleep and wakefulness.   25

a. _____ *rhythm* is a term given to the usual daily pattern of sleep and wakefulness in humans.

b. Whether you are awake or asleep depends upon whether the reticular

formation, also known as the _____ system (RAS) is active. The RAS has two principal parts; the *(thalamic? pons and midbrain?)* portion is responsible for your waking from a deep sleep, and the

_____ portion helps you to maintain a conscious state.

c. List three types of sensory input that can stimulate the RAS.

d. Stimulation of the reticular formation leads to *(increased? decreased?)* activity of the cerebral cortex, as indicated by _____ _____ recordings of brain waves.

e. Continued feedback between the RAS, the _____, and the _____ maintains the state called *consciousness*. Inactivation of the RAS produces a state known as _____ _____.

7. List three factors that are known to alter consciousness.

26 8. Do the following learning activity about sleep.
a. Name two kinds of sleep

b. *(REM? NREM?)* sleep is required first in order for *(REM? NREM?)* sleep to occur.

c. In NREM sleep, a person gradually progresses from stage *(1 to 4? 4 to 1?)* into deep sleep. Blood pressure and temperature are lowest in stage *(1? 4?)*. Alpha waves are present in stage(s) *(1? 3 and 4?)*, while slow delta waves characterize stage(s) *(1? 3 and 4?)* sleep.

d. REM sleep is also called _____ sleep since it follows, but is much more active than, deep stage of sleep. Respirations and pulse are much ____-creased over their levels in stage 4; alpha waves are present as in stage ____ sleep. The name REM indicates that _____ _____ movements can be observed.

e. Dreaming occurs during *(stages 1–4 of NREM? REM?)* sleep.

f. Periods of REM and NREM sleep alternate throughout the night in about ____-minute cycles. During the early part of an 8-hour sleep period REM periods last about *(5–10? 30? 50?)* minutes; they gradually increase in length until the final REM period lasts about _____ _____ minutes. In a normal night REM sleep lasts for a total of about *(0.5? 1.5–2? 4?)* hours.

9. Contrast the functions of EEG and EOG in the study of sleep.

10. Contrast these two terms: *narcolepsy* and *hypersomnia*.

Answers to Numbered Questions in Learning Activities

1. (b) They all contain dendrites of sensory neurons.

2. (a) P. (b) Ad. (c) M. (d) Af.

3. (a) E. (b) V. (c) E. (d) P. (e) V.

4. (a) Nociceptors. (b) Electromagnetic (or photo-). (c) Mechano-. (d) Chemo-. (e) Thermo-.

5. (a) G. (b) S. (c) S. (d) G.

6. (a) Touch, pressure, vibration. (b) Light. (c) Deep. (d) Vibration.

7. (a) RHP. (b) MC. (c) PC. (d) FNE and MD.

8. (a) Medial aspects of left arm, T1, T4. (b) Shoulder and neck (right side), diaphragm, cervical, 3, 4, referred.

9. (b) Enkephalin, substance P.

10. (a) Joint kinesthetic receptor. (b) Tendon organ. (c) Muscle spindle.

11. A C B

12. (a) Cortex, gyri. (b) Longitudinal, foot. (c) Face, hearing. (d) Right, sensation, hand, left. (e) General sensory or somatosensory, are; lips, face, thumb and other fingers (see Figure 15-4, page 346 in the text).

Table LG 15-1.
Principal Sensory and Motor Pathways Controlling Left Side of Body.

Function	Location of Cell Bodies (CB) and Axons		
	First Order	**Second Order**	**Third Order**
a. Sense of pain and temperature in left hand: lateral spinothalamic pathway	CB: Posterior roct ganglion, left side. Axon enters posterior gray of left side of cord	CB: Posterior gray horn, left side. Axon crosses and then ascends through right lateral spinothalamic tract	CB: Thalamus, right side. Axon passes to postcentral gyrus of cerebral cortex
b. Light touch and pressure in left hand: anterior spinothalamic pathway	Same as for (a)	Same as for (a) except axon ascends through right anterior spinothalamic tract	Same as for (a)
c. Two-point discrimination, proprioception and vibration in left hand: posterior column pathway	CB: Posterior roct ganglion, left side. Axon ascends through left posterior column tracts to left side medulla	CB: Nucleus cuneatus or gracilis of medulla, left side. Axon crosses and enters medial lemniscus tract to end in right side of thalamus	Same as for (a) and (b)

	Upper Motor Neuron		**Lower Motor Neuron**
d. Movement of left hand: lateral corticospinal pathway	CB: Right side of precentral gyrus of cerebral cortex. Axon descends through internal capsule and brainstem, crosses in medulla to enter left lateral corticospinal tract. Same axon descends to finally enter left anterior gray of cord at level of spinal nerve that will be stimulated. (Here it may stimulate an association neuron that synapses with lower motor neuron.)		CB: Left anterior gray horn of spinal cord. Axon enters ventral root and passes via left spinal nerve to effector
e. Movement of left hand: anterior corticospinal pathway	CB: Right side of precentral gyrus of cerebral cortex. Axon descends through internal capsule and brainstem, does not cross in medulla, and enters right anterior corticospinal tract and descends. At level of the spinal nerve, the same axon crosses to enter left anterior gray of cord. (Here it may stimulate an association neuron that synapses with lower motor neuron.)		Same as for (d)

|14| Second.

|15| (a) Subconscious. (b) Posture, muscle tone. (c) The left side. (d) Both sides.

|16| (a) CC. (b) B. (c) C.

|17| (a) S. (b) C. (c) CMC.

|18| (a) Frontal, precentral. (b) Movement, left, side of the face, VII. (c) Movement of left foot affected (spasticity).

|19| G D B F E A H C.

|20| (a) A. (b) E. (c) F. (d) F. (e) D. (f) E (except for the 15 percent of fibers in right anterior corticospinal tract). (g) F (since destruction is below origin of brachial plexus). (h) F. (i) C. (j) F. (k) B (and also sensation of temperature is lost by entire left side of body).

|21| (a) 15, 85, lateral. (b) Left. (c) See Table in |13|.

|22| They originate in the motor cortex and terminate in the bulb of the brain (a term pertaining to the medulla which is a slightly bulbous enlargement above the spinal cord) and in the pons.

|23| Memory, sleep/wakefulness, emotions.

|24| (a) Activated, reverberating. (b) Consolidation, very little. (c) Activated, recent.

|25| (a) Circadian. (b) Reticular activating, thalamic, pons and midbrain (mesencephalic). (c) Examples: light, sounds, touch, pressure, impulses from cerebral cortex. (d) Increased, electroencephalograph (EEG). (e) cerebral cortex, spinal cord, sleep.

|26| (a) Rapid eye movement (REM) and nonrapid eye movement (NREM). (b) NREM, REM. (c) 1 to 4, 4, 1, 3 and 4. (d) Paradoxical, in, 1, rapid eye. (e) REM. (f) 90, 5–10, 50, 1.5–2.

MASTERY TEST: Chapter 15

Questions 1–4: Choose the one best answer to each question.

_____ 1. All of these sensations are conveyed by the posterior column pathway EXCEPT:

    A. Pain and temperature   B. Proprioception
    C. Fine touch, two-point discrimination
    D. Vibration   E. Stereognosis

_____ 2. Choose the FALSE statement about REM sleep.

    A. Infant sleep consists of a higher percentage of REM sleep than does adult sleep.
    B. Dreaming occurs during this type of sleep.
    C. The eyes move rapidly behind closed lids during REM sleep.
    D. EEG readings are similar to those of stage 4 of NREM sleep.
    E. REM sleep occurs periodically throughout a typical 8-hour sleep period.

_____ 3. Choose the FALSE statement about receptors.

A. Receptors are very excitable.
B. All receptors contain dendrites of sensory neurons.
C. Some receptors have simple structure, some complex.
D. All receptors except pain have a high threshold to a specific stimulus and a low threshold to all other stimuli.

_____ 4. Choose the FALSE statement about pain receptors.

A. They may be stimulated by any type of stimulus, such as heat, cold, or pressure.
B. They have a simple structure with no capsule.
C. They are characterized by a high level of adaptation.
D. They are found in almost every tissue of the body.
E. They are important in helping to maintain homeostasis.

Questions 5–8: Arrange the answers in correct sequence.

_____ _____ _____ 5. Density of sense receptors, from most dense to least:

A. In tip of tongue
B. In tip of finger
C. In back of neck

_____ _____ _____ _____ 6. In order of occurrence in a sensation:

A. Reception by a sense organ or other receptor
B. Translation of the impulse into a sensation
C. Conduction along a nervous pathway
D. A stimulus

_____ _____ _____ _____ 7. Levels of sensation, from those causing simplest, least precise reflexes to those causing most complex and precise responses:

A. Thalamus
B. Brain stem
C. Cerebral cortex
D. Spinal cord

_____ _____ _____ _____ _____ 8. Pathway for conduction of most of the impulses for voluntary movement of muscles:

A. Anterior gray horn of the spinal cord
B. Precentral gyrus
C. Internal capsule
D. Location where decussation occurs
E. Lateral corticospinal tract

Questions 9–20: Circle T (true) or F (false). If the statement is false, change the underlined word or phrase so that the statement is correct.

T F 9. In general, the <u>left</u> side of the brain controls the right side of the body.

T F 10. Generator (receptor) potentials <u>obey</u> the all-or-none principle.

T F 11. The final common pathway consists of <u>upper motor</u> neurons.

T F 12. Circadian rhythm pertains to the <u>complex feedback circuits involved in producing coordinated movements.</u>

T F 13. <u>Sight, hearing, smell, and pressure</u> are all special senses.

T F 14. The diaphragm seems to refer pain to the <u>shoulder and neck region since the phrenic nerve enters the cord at the same levels (C3 to C5) as cutaneous nerves of the shoulder and neck.</u>

T F 15. The neuron that crosses to the opposite side in sensory pathways is usually the <u>second-order</u> neuron.

T F 16. Cutaneous receptors are <u>evenly distributed, that is, of equal density,</u> over all body surfaces.

T F 17. Damage to the left lateral spinothalamic tract would be most likely to result in loss of awareness of <u>pain and vibration sensations in the left</u> side of the body.

T F 18. Conscious sensations, such as those of sight and touch, can occur only in the <u>cerebral cortex.</u>

T F 19. Pain in the left arm that is experienced during a heart attack is an example of <u>phantom pain.</u>

T F 20. Damage to the final common pathway will result in <u>flaccid paralysis.</u>

Questions 21–25: fill-ins. Complete each sentence with the word or phrase that best fits.

_____ 21. ____ is a neurotransmitter released from axons of posterior root ganglia in pain pathways.

_____ 22. Inactivation of the ____ results in the state of sleep.

_____ 23. Rubrospinal, tectospinal, and vestibulospinal tracts are all classified as ____ tracts.

_____ 24. A CVA affecting the medial portion of the postcentral gyrus of the right hemisphere (area Z in Figure LG 15-1, page LG 306) is most likely to result in symptoms such as ____.

_____ 25. Short-term memory is also referred to as ____ memory.

The Autonomic Nervous System

You have studied the structures and functions of the somatic nervous system in the preceding three chapters. Now you will consider control of the viscera by means of the autonomic nervous system (ANS). You will identify structural components and pathways of this system and contrast the two great divisions of the ANS (Objectives 1–5). You will consider the involvement of the hypothalamus with the ANS (6). You will also discuss relationships between biofeedback and meditation and the ANS (7–8).

Topics Summary

A. Somatic efferent and autonomic nervous systems
B. Structure
C. Physiology
D. Visceral autonomic reflexes
E. Control by higher centers

Objectives

1. Compare the structural and functional differences between the somatic efferent and autonomic portions of the nervous system.
2. Identify the principal structural features of the autonomic nervous system.
3. Compare the sympathetic and parasympathetic divisions of the autonomic nervous system in terms of structure, physiology, and neurotransmitters released.
4. Describe the various postsynaptic receptors involved in autonomic responses.
5. Discuss a visceral autonomic reflex and its components.
6. Explain the role of the hypothalamus and its relationship to the sympathetic and parasympathetic divisions.
7. Explain the relationship between biofeedback and the autonomic nervous system.
8. Describe the relationship between meditation and the autonomic nervous system.

320   **Learning Activities**

(page 357)   **A. Somatic efferent and autonomic nervous systems**

1. Why is the autonomic nervous system so named? Is the autonomic nervous system (ANS) entirely independent of higher control centers? Explain.

☐1 2. Give examples of structures innervated by each type of nerve.
   a. Somatic afferent

   b. Visceral afferent

   c. Somatic efferent

   d. Visceral efferent

☐2 3. Which of the types of nerves in Learning Activity ☐1 above are autonomic?

4. *For extra review.* Study Learning Activity ☐8 in Chapter 12 (page LG 230).
5. Name the two divisions of the ANS.

6. Many visceral organs have dual innervation by the autonomic system.
   a. What does dual innervation mean?

   b. How do the sympathetic and parasympathetic divisions work in harmony to control viscera?

7. Contrast the somatic efferent and autonomic (visceral efferent) nervous systems by identifying characteristics of each in this Learning Activity. ⬛3

_____ a. All neurons produce the transmitter acetylcholine (ACh).

_____ b. Some neurons release ACH while others produce norepinephrine (NE).

_____ c. Involuntary control (effectors are not under conscious control).

_____ d. Effectors are all skeletal muscles.

SE. Somatic efferent
VE. Visceral efferent (ANS)

B. Structure

(page 358–363)

1. Complete this exercise describing structural differences between visceral efferent and somatic efferent pathways. ⬛4
   a. Somatic pathways begin in the *(anterior? lateral?)* gray horns of the cord, whereas visceral pathways begin in the _____ gray horns of the cord or in nuclei within the _____.
   b. Somatic pathways begin at *(all? only certain?)* levels of the cord, whereas visceral routes begin at _____ levels of the cord.
   c. Between the spinal cord and effector, somatic pathways include ____ neuron(s), whereas visceral pathways require ____ neuron(s), known as the pre-_____ and _____ neurons.

2. Contrast preganglionic with postganglionic fibers in this exercise. ⬛5
   a. Consider the sympathetic division first. Its preganglionic cell bodies lie in the lateral gray of segments ____ to ____ of the cord. For this reason, the sympathetic division is also known as the _____ outflow.

b. Preganglionic axons *(are? are not?)* myelinated, and therefore they appear white. These axons may branch to reach a number of different

_____ where they synapse. These ganglia contain *(pre-? post-?)* ganglionic neuron cell bodies whose axons proceed to the appropriate organ. Postganglionic axons are *(gray? white?)* since they *(do? do not?)* have a myelin covering.

c. Parasympathetic preganglionic cell bodies are located in two areas. One

is the _____ from which axons pass out in cranial

nerves ____, ____, ____, and ____. A second location is the *(anterior? lateral?)* gray horns of segments ____ through ____ of the cord.

d. Based on location of *(pre-? post-?)* ganglionic neuron cell bodies, the

parasympathetic division is also known as the _____

_____ outflow. The axons of these neurons appear *(white? gray?)*. They extend to *(the same? different?)* autonomic ganglia

from those in sympathetic pathways (see Activity ⬛7⬛). There post-ganglionic cell bodies send out short *(gray? white?)* axons to innervate viscera.

e. Most viscera *(do? do not?)* receive fibers from both the sympathetic and the parasympathetic divisions. However, the origin of these divisions in the CNS and the pathways taken to reach viscera *(are the same? differ?)*.

⬛6⬛ 3. Contrast these two types of ganglia.

	Posterior Root Ganglia	Autonomic Ganglia
a. Afferent or efferent neurons		
b. Neurons in somatic or visceral pathway or both		
c. Synapsing does or does not occur		

⬛7⬛ 4. Complete the table contrasting types of autonomic ganglia.

	Sympathetic Trunk	Prevertebral	Terminal
a. Sympathetic or parasympathetic		Sympathetic	
b. Alternate name	Paravertebral or lateral or sympathetic chain or trunk		

	Sympathetic Trunk	Prevertebral	Terminal
c. General location			Close to or in walls of effectors

5. Identify descriptions that fit the kinds of ganglia listed below. Answers may ☐8 be used more than once; some questions require more than one answer.

_____ a. Located near first rib
_____ b. Located in solar plexus
_____ c. Site of synapsing of neurons in lesser splanchnic nerves
_____ d. At level of C2 vertebra; supply the head with postganglionic sympathetic fibers
_____ e. Supply heart with nerve fibers
_____ f. Types of prevertebral ganglia
_____ g. Sympathetic ganglia

C. Celiac
IC. Inferior cervical
IM. Inferior mesenteric
MC. Middle cervical
SC. Superior cervical
SM. Superior mesenteric

6. Refer to Figure 16-3a (page 360) in your text. Trace with your finger the ☐9 route of a preganglionic neuron. Now describe the pathway common to all sympathetic preganglionic neurons by listing the structures in correct sequence._____ _____ _____ _____
   A. Ventral root of spinal nerve
   B. Sympathetic trunk ganglion
   C. Lateral gray of spinal cord (between T1 and L2)
   D. White ramus communicantes

7. Now consider the possible routes sympathetic preganglionic neurons may take once they are in the sympathetic trunk ganglion. (It may be helpful to refer again to Figure 16-3a in your text and again trace pathways with your finger.) ☐10
   a. What is the shortest possible path they can take to reach a postganglionic neuron cell body?

b. They may ascend and/or descend the sympathetic chain to reach trunk ganglia at other levels. This is important since sympathetic preganglionic cell bodies are limited in location to ____ to ____ levels of the cord, yet the sympathetic trunk extends from ____ to ____ levels of the vertebral column. Sympathetic fibers must be able to reach these distant areas to provide the entire body with sympathetic innervation.

c. Viscera in skin and extremities (such as sweat glands, blood vessels, hair muscles) receive sympathetic innervation through the following route. Preganglionic neurons synapse with postganglionic cell bodies in sympathetic trunk ganglia (at the level of entry or after ascending or descending). Postganglionic fibers then pass through _____ which connect to _____.

d. Some preganglionic fibers do not synapse as described in (a) or (b), but pass on through trunk ganglia without synapsing there. They course through _____ nerves to _____ ganglia. Here they synapse with postganglionic neurons whose axons form _____ en route to viscera.

e. Prevertebral ganglia are located only in the _____. In the neck, thorax, and pelvis the only sympathetic ganglia are those of the trunk (see Figure 16-2, page 359 in the text). As a result, cardiac nerves, for example, contain only *(preganglionic? postganglionic?)* fibers, as indicated by broken lines in the figure.

f. A given sympathetic preganglionic neuron is likely to have *(few? many?)* branches; these may take any of the paths described. Once a branch synapses, it *(can? cannot?)* synapse again, since any autonomic pathway consists of just _____ neurons (preganglionic and postganglionic).

8. Test your understanding of these routes by completing the sympathetic pathway shown in Figure LG 16-1. A preganglionic neuron located at level T5 of the cord has its axon drawn as far as the white ramus. Finish this pathway by showing how the axon may branch and synapse to eventually innervate these three organs: the heart, a sweat gland in skin of the upper thoracic region, and the stomach. Draw the preganglionic fibers in solid lines and the postganglionic fibers in broken lines.

[11] 9. In the past several Learning Activities, you have been focusing on the *(sympathetic? parasympathetic?)* division of the ANS. Now that you are somewhat familiar with the rather complex sympathetic pathways, you will notice that parasympathetic pathways are much simpler. This is due to the fact that there is no parasympathetic chain of ganglia; the only parasympathetic ganglia are _____, which are in or close to the organs innervated. Also parasympathetic preganglionic fibers synapse with *(more? fewer?)* postganglionic neurons. What is the significance of the structural simplicity of the parasympathetic system?

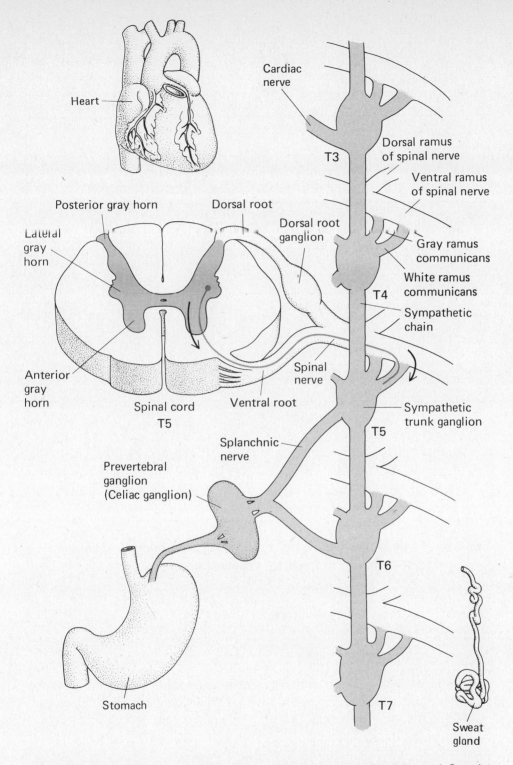

Figure LG 16-1 A typical sympathetic pathway beginning at level T5 of the cord. Complete pathway to the heart, a sweat gland in the skin of the upper thoracic region, and the stomach.

12

10. Complete the table about the cranial portion of the parasympathetic system.

Cranial Nerve		Name of Terminal Ganglion	Structures Innervated
Number	Name		
a.			Iris and ciliary muscle of eye
b.		1. Pterygopalatine	1.
		2. Submandibular	2.
c.	Glossopharyngeal		
d.		Ganglia in cardiac and pulmonary plexuses and in abdominal plexuses	

11. Describe the pathway of neurons in the sacral portion of the parasympathetic system. Tell what organs they innervate.

13

12. *For extra review.* Write S if the description applies to the sympathetic division of the ANS, P if it applies to the parasympathetic division, and P, S if it applies to both.

——— a. Also called thoracolumbar outflow
——— b. Has long preganglionic fibers leading to terminal ganglia and very short postganglionic fibers
——— c. Has relatively short preganglionic fibers and long postganglionic fibers
——— d. Sends some preganglionic fibers through cranial nerves
——— e. Has some preganglionic fibers synapsing in vertebral chain (trunk)
——— f. Has more widespread effect in the body, affecting more organs
——— g. Has some fibers running in gray rami to supply sweat glands, hair muscles, and blood vessels

_____ h. Has fibers in white rami (connecting spinal nerve with vertebral chain)

_____ i. Contains fibers that supply viscera with motor impulses

_____ j. Celiac and superior mesenteric ganglia are sites of postganglionic neuron cell bodies

C. Physiology

(pages 363–366)

1. Refer to Figure LG 16-2 and do this exercise. |14|

   a. In general, which system has long preganglionic fibers and short postganglionic fibers? _(Sympathetic? Parasympathetic?)_ Identify the pathways accordingly.

   b. Axons that release the transmitter acetylcholine (ACh) are known as _(adrenergic? cholinergic?)_. Those that release norepinephrine (NE, also called noradrenalin) are called _____.

   c. Write _C_ next to ends of axons in Figure LG 16-2 that are all cholinergic. Write _C, A_ after the one category of axon which in some cases are cholinergic and in others, adrenergic. In other words, most sympathetic nerves are said to be _(cholinergic? adrenergic?)_, while all parasympathetic nerves are _____.

   d. During stress, the _(sympathetic? parasympathetic?)_ division of the ANS prevails, so stress responses are primarily _(adrenergic? cholinergic?)_ responses.

   e. Another source of NE as well as epinephrine is the gland known as the _____. Therefore chemicals released by the adrenal gland (as in stress) will mimic the action of the _(sympathetic? parasympathetic?)_ division of the ANS.

2. _ACTive learning._ Imagine the following or role play in a class or small group. A miniaturized you sits down by a muscle cell of a human heart (which can be played by a friend). Your goal is to observe how the heart can be regulated. |15|

   a. You look out and notice that there are two kinds of nerve fibers approaching the heart. (Enter two people.) Each person is extending a long arm or leg as a postganglionic axon. One wears a T-shirt with a large S _(sympathetic)_ on it. What letter do you suppose is on the other shirt?

      _____ Which of these is representing the vagus nerve? The person wearing the shirt with _____ on it.

   b. You notice that the vagus person is releasing handfuls of the transmitter _(ACh? NE?)_ (represented by triangular papers with large letters ACh). The transmitter quickly attaches to the heart muscle cell person by means of _____ receptors designed to fit the ACh. (The heart person can have appropriate receptors taped on. See Figure LG

(a) •————<————————————<

• , cell body

< , end of axon

(b) •————————————————<————<

Figure LG 16-2 Comparison of preganglionic and postganglionic neurons. Label as directed in Learning Activity |14|.

16-3.) What is the response of the heart to the ACh? Heart muscle *(gets up and starts contracting rapidly and with great strength? slows down almost falling asleep?).*

c. After a while you note that the sympathetic (postganglionic) axon is also depositing a transmitter at the heart. You look closely and observe that these oddly shaped chemicals are *(ACh? NE?).* These also quickly attach to receptors on the heart; the receptors for NE *(are? are not?)* muscarinic. Instead they are *(alpha? beta?)* since this is the specific receptor for NE on heart cell membranes. When this beta receptor is excited (stimulated) by NE, what is the response of the heart?

d. In actuality there are many ANS nerves stimulating the heart at any given time. The results are additive. At times more of the vagus nerve fibers are active (as if 15 P-shirted students are releasing ACh). Result? The heart slows down. So the vagus is active *(during stress? during rest?).* During exercise and other stressful activities, the many S-shirted students are exerting their *(adrenergic? cholinergic?)* effects on the heart.

3. *A clinical challenge.* Certain medications can mimic the effects of the body's own transmitters. Refer again to Figure LG 16-3 and do this exercise.

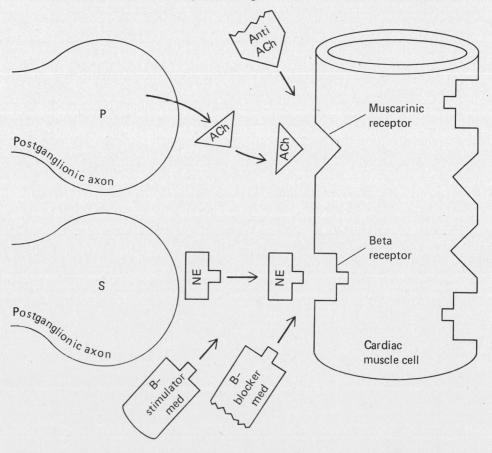

Figure LG 16-3 Diagram of ANS neurotransmitters at neuromuscular junction in the heart. ACh, acetylcholine; NE, norepinephrine; P, parasympathetic; S, sympathetic. Refer to Learning Activities 15 – 16 .

a. A beta-stimulator (or beta-exciter) mimics *(ACh? NE?)*, causing the heart rate and strength of contraction to ____-crease. Name one such drug. (For help, refer to a pharmacology text.) _____ Such a drug has a structure that is close enough to the shape of NE that it can "sit on" the beta receptor as NE would and activate it.

b. A beta-blocker (or beta-inhibitor) also has a shape similar to that of NE and so can sit on _____ receptors. Name one such drug. _____. However this drug cannot activate the receptor, but simply prevents NE from doing so. So a beta-blocker ____-creases heart rate and force of contraction.

c. Other drugs known as anticholinergic (or ACh-blockers) can take up residence on ACh receptors so that ACh being released from vagal nerve stimulation cannot exert its normal effects. A person taking such a medication may exhibit a(n) ____-creased heart rate.

4. Do this exercise on other types of ANS receptors. [17]

a. Some slightly different types of beta receptors are located in smooth muscle of air passageways of lungs. Interaction of NE with these receptors causes the opposite response: relaxation of the smooth muscle. So the effect of the sympathetic stimulation of lungs is to cause *(dilation? constriction?)* of airways, making breathing *(easier? more difficult?)* during stressful times.

b. The effect of NE on the smooth muscle of the stomach, intestines, bladder, and uterus is also *(contraction? relaxation?)*, so that during stress, the body ____-creases activity of these organs and can focus on more vital activities such as heart contractions.

c. Alpha receptors in smooth muscle of blood vessels of the skin respond to NE by contracting. The result is *(dilation? constriction?)* of these vessels due to sympathetic nerve stimulation. Since blood is squeezed out of these skin vessels and forced to enter major blood pathways, blood pressure ____-creases. Name an alpha-stimulator drug: _____ _____

d. Alpha-blocker drugs have just the opposite effect, so tend to *(constrict? dilate?)* blood vessels. Name one of these dilator drugs _____ _____. These can help a person to ____-crease blood pressure.

5. Use arrows to show whether parasympathetic (P) or sympathetic (S) fibers stimulate (↑) or inhibit (↓) each of the following activities. Use a dash (—) to indicate that there is no parasympathetic innervation. The first one is done for you. [18]

a. P ↓ S ↑ Dilation of pupil

b. P ____ S ____ Heart rate and blood flow to coronary (heart muscle) blood vessels

c. P ____ S ____ Constriction of skin blood vessels

d. P _____   S _____   Salivation and digestive organ contractions

e. P _____   S _____   Sweating

f. P _____   S _____   Dilation of bronchioles for easier breathing

g. P _____   S _____   Contraction of bladder and relaxation of internal urethral sphincter causing urination

h. P _____   S _____   Relaxation of gastrointestinal sphincters promoting defecation

i. P _____   S _____   Contraction of pili of hair follicles causing "goose bumps"

j. P _____   S _____   Contraction of spleen which transfers some of its blood to general circulation, causing increase in blood pressure

k. P _____   S _____   Release of epinephrine and norepinephrine from adrenal medulla

l. P _____   S _____   Coping with stress, fight-or-flight response

m. P _____   S _____   Conservation of energy, "rest and repose"

n. P _____   S _____   Erection of genitalia

19 6. Write an explanation based on anatomy of the ANS that tells why skin structures such as sweat glands, hair muscles, and blood vessels lack parasympathetic innervation.

7. Write a paragraph outlining the changes effected by the ANS during a fight-or-flight response, such as running in fear.

8. What is *Horner's syndrome?* Describe symptoms.   **331**

D. **Visceral autonomic reflexes**   (pages 366–367)

1. On Figure LG 16-1 draw an afferent neuron (in contrasting color) to show how the sympathetic neuron could be stimulated.
2. Autonomic neurons can be stimulated by *(somatic afferents? visceral afferents? either somatic or visceral afferents?).*   20
3. Arrange in correct sequence structures in the pathway for a painful stimulus at your fingertip to cause a visceral (autonomic) reflex, such as sweating. Write the letters of the structures in order on the lines provided.   21

___ ___ ___ ___ ___ ___ ___ ___ ___ ___

   A. Association neuron in spinal cord
   B. Nerve fiber in gray ramus
   C. Pain receptor in skin
   D. Cell body of postganglionic neuron in trunk ganglion
   E. Cell body of preganglionic neuron in lateral gray of cord
   F. Nerve fiber in anterior root of spinal nerve
   G. Nerve fiber in white ramus
   H. Fiber in spinal nerve in brachial plexus
   I. Sweat gland
   J. Sensory neuron
4. Most visceral sensations *(do? do not?)* reach the cerebral cortex, so most visceral sensations are at *(conscious? subconscious?)* levels. Hunger and nausea are exceptions.

E. **Control by higher centers**   (pages 367–368)

1. Explain the role of each of these structures in control of the autonomic system.
   a. Thalamus

   b. Hypothalamus (Which part controls sympathetic nerves? Which controls parasympathetic?)

   c. Cerebral cortex (When is it most involved in ANS control? In stress or nonstress situations?)

2. Explain how biofeedback is used to help in each of the following cases.
   a. Heart rate is too fast

   b. Migraine headache

   c. Childbirth

3. Explain how yoga and transcendental meditation *(TM)* may demonstrate whether or not the autonomic nervous system is truly autonomous.

Answers to Numbered Questions in Learning Activities

1 There are many examples possible; here are a few. (a) Skin exteroceptors (for pain, heat, cold, touch) and proprioceptors in muscles, tendons, and joints. (b) Visceroceptors, as in heart, digestive organs, walls of blood vessels. (c) Skeletal muscles, such as gastrocnemius. (d) Smooth muscle (digestive organs, urinary bladder, walls of blood vessels, hair muscles), cardiac muscle (heart), and glands (sweat, salivary, other digestive glands).

2 Visceral efferent.

3 (a) SE. (b) VE. (c) VE. (d) SE.

4 (a) Anterior, lateral, brain stem. (b) All, only certain (sympathetic: T1 to L2; parasympathetic: S2 to S4 plus brain stem). (c) 1, 2, ganglionic, postganglionic.

5 (a) T1, L2, thoracolumbar. (b) Are, ganglia, post-, gray, do not. (c) Brain stem, III, VII, IX, X, lateral, S2, S4. (d) Pre-, craniosacral, white, different, gray. (e) Do, differ.

6 (a) Afferent, efferent. (b) Both (somatic and visceral afferent), somatic efferent only. (c) Does not, does.

7 (a) Sympathetic, parasympathetic. (b) Collateral or prevertebral, intramural (within walls). (c) In vertical chain along either side of bodies of vertebrae from base of skull to coccyx, in three sites anterior to spinal cord close to major abdominal arteries.

8 (a) IC. (b) C. (c) SM. (d) SC. (e) IC, MC. (f) C, IM, SM. (g) All.

9 C A D B.

10 (a) Immediately synapse in trunk ganglion. (b) T1, L2; C3, sacral. (c) Gray rami, spinal nerves. (d) Splanchnic, prevertebral, plexuses. (e) Abdomen, postganglionic. (f) Many, cannot, 2.

11 Sympathetic terminal ganglia, fewer. Parasympathetic effects are much less widespread than sympathetic, distributed to fewer areas of the body, and characterized by a response of perhaps just one organ, rather than an integrated response in which many organs respond *in sympathy* with one another.

12 (a) III, oculomotor, ciliary. (b) VII; facial; 1. nasal mucosa, palate, pharynx, lacrimal gland; 2. submandibular and sublingual salivary glands. (c) IX, otic, parotid salivary gland. (d) X, vagus, thoracic and abdominal viscera.

13 (a) S. (b) P. (c) S. (d) P. (e) S. (f) S. (g) S. (h) S. (i) P, S. (j) S.

14 (a) Parasympathetic; label upper pathway (a) *sympathetic* and lower one (b) *parasympathetic.* (b) Cholinergic, adrenergic. (c) Write C at ends of sympathetic preganglionic axon, and C at end of both parasympathetic axons; write C, A next to end of sympathetic postganglionic axon, adrenergic, cholinergic. (d) Sympathetic, adrenergic. (e) Adrenal medulla, sympathetic.

15 (a) P, P. (b) ACh, muscarinic, slows down, almost falling asleep. (c) NE, are not, beta, gets up and start contracting rapidly and with great strength. (d) During rest, adrenergic.

16 (a) NE, in, isoproterenol. (b) Beta, propranolol, de. (c) In (a double negative: the inhibiting vagus is itself inhibited).

17 (a) Dilation, easier. (b) Relaxation, de. (c) Constriction, in, methoxamine. (d) Dilate, phentolamine, de.

18 (b) P↓, S↑. (c) P—, S↑. (d) P↑, S↓. (e) P—, S↑. (f) P↓, S↑. (g) P↑, S↓. (h) P↑, S↓, (i) P—, S↑. (j) P—, S↑. (k) P—, S↑. (l) P↓, S↑. (m) P↑, S↓. (n) P↑, S↓.

19 Parasympathetic nerves do not enter the sympathetic chain or gray rami to reach individual spinal nerves, so they have no access to skin.

20 Either somatic or visceral afferents.

21 C J A E F G D B H I.

MASTERY TEST: Chapter 16

Questions 1–6: Choose ALL correct answers to each question.

_____ 1. Choose all TRUE statements about the vagus nerve.
- A. Its autonomic fibers are sympathetic.
- B. It supplies ANS fibers to viscera in the thorax and abdomen, but not in the pelvis.
- C. It causes the salivary glands and other digestive glands to increase secretions.
- D. Its ANS fibers are mainly preganglionic.
- E. It is a cranial nerve originating from the medulla.

_____ 2. Which activities are characteristic of the stress response, or fight-or-flight reaction?

    A. The liver breaks down glycogen to glucose.
    B. The heart rate decreases.
    C. Kidneys increase urine production since blood is shunted to kidneys.
    D. There is increased blood flow to genitalia, causing erect state.
    E. Hairs stand on end ("goose pimples") due to contraction of arrector pili muscles.
    F. In general, the sympathetic system is active.

_____ 3. Which fibers are classified as autonomic?

    A. Any visceral efferent nerve fiber
    B. Any visceral afferent nerve fiber
    C. Nerves to salivary glands and sweat glands
    D. Pain fibers from ulcer in stomach wall
    E. Sympathetic fibers carrying impulses to blood vessels
    F. All nerve fibers within cranial nerves
    G. All parasympathetic nerve fibers within cranial nerves

_____ 4. Choose all TRUE statements about gray rami.

    A. They contain only sympathetic nerve fibers.
    B. They contain only postganglionic nerve fibers.
    C. They carry impulses from trunk ganglia to spinal nerves.
    D. They are located at all levels of the vertebral column (from C1 to coccyx).
    E. They carry impulses between lateral ganglia and collateral ganglia.
    F. They carry preganglionic neurons from anterior ramus of spinal nerve to trunk ganglion.

_____ 5. Which of the following structures contain some sympathetic preganglionic nerve fibers?

    A. Splanchnic nerves   B. White rami   C. Sciatic nerve
    D. Cardiac nerves   E. Ventral roots of spinal nerves
    F. The sympathetic chains

_____ 6. Which are structural features of the parasympathetic system?

    A. Ganglia close to the CNS and distant from the effector
    B. Forms the craniosacral outflow
    C. Distributed throughout the body, including extremities
    D. Supplies nerves to blood vessels, sweat glands, and adrenal gland
    E. Has some of its nerve fibers passing through lateral (paravertebral) ganglia

Questions 7–10: Choose the ONE best answer to each question.

_____ 7. All of the following axons are cholinergic EXCEPT:

    A. Parasympathetic preganglionic
    B. Parasympathetic postganglionic
    C. Sympathetic preganglionic
    D. Sympathetic postganglionic to sweat glands
    E. Sympathetic postganglionic to heart muscle

_____ 8. All of the following are collateral ganglia EXCEPT:
    A. Otic ganglion   B. Prevertebral ganglia
    C. Celiac ganglion   D. Superior mesenteric ganglion
    E. Inferior mesenteric ganglion

_____ 9. Which region of the cord contains no preganglionic cell bodies at all?
    A. Sacral   B. Lumbar   C. Cervical   D. Thoracic

_____ 10. Which statement about postganglionic neurons is FALSE?
    A. They all lie entirely outside of the CNS.
    B. Their axons are nonmyelinated.
    C. They terminate in visceral effectors.
    D. Their cell bodies lie in the lateral gray matter of the cord.
    E. They are very short in the parasympathetic system.

Questions 11–20: Circle T (true) or F (false). If the statement is false, change the underlined word or phrase so that the statement is correct.

T F 11. Control of the ANS by the cerebral cortex occurs primarily during times when a person is <u>relaxed (nonstressed).</u>

T F 12. Synapsing occurs in both <u>sympathetic and parasympathetic</u> ganglia.

T F 13. The sciatic, brachial, and femoral nerves all contain <u>sympathetic postganglionic</u> nerve fibers.

T F 14. Biofeedback and meditation confirm that the ANS is <u>independent of</u> higher control centers.

T F 15. Under stress conditions the <u>sympathetic system</u> dominates over the parasympathetic.

T F 16. In a visceral autonomic reflex arc <u>only one efferent neuron</u> is involved.

T F 17. Sympathetic cardiac nerves <u>stimulate</u> heart rate, and the vagus <u>slows down</u> heart rate.

T F 18. Viscera <u>do have sensory nerve fibers, but they are not included in the autonomic nervous system</u>.

T F 19. <u>All</u> viscera have dual innervation by sympathetic and parasympathetic divisions of the ANS.

T F 20. The sympathetic system has a <u>more</u> widespread effect in the body than the parasympathetic does.

Questions 21–25: fill-ins. Complete each sentence with the word or phrase that best fits.

_____ 21. The three types of tissue (effectors) innervated by the ANS nerves are ____.

_____ 22. ____ nerves convey impulses from the sympathetic trunk ganglia to collateral ganglia.

_____ 23. About 80 percent of the craniosacral outflow (parasympathetic nerves) is located in the ____ nerves.

_____ 24. Alpha (α) and beta (β) receptors are stimulated by the transmitter ____.

_____ 25. Drugs that block (inhibit) beta receptors in the heart will cause ____-crease in heart rate and blood pressure.

17

The Special Senses

In this chapter you will learn about receptors and pathways that make you more aware of your external environment, those of the special senses: smell (Objective 1), taste (2), vision (3–8), hearing (9–10), and equilibrium (11). Then you will study disorders and medical terminology associated with the special senses (12–13).

Topics Summary

A. Olfactory sensations
B. Gustatory sensations
C. Visual sensations: anatomy
D. Visual sensations: physiology
E. Auditory sensations and equilibrium
F. Disorders, medical terminology

Objectives

1. Locate the receptors for olfaction and describe the neural pathway for smell.
2. Identify the gustatory receptors and describe the neural pathway for taste.
3. Explain the structure and physiology of the accessory visual structures.
4. List and describe the structural divisions of the eye.
5. Discuss retinal image formation by describing refraction, accommodation, constriction of the pupil, convergence, and inverted image formation.
6. Define emmetropia, myopia, hypermetropia, and astigmatism.
7. Diagram and discuss the rhodopsin cycle responsible for light sensitivity of rods.
8. Identify the afferent pathway of light impulses to the brain.
9. Describe the anatomical subdivisions of the ear.
10. List the principal events in the physiology of hearing.
11. Identify the receptor organs for static and dynamic equilibrium.
12. Contrast the causes and symptoms of cataracts, glaucoma, conjunctivitis, trachoma, deafness, vertigo, labyrinthine disease, Ménière's syndrome, otitis media, and motion sickness.
13. Define medical terminology associated with the sense organs.

Learning Activities

(page 371) **A. Olfactory sensations**

☐1 1. Describe olfactory receptors in this exercise.
   a. Receptors for smell are located in the *(superior? inferior?)* portion of the nasal cavity. Receptors consist of _____-polar neurons which have cell bodies located between _____ cells. The distal end of each olfactory receptor cell contains dendrites, known as olfactory _____. A third type of cell, known as the basal cell, is believed to have the function of _____.

   b. What is the function of olfactory (Bowman's) glands in the nose?

   c. The _____ theory of olfaction suggests that different chemical receptors in olfactory hair membranes respond to different substances, leading to _____
   _____.

   2. Adaptation to smell occurs *(slowly? rapidly?)* at first and then happens at a much slower rate.

☐2 3. Complete this exercise about the nerves associated with the nose.
   a. Upon stimulation of olfactory hairs impulses pass to cell bodies and axons; the axons of these olfactory cells form cranial nerves *(I? II? III?)*, the olfactory nerves. These pass from the nasal cavity to the cranium to terminate in the olfactory _____ located just inferior to the _____ lobes of the cerebrum.

   b. Neurons in the olfactory bulb then convey impulses along the olfactory _____ directly to the _____
   _____. Note that olfaction is one sense that *(does? does not?)* involve the thalamus.

   c. The olfactory pathway just described transmits impulses for the *(general? special?)* sense of smell. General sensations of the nose, such as pain, cold, or tickling, are carried primarily by cranial nerve ____, the trigeminal.

   4. *For extra review.* Study olfactory pathway in Chapter 14 Learning Activity ☐28, page LG 283.

(pages 372–373) **B. Gustatory sensations**

☐3 1. Describe receptors for taste in this exercise.

   a. Taste, or _____ sensations, receptors are located in taste buds. These consist of supporting cells that form a _____
   _____; inside it are 4 to 20 gustatory cells with _____ projecting from a taste pore. The lifespan of each of these gustatory cells is about *(10 years? 10 days?)*.

b. Taste buds are located on elevated projections of the tongue called _____. The largest papillae are located in a V-formation at the back of the tongue. These are _____ _____ papillae. _____ papillae are mushroom-shaped and located on sides and tip of the tongue. Pointed _____ papillae cover the _____ of the tongue, and these *(do? do not?)* contain many taste buds.

2. Briefly describe how a generator potential is produced in gustatory cells.

3. Do the following exercise about taste.

   a. There are only four primary taste sensations. _____, _____, _____, and _____. The taste buds at the tip of the tongue are most sensitive to _____ taste, while the posterior portion is most sensitive to _____ taste.

   b. The four primary tastes are modified by _____ sensations to produce the wide variety of "tastes" experienced. If you have a cold (with a "stuffy nose"), you may not seem to taste them, largely because of loss of the sense of _____.

   c. The threshold for smell is quite *(high? low?)*, meaning that *(a large? only a small?)* amount of odor must be present in order for smell to occur. The threshold varies for the four primary tastes; the threshold for *(sour? sweet and salt? bitter?)* is lowest, while those for _____ are highest.

   d. Taste impulses are conveyed to the brain stem by cranial nerves ____, ____, ____, and ____; the _____ finally relays impulses to the cerebral cortex.

C. Visual sensations: anatomy

(pages 373–380)

1. Examine your own eye structure in a mirror. With the help of Figure 17-3 (page 375) in your text, identify each of the accessory structures of the eye listed below.

   Commissure   Lacrimal caruncle
   Conjunctiva   Tarsal glands
   Eyebrow   Palpebra
   Eyelashes   Tarsal plate

2. *For extra review.* Match the structure listed above with the following descriptions. Write the name of the structure on the line provided.

   a. Eyelid: _____
   b. Thick fold of connective tissue that forms much of the eyelid:

      _____

   c. Covers the orbicularis oculis muscle; provides protection: _____

      _____

340

d. Short hairs; infection of glands at the base of these hairs is *sty:*
_____

e. Angle where eyelids meet medially and laterally: _____
_____

f. Reddish elevation in the medial commissure where whitish material may be produced during sleep: _____

g. Tarsal glands that secrete oil; infection is *chalazion:* _____
_____

h. Mucous membrane lining the eyelids and covering anterior surface of the eye: _____

3. The lacrimal glands produce _____ which contain the bacteriocidal enzyme named _____. State three functions of lacrimal fluid (tears).

4. Write the term that describes each of these major structures in the lacrimal pathway. Numbers in parentheses indicate how many of that structure are associated with one lacrimal apparatus.

   6

   a. Gland (1) at upper outer area of the orbit: _____

   b. Ducts (6–12) which empty out onto conjunctival surface: _____
   _____

   c. Small openings (2) at medial commissure: _____
   d. Duct (1) leading into nose which, if blocked during a cold, can cause "watery" eyes: _____

5. Refer to Figure LG 17-1 and identify parts of the eye in this learning activity.

   7

   a. Name the two parts of the *fibrous tunic:* _____ and _____. Color these green on the figure.
   b. List three structures that form the *vascular tunic* (or *uvea*): _____, _____, and _____. Color these red on the figure.
   c. The innermost layer of the eye is the _____ *tunic,* or _____. Identify this on the figure.

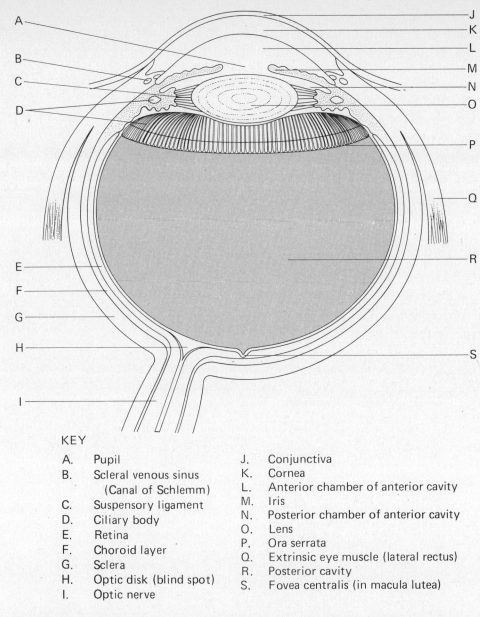

KEY

A. Pupil
B. Scleral venous sinus
(Canal of Schlemm)
C. Suspensory ligament
D. Ciliary body
E. Retina
F. Choroid layer
G. Sclera
H. Optic disk (blind spot)
I. Optic nerve

J. Conjunctiva
K. Cornea
L. Anterior chamber of anterior cavity
M. Iris
N. Posterior chamber of anterior cavity
O. Lens
P. Ora serrata
Q. Extrinsic eye muscle (lateral rectus)
R. Posterior cavity
S. Fovea centralis (in macula lutea)

Figure LG 17-1 Structure of the eyeball in horizontal section. Color as direced in Learning Activity ⬜7 .

6. Contrast rods and cones by completing this table.

	Rods	Cones
a. Quantity	10 or 20 times more than the number of cones	
b. Location		
c. Function		Sense color and acute (sharp) vision

8

7. Do this exercise that describes the retina.
   a. After light strikes the back of the retina, nerve impulses pass anteriorly through three zones of neurons. First is the region of rods and cones, or _____ zone; next is the _____ layer; most anterior is the _____ layer.
   b. Axons of the ganglion layer converge to form the optic _____ _____. This nerve exits from the eye at the point known as the _____ disc, or _____. The name indicates that no image formation can occur here since no rods or cones are present.

8. Briefly explain these medical terms.
   a. Keratoplasty

   b. Detachment of the retina (Where does it occur? How can it be reattached?)

   c. Senile macular degeneration (Which age group does it usually affect? How can you test for it?)

9

9. Describe the lens and the cavities of the eye in this exercise.
   a. The lens is composed of *(lipid? protein?)* arranged in layers, much like an onion. Normally the lens is *(clear? cloudy?)*. A loss of transparency of the lens occurs in the condition called _____.
   b. The lens divides the eye into anterior and posterior *(chambers? cavities?)*. The posterior cavity is filled with _____ humor which has a *(watery? jelly-like?)* consistency. This humor helps to hold the _____ in place. Vitreous humor is formed during embryonic life and *(is? is not?)* replaced in later life.

   c. The anterior cavity is subdivided by the _____ into two chambers. Which is larger? *(Anterior chamber? Posterior chamber?)* A watery fluid called _____ humor is present in this chamber. It is formed by the _____ bodies.
   d. How is the formation and final destination of aqueous fluid similar to that of cerebrospinal fluid (CSF)?

e. State two functions of aqueous humor.

f. Aqueous humor creates pressure within the eye, normally about
    _____ mm Hg. In glaucoma this pressure *(increases? decreases?),* leading
    to destruction of parts of the eye and subsequent blindness.
10. *For extra review,* check your understanding by matching eye structures at
    right with descriptions below. Use each answer once.
    10

_____ a. "White of the eye"

_____ b. A clear structure, composed of
          protein layers arranged like an
          onion

_____ c. Blind spot; area in which there are
          no cones or rods

_____ d. Area of sharpest vision; area of
          densest concentration of cones

_____ e. Nonvascular, transparent, fibrous
          coat; most anterior eye structure

_____ f. Layer containing neurons; if de-
          tached, causes blindness

_____ g. Dark brown layer; prevents reflec-
          tion of light rays; also nourishes
          eyeball since it is vascular

_____ h. A hole; appears black, like a circu-
          lar doorway leading into a dark
          room

_____ i. Regulates the amount of light en-
          tering the eye; colored part of the
          eye

_____ j. Attaches to the lens by means of
          radially arranged fibers called the
          suspensory ligaments

_____ k. Serrated margin of the retina

_____ l. Located at the junction of iris and
          cornea; drains aqueous humor

CF. Central fovea
Cho. Choroid
Cor. Cornea
CM. Ciliary muscle
I.   Iris
L.   Lens
OD. Optic disc
OS. Ora serrata
P.   Pupil
R.   Retina
S.   Sclera
SVS. Scleral venous sinus
     (Canal of Schlemm)

D. Visual sensations: physiology

(pages 380–386)

1. List the three principal processes necessary in order for vision to occur.
   11
   a.

   b.

   c.

2. Define these four processes necessary for formation of an image on the retina.

a. Refraction of light

b. Accommodation of the lens

c. Constriction of the pupil

d. Convergence of the eyes

3. On the lines provided list in order the four clear, colorless eye structures in which light rays are refracted. In the parentheses write *More* or *Less* to indicate relative density. The first one is done for you: Light rays pass from *less* dense air into the *more* dense cornea.

12

Air   a. Cornea   b.   c.   d.

_____ → _____ → _____ → _____ → _____

(Less)   (More)   (   )   (   )   (   )

13

4. Explain how accommodation of the lens enables your eyes to focus clearly.

a. Objects close to your eyes reflect light rays that are *(parallel? divergent?)*. In order to focus these on your retina the rays from these objects must be refracted *(more? less?)* than those from faraway objects.

b. The lens must become *(more? less?)* convex for near vision. This occurs by a thickening and bulging forward of the lens caused by the *(contraction? relaxation?)* of the _____ muscle. In fact, after long periods of close work (such as reading), you may experience eye strain.

c. The normal (or _____ eye) can refract rays from a distance to _____ meters and form a clear image on the retina.

d. In the nearsighted (or _____) eye, images focus *(in front of? behind?)* the retina. This may be due to an eyeball that is too *(long? short?)*. Corrective lenses should be slightly *(concave? convex?)* so they refract rays less and allow them to focus farther back on the retina.

e. The farsighted person can see *(near? far?)* well, but has difficulty seeing *(near? far?)* without the aid of corrective lenses. The farsighted (or

_____) person requires *(concave? convex?)* lenses. After about age 40, most persons lose the ability to see near objects clearly.

This condition, called _____

_____, is due to loss of elasticity of the _____

_____.

5. What is the meaning of *astigmatism?*

6. Choose the correct answer to complete each statement. `14`
   a. Both accommodation of the lens and constriction of the pupil involve *(extrinsic? intrinsic?)* eye muscles.
   b. The pupil becomes smaller when *(radial? circular?)* muscles of the iris contract under the influence of *(sympathetic? parasympathetic?)* innervation.
7. Define *single binocular vision.* Explain why *convergence* is necessary for such vision.

8. Answer these questions about the image focused on the retina. `15`
   a. How would a letter "e" look as it is focused on the retina?

   b. *Clinical application.* An object in your lower field of view (such as the shoe of a person standing in front of you) will have its image formed on the *(superior? inferior?)* portion of your retina. So a person who has the superior portion of the retina detached would be more likely to lose sight of the *(sky? ground?).*
   c. Explain why you see things right side up, even though refraction produces inverted images on your retina.

16 9. Complete this exercise about what happens in the rods of your eyes after you walk into a dark movie theatre. Refer to Figure 17–7 (page 383) in your text.

    a. In order to see in dim light (as in the dark theatre), you use *(rods? cones?)*.

    b. Rods produce a pigment called _____ (or visual purple). But in bright light (as in the lobby) this chemical is broken down very *(rapidly? slowly?)*. So as you enter the theatre your rods lack sufficient rhodopsin to see in the dark.

    c. During the short time that you are stumbling around in the darkness, rhodopsin can form in the following manner. An enzyme converts retinal from its all-trans form to its _____ form. The cis-form is a *(straight? curved?)* molecule which fits well with a protein named _____ to produce rhodopsin.

    d. As soon as you have sufficient rhodopsin present in the rods, stimulation of the rods by the dim light in the theatre will begin breaking down rhodopsin. This is a rapid but stepwise process, ultimately resulting in the cleavage of rhodopsin into two parts: again the protein _____ and the *(cis? all-trans?)* form of retinal.

    e. *For extra review,* write the names of the intermediate chemicals formed in the breakdown of rhodopsin. (Hint: note that the names happen to be in alphabetical order.) Rhodopsin → _____ →

    _____ → _____ →

    _____ → _____ (scotopsin + all-trans-retinal).

    Place an asterisk at the point in this pathway at which a generator impulse develops and a nerve impulse passes to your brain informing you of what you have seen in dim light, such as the images of persons in seats around you.

    f. Your vision becomes increasingly better after a few more minutes in the dim light since you continue to convert all-trans-retinal to cis-retinal (by enzyme action). (See *c* above.) But should you step out in the lobby to buy popcorn, the bright light will *(produce? destroy?)* rhodopsin so rapidly and completely that you will have to start the dark-induced synthesis of rhodopsin all over again once you re-enter the dimly lit theatre.

    g. Night blindness (also known as _____) is associated with vitamin ____ deficiency since that vitamin is a precursor of *(scotopsin? retinal?)*.

10. Explain how cones help you to see in "living color" by completing this paragraph.

17

Cones contain pigments which *(do? do not?)* require bright light for breakdown (and therefore for a generator potential). Three different pigments are present, sensitive to three colors: _____,

_____, and _____. A person who is red-green color-blind lacks some of the cones receptive to two colors

(_____ and _____) and so cannot distinguish between these colors. This condition occurs more often in *(males? females?)*.

11. Describe the conduction pathway for vision by arranging these structures in sequence. Write the letters in correct order on the lines provided. Also, indicate the four points where synapsing occurs by placing an asterisk (*) between the letters. 18

___ ___ ___ ___ ___ ___ ___ ___

   B. Bipolar cells   ON. Optic nerve
   C. Cerebral cortex (visual areas)   OT. Optic tract
   G. Ganglion cells   P. Photoreceptor cells (rods, cones)
   OC. Optic chiasma   T. Thalamus

12. Describe two functions of the horizontal cells in the retina. 19

13. Answer these questions about the visual pathway to the brain. It may be helpful to refer to Figure 17-8 (page 385) in the text. 20
   a. Hold your left hand diagonally up and to the left so you can still see it. Your hand is in the *(temporal? nasal?)* visual field of your left eye, and

   in the _____ visual field of your right eye.
   b. Due to refraction, the image of your hand will be projected on to the *(left? right?) (upper? lower?)* portion of the retinas of your eyes.
   c. All nerve fibers from these areas of your retinas reach the *(left? right?)* side of your thalamus and cerebral cortex.
   d. Damage to the right optic tract, right side of thalamus, or right visual cortex would result in loss of sight of the *(left? right?)* visual fields of each eye.

E. Auditory sensations and equilibrium   (pages 386–393)

1. Refer to Figure LG 17-2 and do the following exercise.
   a. Color the ear as follows: external ear, red; tympanic membrane (eardrum), purple; ossicle (ear bones), blue; inner ear, yellow.
   b. Cover KEY and check your knowledge of ear structures by writing correct labels next to each lettered lead line.

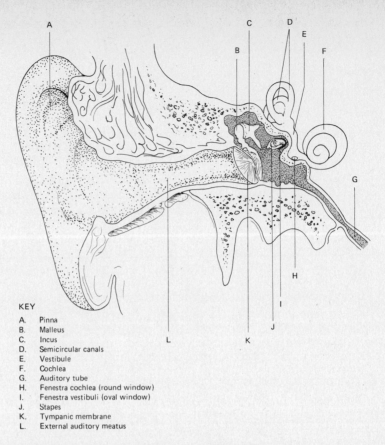

KEY

A. Pinna
B. Malleus
C. Incus
D. Semicircular canals
E. Vestibule
F. Cochlea
G. Auditory tube
H. Fenestra cochlea (round window)
I. Fenestra vestibuli (oval window)
J. Stapes
K. Tympanic membrane
L. External auditory meatus

Figure LG 17-2 Diagram of the right ear in frontal section.

2. *For extra review.* Select the ear structures at right that fit the descriptions below. Not all answers will be used.

[21]

_____ a. Tube used to equalize pressure on either side of tympanic membrane	AT. Auditory tube
	I. Incus
_____ b. Chamber posterior to middle ear; permits middle ear infection to spread to cause mastoiditis	M. Malleus
	OW. Oval window
	P. Pinna
_____ c. Eardrum	RW. Round window
_____ d. Structure on which stapes exerts pistonlike action	S. Stapes
	TA. Tympanic antrum
_____ e. Flared portion of the outer ear	TM. Tympanic membrane
_____ f. Ossicle adjacent to eardrum	
_____ g. Anvil-shaped ear bone	

3. The tensor tympani and stapedius muscles protect the ear from *(loud? soft?)* noises. Explain how these muscles accomplish this function.

4. Describe the internal ear in this exercise.
   a. The internal (inner) ear is a kind of complicated maze or a _____. It consists of a *(bony? membranous?)* outer portion which contains _____-lymph. Tubules and sacs form the inner portion; these membranes enclose _____-lymph.
   b. The inner ear contains three structures named according to their shape. The entranceway, or vestibule, contains two sacs, the _____ _____ and _____.
   c. Just posterior to the vestibule are the three _____ canals. Each has an enlarged end, called an _____.
   d. Anterior to the vestibule is the _____, a snail-shaped structure. A cross section of the cochlea shows three canals. The upper and lower canals (or scala) are filled with _____. The middle canal, called the _____ duct, contains endolymph.
5. Fill in the blanks and circle correct answers about sound waves. 23
   a. Sound waves range from frequencies of 20 to 20,000 cycles/sec (Hz). Humans can hear sounds in the range of _____ Hz.
   b. A musical high note has a *(higher? lower?)* frequency than a low note. So frequency is *(directly? indirectly?)* related to pitch.
   c. How loud a sound is (intensity) is measured in units called _____. Normal conversation is at a level of about _____ dB, while sounds at _____ dB can cause pain.
6. Summarize events in the process of hearing in this activity. It may help to refer to Figure 17-11 (page 391) in the text. 24
   a. Sound waves travel through the _____ and strike the _____ membrane. Sound waves are magnified by the action of the three _____ in the middle ear.
   b. The ear bone named *(malleus? incus? stapes?)* strikes the *(round? oval?)* window, setting up waves in *(endo-? peri-?)*lymph. This pushes on the floor of the upper *scala (vestibuli? tympani?)*. As a result the cochlear duct is moved, and so is the perilymph in the lower canal, the *scala (vestibuli? tympani?)*. The pressure of the perilymph is finally expended by bulging out the *(round? oval?)* window.
   c. So far we have not discussed how nerve impulses are initiated. As the cochlea duct moves, tiny hair cells embedded in the floor of the duct are stimulated. These hair cells are parts of the _____ organ; its name is based on its spiral arrangement on the _____ _____ membrane all the way around the 2¾ coils of the cochlear duct.
   d. As spiral organ hair cells are moved by waves in endolymph, hairs move against the gelatinous _____ membrane. This movement generates a potential in the hair cells which contain dendrites of the _____ branch of cranial nerve _____. The

pathway continues to the brain stem, _____ (relay center), and finally to the _____ lobe of the cerebral cortex.

7. Check your understanding of the pathway of fluid conduction by placing the following structures in correct sequence. Write the letters on the lines provided.

25

___ ___ ___ ___ ___ ___ ___

B. Basilar membrane   ST. Scala tympani (perilymph)
C. Cochlear duct (endolymph)   SV. Scala vestibuli (perilymph)
O. Oval window   VM. Vestibular membrane
RW. Round window

26

8. *For extra review,* refer to Figure LG 17-2 and write in correct sequence the KEY letters of structures in the entire pathway of sound and fluid conduction. The first and last are done for you.

A _ _ _ _ _ _ _ _ H

9. Describe the roles of K^+ and Ca^{2+} in stimulating hair cells and generating nerve impulses related to hearing.

27

10. Contrast receptors for hearing and for equilibrium in this summary of the inner ear.
    a. Receptors for hearing and equilibrium are all located in the *(middle? inner?)* ear. All consist of supporting cells and _____ cells that are covered by a _____ _____ membrane.

    b. In the spiral organ, which senses _____, the gelatinous membrane is called the _____ membrane. Hair cells move against this membrane as a result of *(sound waves? change in body position?).*

    c. In the macula, located in the *(semicircular canals? vestibule?),* the gelatinous membrane is embedded with calcium carbonate crystals called _____. These respond to gravity in such a way that the macula is the receptor for *(static? dynamic?)* equilibrium. An example of such equilibrium occurs as you are aware of your *(position while standing on your head? change in position on a careening roller coaster?).*

d. In the semicircular canals the gelatinous membrane is called the

_____. Its shape is (*flat? like an inverted cup?*). The cupula is part of the (*crista? saccule?*) located in the ampulla. Change in direction (as in a roller coaster) causes _____ to bend hairs in the cupula. Cristae in semicircular canals are therefore receptors for (*static? dynamic?*) equilibrium.

11. Match the structures listed at right with their roles in maintaining equilibrium.

28

_____ a. Termination of most fibers of the vestibular branch of the cranial nerve VIII, carrying impulses from macula

_____ b. Termination of some vestibular nerve fibers; chief coordinator in equilibrium

_____ c. Send impulses for appropriate eye movements

_____ d. Help control head and neck movements; cranial nerve XI

_____ e. Send messages for skeletal muscle responses to achieve balance

AN. Accessory nerve
C. Cerebellum
M. Medulla (vestibular nuclei)
OTAN. Oculomotor, trochlear, and abducens nerves
VT. Vestibulospinal tracts

G. Disorders, medical terminology

(pages 394–395)

1. Match the name of the disorder with the description.

31

_____ a. Inflammation of sebaceous glands at the base of hair follicles of eyelashes

_____ b. Excessive intraocular pressure resulting in blindness; second most common cause of blindness

_____ c. Ringing in the ears

_____ d. Crossed eyes

_____ e. Middle ear infection

_____ f. Associated with vitamin A deficiency

_____ g. Pinkeye

_____ h. Disturbance of the inner ear; cause unknown

_____ i. Loss of transparency of the lens

_____ j. Farsightedness due to loss of elasticity of lens, especially after age 40

_____ k. Condition requiring corrective lenses to focus distant objects

Cat. Cataract
Con. Conjunctivitis
G. Glaucoma
Men. Ménière's syndrome
My. Myopia
NB. Night blindness
OM. Otitis media
P. Presbyopia
Str. Strabismus
Sty. Sty
T. Tinnitus

2. Discuss causes and treatments of:
   a. Cataract

   b. Trachoma

   c. Otitis media

   d. Motion sickness

32 3. Do this exercise related to treatment of glaucoma.
   a. In this disorder intraocular pressure increases due to excess *(aqueous? vitreous?)* humor. One method of treatment is to *(increase? decrease?)* outflow of this humor. Drugs which *(dilate? constrict?)* the pupil help since they tend to thin it out (make it bulge less) so that aqueous humor can more readily pass into the _____. Which division of the autonomic nervous system constricts the pupil? *(Sympathetic? Parasympathetic?)* These nerves are *(cholinergic? adrenergic?)*. Cholinergic drugs can therefore be used to mimic these nerves and constrict the pupil reducing intraocular pressure.
   b. A second method of treatment involves *(increasing? decreasing?)* production of aqueous humor. Name two drugs that act in this manner.

Answers to Numbered Questions in Learning Activities

1 (a) Superior, bi-, supporting, hairs, producing new supporting cells. (b) Produce mucus that acts as a solvent for odoriferous substances. (c) Chemical, generator potential and nerve impulse.

2 (a) I, bulb, frontal. (b) Tracts, cerebral cortex, does not. (c) Special, V.

3 (a) Gustatory, capsule, hairs, 10 days. (b) Papillae, circumvallate, fungiform, filiform, anterior, do not.

4 (a) Sweet, sour, salt and bitter; sweet, bitter. (b) Olfactory, smell. (c) Low, only a small, bitter, sweet and salt. (d) VII, IX and X, thalamus.

5 (a) Palpebra. (b) Tarsal plate. (c) Eyebrow. (d) Eyelashes. (e) Commissure. (f) Lacrimal caruncle. (g) Tarsal glands. (h) Conjunctiva.

6 (a) Lacrimal gland. (b) Excretory lacrimal ducts. (c) Punctae lacrimalia. (d) Nasolacrimal duct.

7 (a) Sclera *(G)*, cornea *(K)*. (b) Choroid *(F)*, ciliary body *(D)*, and iris *(M)*. (c) Nervous, retina *(E)*.

8 (a) Photoreceptor, bipolar, ganglion. (b) Nerve, optic, blind spot.

9 (a) Protein, clear, cataract. (b) Cavities, vitreous, jellylike, retina, is not. (c) Iris, anterior chamber, aqueous, ciliary. (d) Both are formed from blood vessels (choroid plexuses) and the fluid finally returns to venous blood. (e) Provides nutrients and oxygen and removes wastes since cornea and lens are avascular; also provides pressure to separate cornea from lens. (f) 16, increases.

10 (a) S. (b) L. (c) OD. (d) CF. (e) Cor. (f) R. (g) Cho. (h) P. (i) I. (j) CM. (k) OS. (l) SVS.

11 (a) Formation of image on retina. (b) Stimulation of photoreceptor cells. (c) Conduction of nerve impulses to brain.

12 (b) Aqueous humor (less). (c) Lens (more). (d) Vitreous humor (less).

13 (a) Divergent, more. (b) More, contraction, ciliary. (c) Emmetropic, 6 (20 ft). (d) Myopic, in front of, long, concave. (e) Far, near, hypermetropic, convex, presbyopia, lens.

14 (a) Intrinsic. (b) Circular, parasympathetic.

15 (a) "ə" ("e" is inverted 180° and is much smaller). (b) Superior, ground. (c) Brain learns to "turn" images so that you see things right side up.

16 (a) Rods. (b) Rhodopsin, rapidly. (c) Cis, curved, scotopsin. (d) Scotopsin, all-trans. (e) Bathorhodopsin → lumirhodopsin → metarhodopsin I $\overset{*}{\rightarrow}$ metarhodopsin II → pararhodopsin. (f) Destroy. (g) Nyctalopia, A, retinal.

17 Do; red, green, blue; red, green; males.

18 P * B * G ON OC OT * T * C. (Note that no synapsing occurs in the optic chiasma.)

19 They enhance contrast by inhibiting bipolar neurons in areas lateral to excited rods and cones. They also help in the process of differentiating colors.

20 (a) Temporal, nasal. (b) Right, lower. (c) Right. (d) Left.

21 (a) AT. (b) TA. (c) TM. (d) OW. (e) P. (f) M. (g) I.

22 (a) Labyrinth, bony, peri-, endo-. (b) Saccule, utricle. (c) Semicircular, ampulla. (d) Cochlea, perilymph, cochlear.

23 (a) 1,000–4,000. (b) Higher, directly, (c) Decibels (dB), 45, 115–120.

24 (a) External auditory meatus, tympanic, ossicles. (b) Stapes, oval, peri-, vestibuli, tympani, round. (c) Spiral, basilar. (d) Tectorial, cochlear, VIII, thalamus, temporal.

25 O SV VM C B ST RW.

26 L K B C J I E F.

27 (a) Inner, hair, gelatinous. (b) Hearing, tectorial, sound waves. (c) Vestibule, otoliths, static, position while standing on your head. (d) Cupula, like an inverted cup, crista, endolymph, dynamic.

28 (a) M. (b) C. (c) OTAN. (d) AN. (e) VT.

29 (a) ON. (b) R. (c) Ch, S. (d) L. (e) Co.

30 (a) Ecto. (b) Meso. (c) Ecto.

31 (a) Sty. (b) G. (c) T. (d) Str. (e) OM. (f) NB. (g) Con. (h) Men. (i) Cat. (j) P. (k) My.

32 (a) Aqueous, increase, constrict, Scleral venous sinus, parasympathetic, cholinergic. (b) Decreasing, Diamox and Timoptic.

MASTERY TEST: Chapter 17

Questions 1–7: Circle T (true) or F (false). If the statement is false, change the underlined word or phrase so that the statement is correct.

T   F   1. Decomposition of photopigments in rods and cones leads to <u>increased</u> influx of Na$^+$ into the photoreceptor cells and therefore results in <u>depolarization</u>.

T   F   2. As a person walks toward you, the light rays from that person which are directed toward your eyes become more and more <u>parallel</u>.

T F 3. <u>Crista, macula, otilith, and spiral organ</u> are all structures located in the inner ear.

T F 4. <u>Both aqueous and vitreous humors are</u> replaced constantly throughout your life.

T F 5. The receptor organs for special senses are <u>less</u> complex structurally than those for general senses.

T F 6. Mastoiditis is most likely to result from spread of infection of the <u>inner ear.</u>

T F 7. <u>Convergence and accommodation are both results</u> of contraction of smooth muscle of the eye.

Questions 8–14: Arrange the answers in correct sequence.

_____ _____ _____ 8. Layers of the eye, from superficial to deep:

  A. Sclera
  B. Retina
  C. Choroid

_____ _____ _____ _____ 9. From anterior to posterior:

  A. Vitreous humor
  B. Optic nerve
  C. Cornea
  D. Lens

_____ _____ _____ _____ 10. Pathway of aqueous humor, from site of formation to destination:

  A. Anterior chamber
  B. Scleral venous sinus
  C. Ciliary body
  D. Posterior chamber

_____ _____ _____ _____ _____ 11. Pathway of sound waves and resulting mechanical contraction:

  A. External auditory canal
  B. Stapes
  C. Malleus and incus
  D. Oval window
  E. Tympanic membrane

_____ _____ _____ _____ _____ 12. Pathway of tears, from site of formation to entrance to nose:

  A. Lacrimal gland and ducts
  B. Lacrimal sac
  C. Nasolacrimal duct
  D. Surface of conjunctiva
  E. Puncta lacrimalia and lacrimal canals

356

_____ _____ _____ _____ _____ 13. Order of impulses along conduction pathway for smell:

    A. Olfactory bulb
    B. Olfactory hairs
    C. Olfactory nerves
    D. Olfactory tract
    E. Primary olfactory area of
       cortex

_____ _____ _____ _____ _____ 14. From anterior to posterior:

    A. Anterior chamber
    B. Iris
    C. Lens
    D. Posterior cavity
    E. Posterior chamber

Questions 15–20: Choose the one best answer to each question.

_____ 15. Choose the FALSE statement about rods.

    A. There are more rods than cones in the eye.
    B. Rods are concentrated in the fovea and are less dense around the periphery.
    C. Rods enable you to see in dim (not bright) light.
    D. Rods contain rhodopsin
    E. No rods are present at the optic disc.

_____ 16. Choose the FALSE statement about the lens of the eye.

    A. It is biconvex.
    B. It is avascular.
    C. It becomes more rounded (convex) as you look at distant objects.
    D. Its shape is changed by contraction of the ciliary muscle.
    E. Change in the curvature of the lens is called accommodation.

_____ 17. Choose the FALSE statement about the middle ear.

    A. It contains three ear bones called ossicles.
    B. Infection in the middle ear is called otitis media.
    C. It communicates with the nasopharynx by means of the auditory tube.
    D. It functions in conduction of sound from the external ear to the inner ear.
    E. The cochlea is located here.

_____ 18. Choose the FALSE statement about the semicircular canals.

    A. They are located in the inner ear.
    B. They sense acceleration or changes in position.
    C. Nerve impulses begun here are conveyed to the brain by the vestibular branch of cranial nerve VIII.
    D. There are four semicircular canals in each ear.
    E. Each canal has an enlarged portion called an ampulla.

_____ 19. Destruction of the left optic tract would result in:

    A. Blindness in the left eye
    B. Loss of left visual field of each eye
    C. Loss of right visual field of each eye
    D. Loss of lateral field of view of each eye ("tunnel vision")
    E. No effects on the eye

_____ 20. Mr. Frederick has a detached retina of the lower right portion of one eye. As a result he is unable to see objects in which area of which visual field of that eye?

    A. High in the left   B. High in the right
    C. Low in the left   D. Low in the right

Questions 21–25: fill-ins. Complete each sentence with the word or phrase that best fits.

_____ 21. _____ is the most common disorder leading to blindness.

_____ 22. The _____ form of retinal is shaped so that it can combine with scotopsin to form rhodopsin.

_____ 23. _____ is the study of the structure, functions, and diseases of the eye.

_____ 24. The names of three ossicles are _____.

_____ 25. Two functions of the ciliary body are _____.

The Endocrine System

18

You have studied how coordination and integration of many body functions are accomplished by rapid-firing nerve impulses. In this chapter you will learn about a slower acting, yet equally vital, control system involving chemical messengers called hormones. You will first define the glands which produce hormones—endocrine glands (Objectives 1–2). You will distinguish different chemical classes of hormones (3) and their mechanisms of action (4–6). Then you will learn about each of the major endocrine glands: their locations, the hormones they produce, their control, their roles in maintaining homeostasis, and some related disorders (7–25). Next you will study the development of the endocrine system and effects of aging upon this system (26–27). You will consider disturbances to homeostasis evoked by stress, and then look at the body's coping mechanisms involving hormones (28–29). Finally you will study medical terminology related to the endocrine system (30).

Topics Summary

A. Endocrine glands, chemistry of hormones
B. Mechanisms of hormonal action, control of hormone secretion
C. Pituitary (hypophysis)
D. Thyroid and parathyroids
E. Adrenals (suprarenals)
F. Other endocrine glands: pancreas, gonads, pineal (epiphysis cerebri), thymus
G. Developmental anatomy and aging of the endocrine system
H. Stress and homeostasis
I. Summary of hormones, medical terminology

Objectives

1. Discuss the functions of the endocrine system in maintaining homeostasis.

2. Define an endocrine gland and an exocrine gland and list the endocrine glands of the body.
3. Describe how hormones are classified according to their chemistry.
4. Explain the mechanisms of hormonal interaction with plasma membrane receptors and intracellular receptors.
5. Identify the role of prostaglandins in hormonal action.
6. Describe the control of hormonal secretions via feedback cycles and explain several examples.
7. Describe the structural and functional division of the pituitary gland into the adenohypophysis and the neurohypophysis.
8. Discuss how the pituitary gland and hypothalamus are structurally and functionally related.
9. List the hormones of the adenohypophysis, their principal actions, and their associated hypothalamic regulating factors.
10. Describe the release of hormones stored in the neurohypophysis and their principal actions.
11. Discuss the symptoms of pituitary dwarfism, giantism, acromegaly, and diabetes insipidus as pituitary gland disorders.
12. Explain the synthesis, storage, and release of thyroid hormones.
13. Discuss the principal actions and control of thyroid gland hormones.
14. Discuss the symptoms of cretinism, myxedema, exophthalmic goiter, and simple goiter as thyroid gland disorders.
15. Explain the principal actions and control of the parathyroid hormone.
16. Discuss the symptoms of tetany and osteitis fibrosa cystica as parathyroid glands disorders.
17. Describe the subdivision of the adrenal (suprarenal) glands into cortical and medullary portions.
18. Distinguish the effects of adrenal cortical mineralocorticoids, glucocorticoids, and gonadocorticoids on physiological activities, and explain how the hormones are controlled.
19. Discuss the symptoms of aldosteronism, Addison's disease, Cushing's syndrome, and adrenogenital syndrome as adrenal cortical disorders.

20. Identify the function of the adrenal medullary secretions as supplements of sympathetic responses.

21. Discuss the symptoms of pheochromocytomas as an adrenal medullary disorder.

22. Compare the principal actions of the pancreatic hormones and describe how they are controlled.

23. Discuss the symptoms of diabetes mellitus and hyperinsulinism as endocrine disorders of the pancreas.

24. Describe the possible functions of hormones of the pineal gland.

25. Describe the functions of the hormones of the thymus gland in immunity.

26. Describe the effects of aging on the endocrine system.

27. Describe the development of the endocrine system.

28. Define the general adaptation syndrome and compare homeostatic responses and stress responses.

29. Identify the body reactions during the alarm, resistance, and exhaustion stages of stress.

30. Define medical terminology associated with the endocrine system.

Learning Activities

(pages 398–399) **A. Endocrine glands, chemistry of hormones**

1. The two control systems of the body are the nervous system and the endocrine system.

   a. Compare the ways in which the nervous system and the endocrine system exert control over the body.

   b. List four categories of effects of hormones.
      (1)

      (2)

      (3)

      (4)

   1 c. Stated very simply, what is the primary function of both systems?

   2 2. Referring to Figure 18-1 (page 398) in the text, identify the major endocrine glands. Approximate their locations in your own body.

a. Compare the arrangement of the endocrine system with that of the nervous system.

b. How does the endocrine system achieve such widespread effects?

3. Complete the table contrasting exocrine and endocrine glands. (For help refer to Chapter 4, Learning Activity [6] page LG 69.)

	Products Secreted into	Examples
a. Exocrine		
b. Endocrine		

4. Contrast the principal classes of hormones based on chemistry. [3]
   a. Hormones that are water soluble include (*amines, peptides and proteins? steroids?*). They are produced by endocrine glands derived from the *(endoderm* or *ectoderm? mesoderm?)* of the embryo. List several endocrine glands that secrete water-soluble hormones.

   b. _____-soluble hormones are produced by endocrine glands derived from embryonic _____. These hormones are synthesized in organelles named _____ and _____. Name three glands that produce such hormones.

5. Before you begin your study of different hormones, look at Table LG 18-1. As you proceed through the chapter, test your knowledge by filling in names, abbreviations, sources, and functions of these hormones. By the end of the chapter, you should be able to complete the table.

Table LG 18-1
Summary of Hormones and Regulating Factors

Abbreviation	Name	Source	Principal Functions
1.	Prostaglandins		
2. GHRF		Hypothalamus	
3.	Somatostatin	Hypothalamus and pancreas	
4.			Stimulates release of TSH
5. CRF			
6. GnRF			Stimulates release of FSH and LH
7.	Prolactin inhibiting factor		
8. PRF			
9. MRF			
10. MIF			
11. GH or STH		Anterior lobe of pituitary	
12.	Thyroid-stimulating hormone		
13. ACTH			
14. FSH			
15. LH			
16.	Prolactin		
17.			Increases skin pigmentation

Abbreviation	Name	Source	Principal Functions
18.	Oxytocin		
19. ADH			
20. —	Thyroxin		
21.			Decreases blood calcium
22.		Parathyroids	
23. —	Aldosterone		
24. —	Cortisol, cortisone		
25. —	Sex hormones (2)	(4)	
26.	(2)	Adrenal medulla	
27. —		Pancreas, alpha cells	
28. —	Insulin		
29. —		Pineal gland	
30. —		Thymus	

(pages 399–
404) **B.** Mechanisms of hormonal action, control of hormonal secretion

1. Study Figure 18-3 (page 401) in your text. Write a paragraph describing how hormones travel in blood, effects on target areas, feedback regulation of further release of hormone, and ultimate disposal of the hormone.

2. Study the cyclic AMP mechanism in Figure 18-4a (page 402) in your text.

| 4 | Then complete this exercise.

   a. A hormone such as antidiuretic hormone (ADH) acts as the

   _____ messenger, as it carries a message from the

   _____ where it is secreted (in this case the pituitary

   gland) to the _____ where it acts (in this case kidney cells).

   b. All hormones travel in the _____, with the result that all body cells are exposed to all hormones. Only certain cells are affected by particular hormones because a hormone attaches only to cells that have

   specific _____ located in the cell *(membrane? cytoplasm? nucleus?).* It is estimated that each target cell may have up to *(100? 1,500? 100,000?)* receptors.

   c. The attachment of the hormone increases activity of the enzyme

   _____, located in the plasma membrane. This enzyme catalyzes conversion of ATP to _____, known

   as the _____ messenger.

   d. Cyclic AMP then activates one or more enzymes known as

   _____. Its role is to enhance the addition of a *(calcium? phosphate?)* from ATP to a protein. The protein–phosphate can then finally set off this target cell's response.

   e. For example, when the hormone ADH stimulates kidney cells, it does so in just the manner described, leading to production of cyclic

   _____, activated protein _____

   enzyme, and then protein _____. This regulating chemical ultimately alters permeability of kidney cells so that kidneys

   retain water, thereby decreasing production of _____.

   (More about this mechanism in Learning Activity | 13 |.)

f. Effects of cyclic AMP are *(short? long?)*-lived. This chemical is rapidly degraded by an enzyme named _____.

g. A number of hormones are known to act by means of the cyclic AMP mechanism. Included are most of the _____-soluble hormones. Name several.

h. Name two other chemicals that may, like cyclic AMP, act as second messengers.

i. What is *calmodulin?*

3. Thyroid hormones, as well as the steroid hormones, use a different mechanism for activating their target cells. Summarize these *gene activation* mechanisms.

4. Describe prostaglandins in this exercise.

   a. Prostaglandins (PGs) are considered *(local or tissue? circulatory?)* hormones. They are synthesized by *(one specific endocrine gland? most tissues of the body?)*.

b. Chemically, prostaglandins are *(proteins? lipids?)*. Each contains a 20-carbon _____ acid and a _____-carbon ring. Name three prostaglandins.

c. Prostaglandins seem to alter the amount of _____ production. Since cyclic AMP activates so many responses, PGs have *(few? a wide range of?)* effects on the body. PGs are considered *(extremely? only slightly?)* potent chemicals.

d. List four functions of prostaglandins.

e. Prostaglandin synthesis has been linked to increase in temperature, as in fever. Aspirin and acetominophen (Tylenol) ____-crease PG synthesis and can therefore ____-crease fever.

5. Complete line 1 (on prostaglandins) in Table LG 18-1.

6. It is vital to maintenance of homeostasis that the appropriate amount of the appropriate hormones be released at any given moment. Describe three general mechanisms of regulation of hormones by completing this exercise.

$\boxed{6}$

a. Release of parathyroid hormone (PTH) is signaled by alteration in the *(nerve stimulation to? calcium level of blood passing through?)* the parathyroid gland. A low blood calcium level triggers PTH release which then

____-creases blood calcium level. This is an example of a *(positive? negative?)* feedback mechanism.

b. Name two hormones that are released in response to sympathetic nerve stimulation. _____ and _____
Name one other hormone that is regulated by nervous stimulation.

_____

c. A third type of control mechanism involves regulating factors synthesized by the _____. Some of these, called releasing factors, cause release of hormones in the anterior pituitary; others, called

_____ factors, have the opposite effect. (More about regulating factors in Learning Activity $\boxed{10}$.)

(pages 404-412) **C. Pituitary (hypophysis)**

1. Complete this exercise about the pituitary gland.

a. The pituitary is also known as the _____.

b. Why is it sometimes called the "master gland"?

c. Where is it located? What protects it?

d. Seventy-five percent of the gland consists of the *(anterior? posterior?)* lobe, called the _____-hypophysis. The posterior lobe (or _____) lobe is somewhat smaller.

7

2. The blood supply to the pituitary is designed in such a way that regulating factors can be transported from the _____ to the anterior pituitary *(after? without?)* first traveling through the heart and general circulation. This anatomical arrangement ____-creases the effectiveness of the regulating factors.

   Arrange in order the names of vessels which supply blood to the adenohypophysis. Use lines provided. ____ ____ ____ ____

   HPV. Hypophyseal portal veins (in infundibulum)
   PP. Primary plexus (near base of hypothalamus)
   SHA. Superior hypophyseal arteries
   SP. Secondary plexus (in anterior pituitary)

   Now place an asterisk (*) next to the vessels into which the hypothalamus secretes regulating factors. Place a double asterisk (**) next to vessels into which the adenohypophysis delivers its hormones.

3. Refer to Table LG 18-1. All hormones numbered 11 to 17 have the same source: the _____. To indicate this, simply draw a long arrow down the *source* column through parts 11 to 17. Now fill in names and abbreviations of these seven hormones. (Omit functions for now.)

4. Define *tropic hormone*.

8

Circle the tropic hormones in Table LG 18-1.

368

9

5. Refer to Figure LG 18-1 as you do this learning activity about growth hormone.

a. The main function of growth hormone (＿＿) is to stimulate growth and maintain size of ＿＿＿＿＿＿＿ and ＿＿＿＿＿＿＿＿＿＿＿＿＿＿＿＿＿＿. The alternate name for GH, ＿＿＿＿＿＿＿＿＿＿＿＿＿＿＿＿＿＿＿ hormone (STH), indicates its function: to "turn" *(trop-)* the body *(soma-)* to growth. Note that somatotropic hormone *(is? is not?)* considered a true tropic hormone since it *(does? does not?)* stimulate another endocrine gland to produce a hormone.

b. GH stimulates growth by *(accelerating? inhibiting?)* entrance of amino acids into cells where they can be used for ＿＿＿＿＿＿＿＿＿＿＿ synthesis. Therefore, growth hormone stimulates protein *(anabolism? catabolism?)*. Indicate this on Figure LG 18-2.

c. GH promotes breakdown to glycogen in liver to ＿＿＿＿＿＿＿＿＿＿＿ ＿＿＿＿＿＿＿＿＿＿＿＿＿. It also promotes release of stored body fats, as well as stimulates ＿＿＿＿＿＿＿＿＿＿＿-bolism of fats. Show this on the figure.

d. Since cells now turn to fats for energy (by means of the Krebs cycle), and liver glycogen is broken down, the level of glucose in cells and blood ＿＿＿-creases, a condition known as *(hyper? hypo?)*-glycemia. In fact, excess GH can mimic diabetes mellitus, and for this reason GH is said to be diabeto-＿＿＿＿＿＿＿＿＿＿＿.

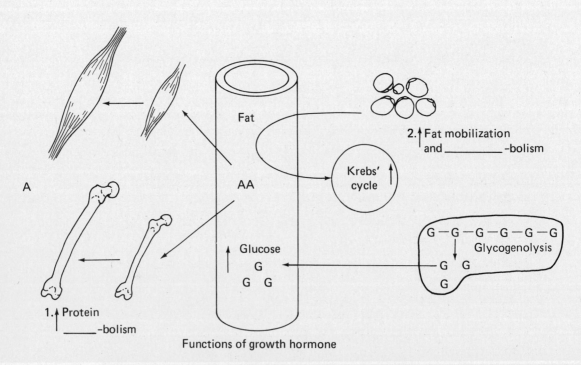

Functions of growth hormone

Figure LG 18-1 Functions of growth hormone in regulating metabolism. AA, amino acid; G-G-G-G-G-G, glycogen; G, glucose. Complete as directed in Learning Activity 9 .

e. GH secretion is controlled by two regulating factors: _____

_____ and _____. One stimulus that promotes release of GH is low blood sugar (for example, during growth when cells require much energy). This condition, *(hyper-? hypo-?)* glycemia, stimulates secretion of *(GHRF? GHIF?)*, which then causes release of *(high? low?)* levels of GH.

f. State the function of somatomedins and insulin-like growth factors (IGF).

g. Hypersecretion of GH in early years leads to *(dwarfism? giantism? acromegaly?)* while deficiency causes _____. Excess GH after closure of epiphyseal plates causes _____, characterized by long bones in three areas: _____, _____, and _____.

6. Now you should be able to complete parts 2, 3, and 11 on Table LG 18-1.
7. Recall that release of certain hormones, like GH, can be decreased by inhibiting factors (IFs). The four tropic hormones lack inhibiting factors, but are instead controlled by negative feedback mechanisms. as shown in Figure LG 18-2. Complete this exercise about them. |10|

a. Releasing factors (RFs) are produced by the _____

_____. Figure LG 18-2a shows the RF named _____ which stimulates release of the hormone ACTH.

b. All tropic hormones are produced by the _____. Tropic hormones *(trop* = turn) literally "turn on" their target glands. Note in Figure LG 18-1a that a high level of the tropic hormone ACTH will *(stimulate? inhibit?)* the adrenal cortex, so that a *(high? low?)* level of the target hormone, adrenocorticoid, will enter the blood.

c. When the blood level of the target hormone is sufficiently high, it will exert a *(positive? negative?)* feedback effect by *(stimulating? inhibiting?)* releasing factor production. This, in turn, leads to a *(high? low?)* level of ACTH.

d. A low level of ACTH will have a direct effect on its target gland, causing a *(high? low?)* level of adrenocorticoid.

c. Note that a low level of target hormone signals the need to increase itself by increasing the level of _____, and consequently the level of _____, which will lead to increased adrenocorticoid production.

f. Now show relationships among the other tropic hormones and their releasing factors and target hormones by completing Figure LG 18-2b–d. Write the name of one hormone or regulating factor on each blank line.

8. Fill in parts 4 to 6 and 12 to 15 of Table LG 18-1.

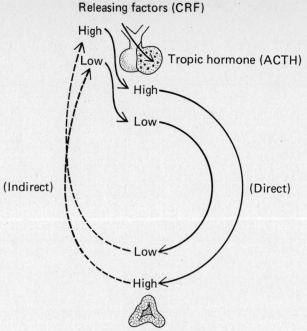

(a)   Target hormone (adrenocorticoid)

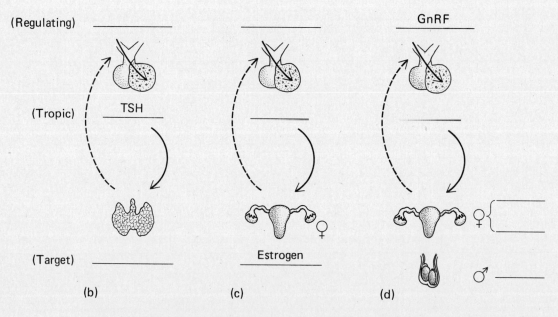

Figure LG 18-2 Control of hormone secretion. (a) CRF, ACTH, and adrenocorticoids. (b–d). Other examples of regulating factors, tropic hormones, and target hormones. Complete as directed in Learning Activity 10 .

11 9. Do this exercise about hormonal control of the mammary glands.

a. Milk is produced in mammary glands following stimulation by the hormone

_____ (_____) which is secreted by the *(anterior? posterior?)* lobe of the pituitary. A different hormone causes ejection of milk from the glands at the time of the baby's suckling. This hormone,

named _____, is released by the *(anterior? posterior?)* pituitary.

b. During pregnancy, prolactin levels *(increase? decrease?)* due to increased levels of *(PIF? PRF?)*. During most of a woman's lifetime prolactin levels are low as a result of high levels of *(PIF? PRF?)*. An exception is the time just prior to each menstrual flow when PIF levels *(increase? decrease?)*. Due to this lack of inhibition prolactin secretion increases, activating (and causing tenderness in) breast tissues.

10. Complete parts 7 to 10 and 16 to 17 of Table LG 18-1.

11. Posterior pituitary hormones *(are? are not?)* controlled by hypothalamic regulating factors as are the anterior pituitary hormones. Instead, the _____ system detects the need for oxytocin and ADH, and neurons actually synthesize these hormones. It is for this reason that the posterior pituitary is called the _____-hypophysis. `12`

12. Describe the production and pathway of hormones from the hypothalamus to the posterior lobe of the pituitary. Include the role of *neurophysins*.

State two advantages of storage and secretion of these hormones by the posterior pituitary (rather than by the hypothalamus).

13. Describe the hormones of the neurohypophysis in this learning activity. `13`

a. The posterior pituitary *(does? does not?)* synthesize hormones, but it does secrete two hormones: _____ and _____ _____. These compounds are synthesized in cell bodies of neurons located in the _____ and then passed down *(axons? dendrites? portal veins?)* to the neurohypophysis.

b. Oxytocin has two effects in the body. It stimulates contraction of the _____ and it causes *(production? ejection?)* of milk from mammary glands. In both cases the hypothalamus releases oxytocin in response to afferent _____ impulses. For example, during labor uterine distension activates nerve fibers which notify the hypothalamus that oxytocin is needed for uterine contraction.

372

   c. The time period from the start of a baby's suckling at the breast to the delivery of milk to the baby is about ＿＿ seconds. Although afferent impulses (from breast to hypothalamus) require only *(a fraction of a second? 45 seconds?),* passage of the oxytocin through the

＿＿＿＿＿＿＿＿＿＿＿＿＿＿ to the breasts requires about ＿＿ seconds.

   d. The second posterior pituitary hormone is ADH. Since *diuresis* means "a large volume of urine," antidiuresis means the opposite. The effect of ADH (or antidiuretic drugs) is to *(increase? decrease?)* urine production. Think of situations in which the body needs to conserve water. Suppose

you run a long distance and lose body fluids in ＿＿＿＿＿＿＿＿＿＿; your body could become dehydrated. A(n) *(increase? decrease?)* of ADH production helps by conserving body fluids. Similarly low blood pressure can be elevated by *(increasing? decreasing?)* ADH, again because ADH causes kidneys to produce *(more? less?)* urine.

   e. How does the hypothalamus "know" to release ADH? The mechanism *(does? does not?)* involve long afferent neurons as in oxytocin release.

Instead, neurons (called ＿＿＿＿＿＿＿＿＿＿＿-receptors) located in

the hypothalamus itself are activated by a "salty" (or ＿＿＿＿＿＿＿

＿＿＿＿＿＿＿＿＿-tonic) extracellular fluid around them. In this way they sense that the body needs to conserve fluids, and they release ADH.

   f. Inadequate ADH production, as occurs in diabetes *(mellitus? insipidus?)* results in *(enormous? minute?)* daily volumes of urine. Unlike the urine produced in diabetes mellitus, this urine *(does? does not?)* contain sugar. It is bland or *insipid.*

  14. Fill in parts 18 and 19 of Table LG 18-1. (Be sure that parts 1 to 19 are complete now.)

(pages 412–419) **D. Thyroid and parathyroids**

  1. Draw a diagram of the thyroid gland. Label the *lobes* and the *isthmus.*

The parathyroids are embedded in the *(anterior? posterior?)* of the thyroid.

  2. Identify the location of your own thyroid gland. Verify this by referring to Figure 18-1 (page 398) of the text.

3. Describe the histology and hormone secretion of the thyroid gland in this [14] exercise.

   a. The thyroid is composed of two types of glandular cells. *(Follicular? Parafollicular?)* cells produce the two hormones _____ and _____. Parafollicular cells manufacture the hormone _____.

   b. The ion *(Cl$^-$? I$^-$? Br?)* is highly concentrated in the thyroid since it is an essential component of T$_3$ and T$_4$. Each molecule of thyroxine (T$_4$) contains four of the I$^-$ ions, while T$_3$ contains _____ ions of I$^-$.

   c. These hormones are combined with a *(lipid? protein?)* to form thyroglobulin (TGB) and *(released immediately? may be stored for months?)*.

   d. Thyroxine travels in the bloodstream bound to plasma proteins, and in this form the hormone is often called PBI, or _____ _____.

4. List three categories of effects of thyroid hormones.

5. Fill in blanks and circle answers about functions of thyroid hormone contrasted to those of growth hormone. [15]

   a. Like growth hormone, thyroid hormone increases protein _____-bolism.

   b. Unlike GH, thyroid hormone increases most aspects of carbohydrate _____-bolism. Therefore, thyroid hormone *(increases? decreases?)* basal metabolic rate (BMR), releasing heat and *(increasing? decreasing?)* body temperature.

   c. Thyroid hormone *(increases? decreases?)* heart rate and blood pressure and *(increases? decreases?)* nervousness.

   d. This hormone is important for development of normal bones and muscles, as well as brain tissue. Therefore, deficiency in thyroid hormone during childhood (a condition called _____) *(does? does not?)* lead to retardation as well as small stature. Note that in childhood GH deficiency, retardation of mental development *(occurs also? does not occur?)*.

6. Release of thyroid hormone is under influence of releasing factor _____ and tropic hormone _____.
   Review Figure LG 18-2b.
   List three factors that stimulate TRF and TSH production.

7. Contrast *myxedema* and *exophthalmic goiter* with regard to thyroid hormone levels and symptoms.

8. A simple goiter is *(an enlargement? a decrease in size?)* of the thyroid gland. [16] It results from the futile attempts of the gland to produce thyroid hormone when iodine is present in *(excessive? deficient?)* quantities.

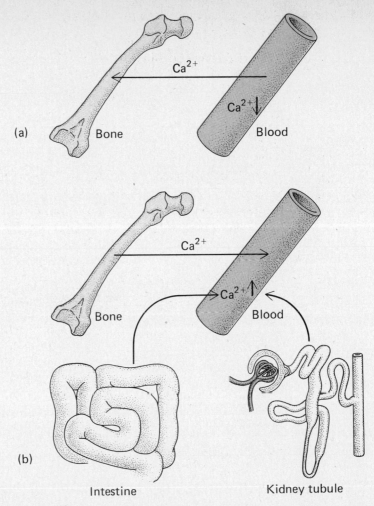

Figure LG 18-3 Hormonal control of calcium ion (Ca²⁺). (a) Effects of calcitonin (CT): increased calcium storage in bones lowers blood calcium level. (b) Parathyroid hormone (PTH) causes increased blood calcium by drawing calcium into blood from three sources. Arrows show direction of calcium flow. Refer to Learning Activity 17 .

17 9. Refer to Figure LG 18-3 and do this exercise.

a. Another hormone produced by the thyroid is _____ _____, which regulates distribution of _____ _____ ion. High levels of calcitonin result in *(high? low?)* calcium levels in blood.

b. How does CT lower blood calcium?

Because of its effects on bone, CT is helpful in treatment of post-menopausal _____.

c. The main action of parathyroid hormone (PTH) is to *(increase? decrease?)* blood levels of calcium. Three sources from which calcium is drawn in order to increase the blood level of calcium are _____ _____, _____, and _____. Regulation of PTH occurs by a *(positive? negative?)* feedback mechanism involving calcium. So if blood level of calcium is low, this causes parathyroids to produce a *(high? low?)* level of PTH.

d. Figure LG 18-3 emphasizes that calcitonin and PTH are *(antagonists? synergists?)* with regard to effects on calcium. Both hormones affect blood *(potassium? phosphate?)* levels too. Several mechanisms are involved, but the overall effects are that calcitonin and PTH act synergistically, both *(raising? lowering?)* blood phosphate levels.

10. Consider effects of calcium imbalances by doing this exercise. [18]

a. One role of calcium is to decrease excessive nerve impulses by blocking sodium ion pores of membranes. Therefore a *de*crease in blood calcium causes ___-crease in nerve impulses. Although calcium *is* necessary for muscle _____, muscles can still contract if calcium levels are somewhat decreased. In fact, they contract convulsively and spasmodically (the condition called _____), if nerves are activated by slightly lowered calcium levels (_____-calcemia).

b. Hypocalcemia results from *(hypo-? hyper-?)* parathyroid conditions, such as following removal of these glands. Without this hormone tetany can occur.

c. Hyperparathyroidism is likely to lead to *(demineralization? excessive deposit of minerals?)* in bones.

11. Complete parts 20–22 of Table LG 18-1.

E. Adrenals (suprarenals)   (pages 419–424)

1. Each adrenal gland has two parts: the adrenal _____ and the adrenal _____.

2. Draw a diagram of the adrenal gland (actual size). Label *cortex* and *medulla*. Then label the three zones of the cortex, and write next to each the types of hormones synthesized there.

3. Refer to Figure LG 18-4 and do the following learning activity about mechanisms controlling fluid balance.

   |19|

   a. Figure LG 18-4a reviews the role of ADH, released by the

   _____. ADH causes kid-

   neys to move _____ out of the urine that is forming, and return it to blood, a process called *reabsorption.* Show this on Figure LG 18-4a.

   b. Complete Figure LG 18-4b to show effects of aldosterone. This is the main

   _____-corticoid produced by the adrenal *(cortex?*

   medulla?). It causes kidney tubules to reabsorb _____ and _____

   back into blood (so they are retained by the body), while _____ and

   _____ are lost in urine. Effects of aldosterone on fluid level of the body can be considered *(similar to? opposite?)* effects of ADH.

   c. Next to the top arrow in Figure LG 18-4b, draw symbols of two anions that may accompany Na^+ back into the blood.

   d. Retention of fluids tends to increase blood pressure (BP). One important mechanism for regulating BP is shown in Figure LG 18-4c. This is the

   _____-_____ mechanism. When BP of kidney blood vessels is low, certain kidney cells, called

   _____ cells, sense the decreased pressure. As a result,

   they secrete an enzyme named _____. Renin acts on

   a blood protein named _____, changing it to

   _____. A lung converting enzyme (CE) finally con-

   verts this to _____.

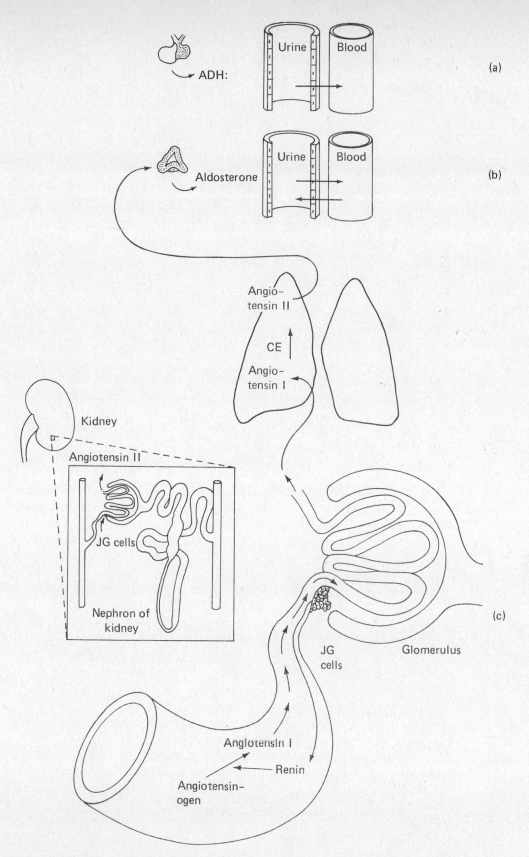

Figure LG 18-4 Mechanisms of fluid and blood pressure control. (a) ADH (from posterior pituitary). (b) Aldosterone (from adrenal cortex). (c) Renin-angiotensin mechanism. JG cells, juxtaglomerular cells; CE, converting enzyme. Complete as directed in Learning Activity 19 .

e. Angiotensin II present in blood passing through the adrenal cortex will stimulate it to produce _____. As described above, aldosterone leads to Na^+ and water *(retention? loss?)* and therefore to a(n) *(increased? decreased?)* BP. In addition, angiotensin II is a powerful vasoconstrictor, which also *(increases? decreases?)* BP.

4. *ACTive learning.* In a study group or lab group, role play the renin-angiotensin-mechanism, using the entire room as the interior of the human body. Determine which parts of the room are the liver, kidneys, lungs, adrenal glands, and blood vessels. Start a foldable item (like a sweatshirt) as an angiotensinogen molecule in the liver. One person can transport this molecule through the bloodstream to the kidney. Here renin (another student) alters the shape of the sweatshirt to change it to angiotensin I. And so on through lungs until aldosterone exerts its final effects upon kidneys, where people named Na^+ and H_2O stay in blood while others named K^+ or H^+ exit from the room via urine.

|20| 5. Excess aldosterone production leads to too *(much? little?)* water retention, with resulting edema. Another effect is *(excess? deficient?)* blood potassium which affects depolarization of nerves and muscles.

6. Name three glucocorticoids. Circle the one that is most abundant in the body.

|21| 7. In general, glucocorticoids are involved with metabolism and resistance to stress. Describe their effects in this exercise.

a. These hormones promote ____-bolism of proteins. In this respect their effects are *(similar to? opposite?)* those of both growth hormone and thyroid hormone.

b. However, glucocorticoids may stimulate conversion of amino acids and fats to glucose, a process known as _____. In this way glucocorticoids (like GH) *(increase? decrease?)* blood glucose.

c. Refer to Figure 18-25 (page 431) in your text. Note that glucocorticoid (and also GH) production *(increases? decreases?)* during times of stress. By raising blood glucose levels, these hormones make energy available to cells most vital for a response to stress. During stress glucocorticoids also act to *(increase? decrease?)* blood pressure and to *(enhance? inhibit?)* the inflammatory process. (They are *anti-inflammatory.*)

8. How are glucocorticoid hormones regulated? (For help, review Figure LG 18-2a, page LG 370.)

9. A third category of adrenal cortex hormones are the _____ 22 _____ or _____ hormones. Name two.

10. Contrast *virilizing adenoma/feminizing adenoma.*

11. Explain the relationship between the adrenal medulla and the sympathetic nervous system. Include the term *chromaffin cells* in your explanation.

12. Name the two principal hormones secreted by the adrenal medulla. Circle 23 the name of the hormone which accounts for 80 percent of adrenal medulla secretion.

13. Describe the effects of epinephrine and norepinephrine.

14. You should now be able to complete parts 23 to 26 of Table LG 18-1.

F. Other endocrine glands: pancreas, gonads, pineal (epiphysis cerebri), thymus (pages 424–428)

1. The islets of Langerhans are located in the _____. Do this exercise describing hormones produced there. 24

a. The name *islets of Langerhans* suggests that these clusters of _____-crine cells lie amidst a "sea" of exocrine cells within the pancreas. Name the three types of islet cells.

b. Refer to Figure LG 18-5a. Glucagon, produced by *(alpha? beta? delta?)* cells, *(increases? decreases?)* blood sugar in two ways. First, it stimulates the breakdown of _____ to glucose, a process known as *(glycogenesis? glycogenolysis? gluconeogenesis?)*. Second, it stimulates the conversion of amino acids and other compounds to glucose. This process is called _____.

For extra review. On separate paper define the terms glycogen and glucagon, noting their differences in spelling and pronunciation.

c. Is glucagon controlled by an anterior pituitary *tropic* hormone? *(Yes? No?)* In fact, control is by effect of _____ directly upon the pancreas. When blood glucose is low, then a *(high? low?)* level of glucagon will be produced. This will raise blood sugar.

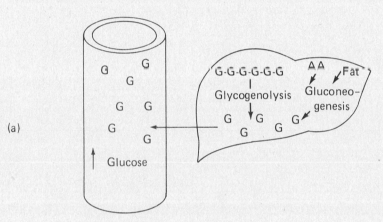

(a)

Functions of glucagon and epinephrine

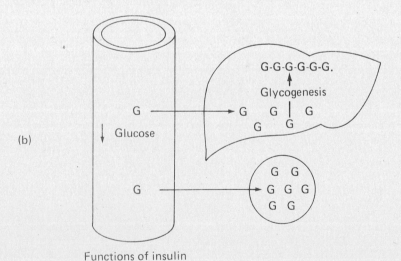

(b)

Functions of insulin

Figure LG 18-5 Functions of glucagon and insulin in regulating blood glucose. AA, amino acid; G, glucose; G-G-G-G-G-G, glycogen (chain of glucoses). Refer to Learning Activity 24.

d. Identify two other factors that can stimulate glucagon release.

_____ and _____ Now write one factor that can inhibit glucagon secretion.

e. Now look at Figure LG 18-5b. The action of insulin is *(the same as? opposite?)* that of glucagon. In other words, *insulin decreases blood sugar.* (Repeat that statement three times; it is important!) Insulin acts by several mechanisms. Two are shown in the figure. First, insulin *(helps? hinders?)* transport of glucose from blood into cells. Second, it accelerates conversion of glucose to _____, the process called

_____.

f. High blood sugar *(triggers? inhibits?)* insulin production. Since GH and ACTH lead to *(hyper? hypo?)*-glycemia, these two hormones can be said to indirectly *(trigger? inhibit?)* insulin production.

2. Write three primary characteristics that distinguish Type I (juvenile-onset) diabetes mellitus from Type 2 (maturity-onset).

3. *A clinical challenge.* Mrs. Marshall has diabetes mellitus. In the absence of sufficient insulin, her cells are deprived of glucose. Explain why she experiences the following symptoms.

   a. Hyperglycemia and glycosuria

   b. Increased urine production (poly-_____) and increased water intake (poly-_____)

382

c. Ketosis, a form of _____-osis

d. Atherosclerosis

26 4. Note on Table LG 18-1 that sex hormones are produced in several locations in the body. List four in box 25 of the table. (More about these in Chapter 28.)

5. Where is the pineal gland located?

Does this gland appear to atrophy with age?

Describe the probable action of melatonin.

6. What is the general function of the thymus?

7. Complete parts 27 to 30 of Table LG 18-1.

27 8. As you continue your study of anatomy and physiology, you will consider hormones produced in three other parts of the body: those from the _____ tract (Chapter 24), hormones made in the _____ during pregnancy (Chapter 28), and also the hormone erythropoietin, which affects _____ production (Chapter 19).

G. **Developmental anatomy and aging of the endocrine system**

1. Categorize the major endocrine glands according to their primary germ 28
   layer origins. Write one endocrine gland on each line provided. One is done
   for you.
   a. Endoderm: _____pancreas_____   _____

   _____   _____

   b. Mesoderm: _____   _____

   c. Ectoderm: _____   _____

   _____   _____

2. Now match the endocrine gland with the correct description of its embryo- 29
   logical origin.

   _____ a. Develops from the foregut area   AC. Adrenal cortex
            that later becomes part of small   AM. Adrenal medulla
            intestine   AP. Anterior pituitary
   _____ b. Derived from the roof of the sto-   P. Pancreas
            modeum (mouth) called the hypo-   PT. Parathyroid
            physeal (Rathke's) pouch   T. Thymus
   _____ c. Originates from the neural crest
            which also produces sympathetic
            ganglia
   _____ d. Derived from tissue from the same
            region that forms gonads
   _____ e. Arise from the third and fourth
            pharyngeal pouches (2 answers)

3. Name two endocrine organs that are perhaps most likely to cause problems
   among the elderly.   30

H. **Stress and homeostasis**   (pages 430–433)

1. Define and give two examples of *stressors.*

[31] 2. Summarize the body's response to stress in this exercise.

a. The part of the brain that senses the stress and initiates response is the

_____. It responds by two main mechanisms.

b. First is the *(alarm? resistance?)* reaction involving the adrenal *(medulla? cortex?)* and the _____ division of the autonomic nervous system.

c. Second is the _____ reaction. Playing key roles in this response are three anterior pituitary hormones: _____,

_____, and _____ and their target hormones. Therefore the adrenal *(medulla? cortex?)* hormones are activated.

[32] 3. Now describe events that characterize the first (or alarm) stage of response to stress. Complete this exercise.

a. The alarm reaction is sometimes called the _____ response.

b. During this stage blood glucose *(increases? decreases?)* by a number of hormonal mechanisms, including those of adrenal medulla hormones

_____ and _____. Glucose must be available for cells to have energy for the stress response.

c. Oxygen must also be available to tissues; the respiratory system *(increases? decreases?)* its activity. Heart rate and blood pressure *(increase? decrease?)*. Blood is shunted to vital tissues such as the _____

_____ and _____ and is directed away from reservoir organs such as the _____

and _____.

d. Nonessential activities such as digestion *(increase? decrease?)*. Sweating *(increases? decreases?)* in order to control body temperature and eliminate wastes.

4. Compare the alarm stage with the resistance stage of stress response by completing this table.

	Alarm	Resistance
a. Order of occurrence (first or second)		
b. Duration (short-lived or long-lived)		
c. Chief regulating structures	Sympathetic organs and adrenal medulla	
d. Purpose and results		

5. Describe the effects of each of these hormones in the resistance stage.
   a. Mineralocorticoids

   b. Glucocorticoids

   c. Thyroxin

   d. GH

6. When the resistance stage fails to combat the stressor, the body moves into the _____ stage. Explain how this is related to:
   a. Loss of potassium

   b. Depletion of glucocorticoids

   c. Weakening of organs

1. *For extra review.* Test your understanding of these hormones by writing the name (or abbreviation where possible) of the correct hormone after each of the following descriptions.

[33]

a. Stimulates release of growth hormone: _____

b. Stimulates testes to produce testosterone:_____

c. Promotes protein anabolism, especially of bones and muscles:

_____

d. Tropic hormone for thyroxin: _____

e. Stimulates development of ova in females and sperm in males:

_____

f. Present in bloodstream of nonpregnant women to prevent lactation:

_____

g. Stimulates ovary to release egg and change follicle cells into corpus

luteum: _____

h. Stimulates uterine contractions and also milk ejection: _____

_____

i. Inhibited by MIF: _____

j. Stimulates kidney tubules to produce small volume of concentrated urine:

_____

k. Target hormone of ACTH: _____

l. Increases blood calcium: _____

m. Decreases blood calcium and phosphate: _____

n. Contains iodine as an important component: _____

o. Raises blood glucose (3): _____,

_____, _____

p. Lowers blood glucose: _____

q. Serves antiinflammatory functions: _____

r. Mimics many effects of sympathetic nerves:_____

s. Leads to relaxation of pubic symphysis and dilation of the cervix of the

uterus close to the time of delivery of baby: _____

2. *For extra review.* Match the disorder with the hormonal imbalance.

_____ a. Deficiency of GH in child; slow bone growth

_____ b. Excess of GH in adult; enlargement of hands, feet, and jawbones

_____ c. Deficiency of ADH; production of enormous quantities of "insipid" (nonsugary) urine

_____ d. Deficiency of insulin; hyperglycemia and glycosuria (surgary urine)

_____ e. Deficiency of thyroxin in child; short stature and mental retardation

_____ f. Deficiency of thyroxin in adult; edematous facial tissues, lethargy

_____ g. Excess of thyroxin; protruding eyes, "nervousness," weight loss

_____ h. Deficiency of thyroxin due to lack of iodine; most common reason for enlarged thyroid gland

_____ i. Deficiency of PTH; decreased calcium in blood and fluids around muscles resulting in abnormal muscle contraction

_____ j. Deficiency of adrenocorticoids; increased K^+ and decreased Na^+ resulting in low blood pressure and dehydration

_____ k. Tumor of adrenal medulla causing sympathetic-like responses, such as increased pulse and blood pressure, hyperglycemia, and sweating

Ac. Acromegaly
Ad. Addison's disease
C. Cretinism
DI. Diabetes insipidus
DM. Diabetes mellitus
E. Exophthalmic goiter
M. Myxedema
PC. Pheochromocytoma
PD. Pituitary dwarfism
S. Simple goiter
T. Tetany

Answers to Numbered Questions in Learning Activities

1 Maintain homeostasis

2 (a) The brain and cord are centrally related but nerves extend out to virtually every part of the body; endocrine glands are found in relatively few locations. (b) Hormones travel through blood, and blood reaches essentially every part of the body.

3 (a) Amines, peptides and proteins; endoderm or ectoderm; thyroid, parathyroid and pancreas (of endodermal origin) and pituitary and adrenal medulla (from ectoderm). (b) Lipid; mesoderm; mitochondria, smooth ER; adrenal cortex, ovaries and testes.

4 (a) First; endocrine gland; target cell. (b) Bloodstream; receptors; membrane, cytoplasm or nucleus; 100,000. (c) Adenylate cyclase, cyclic AMP, second. (d) Protein kinase, phosphate. (e) AMP; kinase; phosphate; urine. (f) Short; phosphodiesterase. (g) Water; ADH, OT, FSH, LH, TSH, ACTH, CT, epinephrine and NE. (h) Calcium ions and cyclic GMP. (i) A protein to which the second messenger Ca^{2+} ions bind and which can then activate or inhibit cell enzymes to determine the final hormonal effect.

5 (a) Local or tissue, most tissues of the body. (b) Lipids, fatty, 5, examples: PGA, PGB, PGE_1, PGE_2, PGI. (c) Cyclic AMP, a wide range of, extremely. (d) See list on page 403 of text. (e) De, de.

6 (a) Calcium level of blood passing through, in, negative. (b) Epinephrine and norepinephrine (NE), ADH. (c) Hypothalamus, inhibiting.

7 Hypothalamus, without, in; SHA PP* HPV SP**.

8 A hormone that stimulates another endocrine gland (a target gland), causing it to secrete its own hormone (a target hormone); numbers 12–15 in table (TSH, ACTH, FSH, LH).

9 (a) GH, bones, skeletal muscles, somatotropic, is not, does not. (b) Accelerating, protein, anabolism. (c) Glucose, cata. (d) In, hyper-, -genic. (e) GHRF, GHIF, hypo-, GHRF, high. (f) They mediate effects of GH, such as conversion of glycogen to glucose. (g) Giantism; dwarfism; acromegaly; hands, feet, face.

10 (a) Hypothalamus, CRF. (b) Anterior lobe of pituitary, stimulate, high. (c) Negative, inhibiting, low. (d) Low. (e) CRF, ACTH. (f) *Figure LG 18-1b:* TRF causes release of tropic hormone TSH, which stimulates thyroid hormone production. *Figure LG 18-1c:* GnRF causes release of tropic hormone FSH which stimulates the target hormone estrogen. *Figure LG 18-1d:* Also as a result of GnRF, tropic hormone LH is released. It stimulates target hormones estrogen and progesterone in females and testosterone in males.

11 (a) Prolactin (or lactogenic hormone) (PRL), anterior, oxytocin, posterior. (b) Increase, PRF, PIF, decrease.

12 Are not, nervous, neuro.

13 (a) Does not, antidiuretic hormone (ADH), oxytocin (OT), hypothalamus, axons. (b) Uterus, ejection, nerve. (c) 30–60, a fraction of a second, bloodstream, 30–60. (d) Decrease, sweat, increase, increasing, less. (e) Does not, osmo-, hyper-. (f) Insipidus, enormous, does not.

14 (a) Follicular; thyroxine (T_4) and triiodothyronine (T_3); calcitonin. (b) I^-; 3. (c) Protein; may be stored for months. (d) protein-bound iodine.

15 (a) Ana. (b) Cata, increases, increasing. (c) Increases, increases. (d) Cretinism, does, does not occur.

16 An enlargement, deficient.

17 (a) Calcitonin (CT), calcium, low. (b) CT inhibits osteoclasts, so they do not break down bone; it also inhibits PTH activity; osteoporosis. (c) Increase, bone, ingested food (into intestinal blood vessels), kidney tubules, negative, high. (d) Antagonists, phosphate, lowering.

18 (a) In-, contraction, tetany, hypo-. (b) Hypo-. (c) Demineralization.

19 (a) Neurohypophysis, water. (b) Mineralo-, cortex; Na^+, water, K^+, H^+, similar to. (c) Cl^- (chloride) or HCO^- (bicarbonate). (d) Renin-angiotensin, juxtaglomerular, renin, angiotensinogen, angiotensin I, angiotensin II. (e) Aldosterone, retention, increased, increases.

20 Much, deficient.

21 (a) Cata-, opposite. (b) Gluconeogenesis, increase. (c) Increases, increase, inhibit.

22 Gonadocorticoids or sex hormones; estrogens and androgens.

23 Epinephrine (80%) and norepinephrine (NE).

24 Pancreas. (a) Endo-; alpha, beta, delta. (b) Alpha, increases, glycogen, glycogenolysis, gluconeogenesis. (c) No, glucose, high. (d) Exercise (since lowers glucose level of blood) and meals rich in protein content, GHIF. (e) Opposite, helps, glycogen, glycogenesis. (f) Triggers, hyper, trigger.

25 (a) Lack of insulin prevents glucose from entering cells, so more is left in blood and spills into urine. (b) -Uria, -dipsia; glucose in urine draws water (osmotically), and as result of water loss, patient is thirsty. (c) Acid-; since cells have little glucose to use, they turn to excessive fat catabolism which leads to keto-acid formation (discussed in Chapter 25). (d) As fats are transported in extra amounts (for catabolism), some fats are deposited in walls of vessels; fatty (= *athero*) thickening (= *sclerosis*) results.

26 Ovaries, testes, adrenal cortex and placenta.

27 Gastrointestinal, placenta, red blood cell.

28 (a) Parathyroids, thymus, thyroid. (b) Adrenal cortex, gonads (ovaries and testes). (c) Anterior and posterior pituitary, pineal gland and adrenal medulla.

29 (a) P. (b) AP. (c) AM. (d) AC. (e) PT, T.

30 Pancreas (diabetes mellitus) and thyroid.

31 (a) Hypothalamus. (b) Alarm, medulla, sympathetic. (c) Resistance; ACTH, GH, TSH, cortex.

32 (a) Fight-or-flight. (b) Increases, norepinephrine, epinephrine. (c) Increases; increase; skeletal muscles, heart, or brain; spleen, skin. (d) Decrease, increases.

33 (a) GHRF. (b) LH. (c) GH (also testosterone). (d) TSH. (e) FSH. (f) PIF. (g) LH. (h) OT. (i) MSH. (j) ADH. (k) Adrenocorticoid (or glucocorticoid, such as cortisol). (l) PTH. (m) CT. (n) Thyroid hormone. (o) GH, glucagon, epinephrine, norepinephrine, adrenocorticoids (or ACTH indirectly). (p) Insulin. (q) Adrenocorticoid. (r) Epinephrine, norepinephrine. (s) Relaxin.

34 (a) PD. (b) Ac. (c) DI. (d) DM. (e) C. (f) M. (g) E. (h) S. (i) T. (j) Ad. (k) PC.

Questions 1–2: Arrange the answers in correct sequence.

_____ _____ _____ _____ 1. Steps in action of TSH upon a target cell:

A. Adenylate cyclase breaks down ATP.
B. TSH attaches to receptor site.
C. Cyclic AMP is produced.
D. Second messenger causes the specific effect mediated by TSH, namely, stimulation of thyroid hormone production.

_____ _____ _____ _____ 2. Steps in renin-angiotensin mechanism to increase blood pressure:

A. Angiotensin I is converted to angiotensin II.
B. Renin converts angiotensinogen, a plasma protein, to angiotensin I.
C. Angiotensin II stimulates aldosterone production which causes water conservation and raises blood pressure.
D. Low blood pressure causes secretion of enzyme renin from cells in kidneys.

Questions 3–9: Choose the one best answer to each question.

_____ 3. All of these compounds are synthesized in the hypothalamus EXCEPT:

A. ADH   B. GHRF   C. CT   D. PIF
E. Oxytocin

_____ 4. Choose the FALSE statement about endocrine glands.

A. They secrete chemicals called hormones.
B. Their secretions enter extracellular spaces and then pass into blood.
C. Sweat and sebaceous glands are endocrine glands.
D. Endocrine glands are ductless.

_____ 5. Choose the FALSE statement about prostaglandins.

A. They are secreted in large quantities in the body.
B. They are a group of lipids which are associated with membranes.
C. They appear to influence formation of cyclic AMP in cells.
D. They have a variety of functions.

_____ 6. All of the following hormones are secreted by the adenohypophysis EXCEPT:

A. ACTH   B. FSH   C. MSH
D. Oxytocin   E. GH

_____ 7. Which one of the following is a function of adrenocorticoid hormones?

        A. Lower blood pressure   B. Raise blood level of calcium
        C. Convert glucose to amino acids
        D. Lower blood level of sodium   E. Antiinflammatory

_____ 8. All of these hormones lead to increased blood glucose EXCEPT:

        A. ACTH   B. Insulin   C. Glucagon
        D. Growth hormone   E. Epinephrine

_____ 9. Choose the FALSE statement about the anterior pituitary.

        A. It secretes prolactin (PRL).
        B. It develops from mesoderm.
        C. It secretes tropic hormones.
        D. It is stimulated by releasing factors from the hypothalamus.
        E. It is also known as the adenohypophysis.

_____ 10. Choose the FALSE statement.

        A. Both ADH and aldosterone tend to lead to retention of water and increase in blood pressure.
        B. PTH activates vitamin D which enhances calcium absorption.
        C. PTH activates osteoclasts, leading to bone destruction.
        D. Positive Trousseau and Chvostek signs are likely to occur in hypocalcemia.
        E. ADH and OT pass from hypothalamus to posterior pituitary via pituitary portal veins.

Questions 11–20: Circle T (true) or F (false). If the statement is false, change the underlined word or phrase so that the statement is correct.

T   F   11. Polyphagia, polyuria, and hypoglycemia are all symptoms associated with insulin deficiency.

T   F   12. Regulating factors are all secreted by the anterior pituitary and affect the hypothalamus.

T   F   13. The hormones of the adrenal medulla mimic the action of parasympathetic nerves.

T   F   14. Excess PTH will cause bones to become demineralized.

T   F   15. Epinephrine and cortisol are both adrenocorticoids.

T   F   16. Secretion of growth hormone, insulin, and glucagon is controlled (directly or indirectly) by blood glucose level.

T   F   17. The general adaptation syndrome permits a person to maintain a constant internal environment under stressful conditions.

T   F   18. The alarm reaction occurs before the resistance reaction to stress.

T   F   19. Most hormones studied so far appear to act by a mechanism involving cyclic AMP rather than by the gene activation mechanism.

T   F   20. A high level of thyroxine circulating in the blood will tend to lead to a <u>low</u> level of TRF and TSH.

Questions 21–25: fill-ins. Complete each sentence with the word or phrase that best fits.

_____ 21. ____ is a hormone used to induce labor since it stimulates contractions of smooth muscle of the uterus.

_____ 22. ____ and ____ are two hormones that are classified chemically as amines.

_____ 23. Pro-opiomelanocortin is a molecule that can give rise to the chemicals named ____.

_____ 24. Type ____ diabetes mellitus is a non-insulin-dependent condition in which insulin level is adequate but cells have decreased sensitivity to insulin.

_____ 25. Three factors that stimulate release of ADH are ____.

UNIT IV
Maintenance of the Human Body

19

The Cardiovascular System: The Blood

In Unit IV you will learn about a number of systems which help the human body to maintain homeostasis on a day-to-day basis. In Chapter 19 you will begin the study of the cardiovascular system with an examination of the blood. You will contrast roles of blood with those of lymph and interstitial fluid (Objectives 1–2, 19). You will identify components of blood and describe their origins, structural features, functions, and counts (3–10). You will study the clotting process (11–14). You will contrast blood groups and relate these to antigen-antibody reactions (15–18). You will study some important blood disorders and the medical terminology associated with blood (20–23).

Topics Summary

A. Types of body fluids, characteristics and functions of blood
B. Components: types, origin
C. Formed elements: erythrocytes
D. Formed elements: leucocytes
E. Formed elements: thrombocytes
F. Plasma
G. Hemostasis
H. Grouping (typing)
I. Interstitial fluid and lymph
J. Disorders, medical terminology

Objectives

1. Contrast the general roles of blood, lymph, and interstitial fluid in maintaining homeostasis.

2. Define the principal physical characteristics of blood and its functions in the body.
3. Compare the origins of the formed elements in blood.
4. Discuss the structure of erythrocytes and their function in the transport of oxygen and carbon dioxide.
5. Explain the importance of a reticulocyte count and hematocrit.
6. List the structural features and types of leucocytes.
7. Explain the significance of a differential count.
8. Discuss the role of leucocytes in phagocytosis and antibody production.
9. Discuss the structure of thrombocytes and explain their role in blood clotting.
10. List the components of plasma and explain their importance.
11. Explain how the body attempts to prevent blood loss.
12. Identify the stages involved in blood clotting.
13. Explain the various factors that promote and inhibit blood clotting.
14. Define clotting time, bleeding time, and prothrombin time.
15. Explain ABO and Rh blood grouping.
16. Define the antigen-antibody reaction as the basis for ABO blood grouping.
17. Define the antigen-antibody reaction of the Rh blood grouping system.
18. Define erythroblastosis fetalis as a harmful antigen-antibody reaction.
19. Compare the location, composition, and function of interstitial fluid and lymph.
20. Contrast the causes of nutritional, pernicious, hemorrhagic, hemolytic, aplastic, and sickle cell anemia.
21. Define polycythemia and describe the importance of hematocrit in its diagnosis.
22. Identify the clinical symptoms of infectious mononucleosis (IM) and leukemia.
23. Define medical terminology associated with blood.

Learning Activities

(page 440) **A. Types of body fluids, characteristics and functions of blood**

1. Explain why specialized cells such as muscle or gland cells are dependent on the fluids of their internal environment for the maintenance of homeostasis.

2. Describe relationships among these three body fluids: blood, interstitial fluid, and lymph. (It may help to refer to Figure LG 1-2, page LG 11.)

3. Name the structures that comprise the:
   a. Cardiovascular system

   b. Lymphatic system

4. Describe these characteristics of blood. State normal values of each.
   a. Viscosity

   b. pH

   c. Temperature

5. Answer these questions about blood volume.

    a. Blood constitutes about _____ percent of body weight. Mr. Jacob is a healthy 150-lb man. His body contains about _____ lb of blood.

    b. Mr. Jacob's body contains about _____ liters of blood. One liter equals about _____ pints. So, when Mr. Jacob donates a pint of blood, he is giving about _____ percent of his blood.

6. Name six substances carried by blood.

7. List five functions of blood other than transport.

B. Components: types, origin

(pages 440–441)

1. Blood is a connective tissue that consists of about _____ percent intercellular material and about _____ percent cells or formed elements. The intercellular material is a liquid named _____ _____. List the types of cells and other formed elements in blood.

2. The process of blood formation is called _____.

3. Name the locations of blood formation in the fetus.

4. Contrast *myeloid tissue* and *lymphoid tissue.*

5. Explain why hemocytoblasts might be called ancestors of all blood cells.

(pages 442–445) **C. Formed elements: erythrocytes**

1. Erythrocytes are also called *(red? white?)* blood cells.
2. Draw a mature red blood cell. Explain the functional advantage offered by the shape of a red blood cell.

3. What accounts for the red color of erythrocytes?

2

4. Write ranges of normal hemoglobin values for each of the following.

   a. Adult male: _____ g/100 ml blood

   b. Adult female: _____ g/100 ml blood

   c. Infant: _____ g/100 ml blood

3

5. Describe functions of hemoglobin in this exercise.

   a. The hemoglobin molecule consists of a central portion, which is the protein *(heme? globin?)* with four side groups called *(hemes? globins?)*.

   b. Each heme contains one _____ atom on which a molecule of *(oxygen? carbon dioxide?)* can be transported. So each hemoglobin molecule can carry *(1? 4?)* oxygen molecule(s). Almost all oxygen is transported in this manner.

   c. Carbon dioxide has one combining site on hemoglobin; it is the amino acid portion of the *(heme? globin?)*, thereby forming _____

   _____-hemoglobin. About *(23? 70? 97?)* percent of carbon dioxide is carried in this state. Most carbon dioxide is transported *(in blood cells? in plasma?)* as the _____ (HCO_3^-) ion.

6. The average life of a red blood cell is about _____ days. Briefly explain reasons for death of erythrocytes. Describe the ultimate fates of hemosiderin, bilirubin, and globin.

7. Draw a square about 1 mm in size on this line: _____. In a cube the $\boxed{4}$ size of that square (1 mm³), about how many red blood cells are present in a healthy person? *(5? 500? 5,000? 500,000? 5 million?)*

8. *For extra review.* Do the following exercise. $\boxed{5}$

   a. A healthy female generally has a *(higher? lower?)* erythrocyte count than a healthy male. This is related to the *(higher? lower?)* metabolic rate of females.

   b. Draw a square 1 cm on a side. Calculate the number of red blood cells in a cubic volume that size (1 cubic centimeter = 1 cc = 1 ml):___

   _____/cc. Note that the body contains

   _____ ml of blood. The body does replace ery-

   throcytes at the rate of about _____/sec.

   Although this is a large number it is a fraction of perhaps

   _____ red blood cells in the entire body!

9. Arrange in order these stages in erythropoiesis. Write letters on the lines $\boxed{6}$ provided._____ _____ _____ _____

   A. Metarubricyte formed.

   B. Cell loses nucleus. _____

   C. Cell starts to synthesize hemoglobin. _____
   D. Cell moves out of marrow into general circulation.

   For extra review, identify the names of the cell at stages *B* and *C* above. Write cell names on long lines provided.

10. What percentage of circulating red blood cells are normally reticulocytes? *(0.1? 0.5? 1.5? 10?)* If the reticulocyte count is higher than 1.5 percent, then erythropoiseis is occurring at a *(slower? more rapid?)* rate than normal. List three reasons why the reticulocyte count might be elevated.

$\boxed{7}$

11. Explain the roles of the terms at right in red blood cell production by matching terms with descriptions.

   _____ a. Decrease of oxygen in cells; serves as a signal that erythropoiesis is needed (to help provide more oxygen to tissues)

   _____ b. Enzyme produced by kidneys when they are hypoxic; causes production of erythropoietin

   _____ c. Hormone that travels to marrow to stimulate red blood cell production

   _____ d. Component of the heme of hemoglobin

   _____ e. Vitamin necessary for normal erythropoiesis

   _____ f. Substance produced by the stomach lining and necessary for normal vitamin B_{12} absorption

   B_{12}. Vitamin B_{12}
   E. Erythropoietin
   Fe. Iron
   H. Hypoxia
   IF. Intrinsic factor
   REF. Renal erythropoietic factor

8   12. Do this exercise about hematocrits. Match numbers at right with descriptions below. Not all numbers will be used.

      _____ a. Hematocrit could be described as   25
              the number of mls of erythrocytes   42
              in _____ ml of blood.   47
      _____ b. Average hematocrit for females   65
      _____ c. Average hematocrit for males   100
      _____ d. Polycythemic hematocrit   1,000
      _____ e. Anemic hematocrit

(pages 445–447)   **D. Formed elements: leucocytes**

9   1. State three structural characteristics of leucocytes which distinguish them from erythrocytes.

10   2. State one general function of leucocytes.

Now write a short paragraph summarizing the process of inflammation. Include roles of different leucocytes, as well as the terms *chemotaxis* and *diapedesis.* For extra help refer to Chaper 22, Learning Activity 16–17, pages LG 488–490.

3. Contrast the five types of white blood cells (WBCs) by completing the table.

401

Type of WBC	Percent of Total WBCs	Granular or Agranular	Functions	Diagram of Cell Type
a.	60–70			
b.			Become plasma cells which produce antibodies	
c.				
d. Eosinophils				
e.			Become mast cells	

4. Check your understanding of types of leucocytes by matching names of leucocytes with descriptions. Answers may be used more than once. ｜11｜

_____ a. Constitute the largest percentage (about 70 percent) of leucocytes

_____ b. 20 to 25 percent of leucocytes

_____ c. Involved in immunity; may form plasma cells for antibody production

_____ d. Involved in allergic response; become mast cells that release serotonin, heparin and histamine

_____ e. Involved in allergic response; release antihistamines

_____ f. Form wandering macrophages that clean up sites of infection

_____ g. Important in phagocytosis (2 answers)

_____ h. Classified as granular leucocytes (3 answers)

B. Basophils
E. Eosinophils
L. Lymphocytes
M. Monocytes
N. Neutrophils

12 5. Complete this exercise about antigens and antibodies.

a. Antigens are defined as substances that _____. Most are composed of *(carbohydrate? lipid? protein?)*. Most *(are? are not?)* synthesized by the body. Several examples of antigens are:

b. Antigens stimulate _____ cells to become plasma cells which produce _____. Chemically, these are all globulin-type _____. They function to *(activate? inactivate?)* antigens.

c. Other lymphocytes are known as *(A? L? T?)*-lymphocytes. One group of T-cells are activated by antigens to kill these antigens either directly or with help of other lymphocytes and macrophages. Such T-cells are appropriately called _____ T-cells.

6. *A clinical challenge.* Explain why multiple myeloma, a cancer of the bone
13 marrow, can lead to the following signs and symptoms.

a. Osteoporosis and pain

b. Hypercalcemia and kidney damage

c. Anemia, leukopenia, and thrombocytopenia

7. State several examples of antigen-antibody responses.

8. Again, draw a square 1 mm on a side: _____ (1 mm³).
   Answer these questions about the number of cells in a cubic space that size
   within the human bloodstream.   14

   a. There are normally about _____ red blood cells in the
      blood within a space this size.

   b. There are normally about _____ white blood cells per
      cubic millimeter.

   c. The ratio of red blood cells to white blood cells is normally about

      _____ to 1.

9. *A clinical challenge.* Ms. Clifton arrives at a health clinic with a suspected
   acute infection. Complete this Learning Activity about her.   15

   a. During infection, it is likely that Ms. Clifton's leucocyte count will *(in?
      de?)* -crease. A count of *(4,000? 8,000? 15,000?)* leucocytes/mm³ blood is
      most likely. This condition is known as *(leucocytosis? leucopenia?).*

   b. In order to confirm changes in counts of specific types of leucocytes, a

      _____ count may be taken. This procedure will indi-
      cate the number of each type of *(RBC? WBC?)* in 100 ml of blood.

   c. An increase in the number of WBCs named _____ is
      most indicative of enhanced phagocytic activity during infection. Neutro-
      phils are most likely to account for *(48? 62? 76?)* percent of the total white
      count in Ms. Clifton's blood. A sign of a chronic infection is increase in
      the percent of cells that can become macrophages; these are

      _____.

10. What accounts for the short life span of white blood cells?

404

11. Discuss advances in bone marrow transplantation by answering these questions.

   a. If donor and recipient bone marrows are not identical, what is likely to happen?

   b. How may donor marrow be treated in order to reduce such undesirable immune responses?

(pages 447–448) **E. Formed elements: thrombocytes**

1. Give the following information about thrombocytes.

   a. State another name for thrombocytes: _____

   b. Tell how they form.

   c. Describe their structure and size.

   d. What is their average life span? _____ days

   e. How many are usually present in 1 mm³? _____ /mm³

2. What is the major function of platelets?

3. Now that you have studied all of the major formed elements of blood, review their development in Figure 19-2a (page 442) of the text. Then match each mature formed element listed at right with a cell in its developmental pathway.

17

_____ a. Megakaryocyte	B. Basophil
_____ b. Lymphoblast	E. Erythrocyte
_____ c. Monoblast	L. Lymphocyte
_____ d. Myeloblast (two answers)	M. Monocyte
_____ e. Rubriblast and reticulocyte	N. Neutrophil
	T. Thrombocyte

4. Describe components of a complete blood count (CBC).

F. Plasma

(pages 448–449)

1. Define *plasma*.

2. Match the names of components of plasma with their descriptions.

18

_____ a. Makes up about 92 percent of plasma	A. Albumin
	E. Electrolytes
_____ b. Regulatory substances carried in blood	F. Fibrinogen
	G. Globulin
_____ c. Cations and anions carried in plasma	GAF. Glucose, amino acids, and fats
_____ d. Constitutes about 55 percent of plasma protein	HE. Hormones and enzymes
_____ e. Made by liver; a protein used in clotting	N. Nonprotein nitrogen (NPN) substances
_____ f. Antibody protein	W. Water
_____ g. Wastes carried to kidneys or sweat glands	
_____ h. Food substances carried in blood	

3. Define *apheresis*.

_____ refers to temporary withdrawal of plasma with selective removal of unwanted components, such as certain antibodies.

(pages 449–453) **G. Hemostasis**

1. Define _hemostasis._

19 2. List the three basic mechanisms of hemostasis.
   a.

   b.

   c.

   Now describe how each of these responses to blood vessel injury helps maintain homeostasis. In the spaces above write a sentence or two about each of the three mechanisms.

3. Contrast _serum/plasma._

4. Three sources of coagulation factors, that is, substances that enhance the coagulation process, are _____, _____ _____, and _____. $\boxed{20}$

5. Clot formation (or coagulation) is *(the same as? one part of?)* the process of hemostasis. (*Hint:* Refer to Learning Activity $\boxed{19}$).

6. Describe the steps in the pathway of clot formation by filling in empty boxes in Figure LG 19-1. $\boxed{21}$

7. Point out major differences in stage I of extrinsic and intrinsic pathways by identifying what prompts formation of thromboplastin in each case. Match the pathway with its description. $\boxed{22}$

_____ a. Begins with damage of tissue cells external (extrinsic) to a blood vessel. Cells release chemicals which stimulate thromboplastin formation.

_____ b. Begins with adherence of platelets to rough edges of damaged blood vessel. Platelets release platelet coagulation factors (Pf_{1-4}).

E. Extrinsic
I. Intrinsic

8. Write plasma coagulation factor numbers next to each of the following. Note that these are listed in the sequence in which they are used in the three stages of clot formation. $\boxed{23}$

_____ a. Thromboplastin: factor _____

_____ b. Prothrombin: factor _____

_____ c. Fibrinogen: factor _____

_____ d. Calcium (used in all three stages): factor _____

A number of steps; these differ in intrinsic and extrinsic pathways $\Big\}$ $\xrightarrow{1}$ Thromboplastin

Prothrombin $\xrightarrow{2}$ a.

Fibrinogen $\xrightarrow{3}$ b.

Figure LG 19-1 Three stages of clot formation. Complete as directed in Learning Activity $\boxed{21}$.

9. State the roles of each of the following in clot formation.

[24]
    a. Vitamin K (Explain why persons with certain digestive disorders may bleed uncontrollably.)

    b. Antihemophilic factor VIII

    c. Christmas factor

    d. Pf_4

10. Although there are several forms of hemophilia, all involve deficiency of

[25]
    _____.

State three symptoms of hemophilia.

Briefly describe two treatments for hemophilia.

11. Describe these two events following clot formation.
    a. Clot retraction *(syneresis)*

    b. Fibrinolysis (Name two enzymes involved.)

12. Explain how certain chemicals may be helpful to persons with undesirable clot formation. Answer these questions. 26

   a. How do streptokinase and urokinase help?

   b. What is t-PA and how does it help?

   c. What advantages does t-PA offer over streptokinase and urokinase?

   d. How do these three chemicals prevent clot formation?

13. In what circumstances is clotting:
   a. Helpful in maintaining homeostasis

   b. Harmful

14. Explain how plaques lead to unwanted clot formation.

15. Contrast these two pairs of terms.
    a. Thrombus/thrombosis

    b. Embolus/embolism

27 | 16. Match anticoagulants with their mechanisms of action.

    _____ a. Reacts with calcium ions, so calcium is not available to convert prothrombin to thrombin; used to preserve blood in blood banks.

    _____ b. Also prevents formation of thrombin from prothrombin; fast-acting, so may be used in open heart surgery.

    _____ c. Acts as antagonist to vitamin K, so lowers prothrombin level; slower acting.

EDTA. EDTA
H. Heparin
W. Warfarin (coumadin)

17. Give average time for the following blood tests. Then write a brief description of the tests.
    a. Bleeding time: ____ min

    b. Clotting time: ____ min

    c. Prothrombin time: ____ sec

18. *A clinical challenge.* Mr. B. is a chronic alcoholic. His clotting mechanisms are altered as indicated by abnormal clotting time and prothrombin time. Write a rationale for those alterations. 28

19. *ACTive learning.* In a study group or lab, role play the clotting mechanisms. Each person can assume one or more roles and can carry a large paper sign indicating the role played, such as prothrombin or vitamin K. Go through the three main steps of clotting, and note how some chemicals (like calcium) appear in more than one step. Designate a part of the room as the liver and note which factors emerge from there. Some persons can play roles of chemicals (like heparin or warfarin), demonstrating which part of the clotting process is inhibited. Finally a person playing plasminogen can be activated (as by t-PA) to dissolve the string of fibrin-persons showing fibrinolysis.

H. Grouping (typing)

1. Contrast *agglutination* with *coagulation*. 29

    a. Coagulation (or _____) involves coagulation factors found in platelets, plasma, or other tissue fluids. The process *(requires? can occur in absence of?)* red blood cells.

    b. Agglutination (or "clumping") of erythrocytes is an antigen- _____ process that *(does? does not?)* require red blood cells since these are sites of _____ used in the agglutination reaction.

2. Do this exercise about factors responsible for blood groups. 30

    a. Agglutinogens are *(antigens? antibodies?)* located *(in plasma? on surface of red blood cells?)*.

    b. Type A blood has *(A? B? both? neither?)* antigen on RBCs.

    c. Type A has *(a? b? both? neither?)* antibody or _____

    _____. These are also called *anti-B* antibodies since they attack agglutinogen B.

    d. If a person with type A blood receives type B blood, the recipient's *(anti-A? anti-B?)* antibodies will attack the *(A? B?)* antigens on the donor's cells.

3. Complete the table contrasting ABO blood types.

Blood Group	Percent of White Population	Percent of Black Population	Sketch of Blood Showing Correct Antigens and Antibodies	Can Donate Safely to	Can Receive Blood Safely From
a. Type A					A, O
b.	10				
c. Type AB		7			
d.				A, B, AB, O	

31 4. Type O is known as the universal *(donor? recipient?)* with regard to the ABO group since type O blood lacks _____ of the ABO group. Type _____ is known as the universal recipient. Explain why.

5. Explain what problems can result from transfusion of an incompatible blood type.

6. State the significance of a recently developed method for converting type B blood into type O.

32

7 Complete this exercise about the Rh group.
   a. The Rh *(positive? negative?)* group is more common. Rh *(positive? negative?)* blood has Rh antigens on the surfaces of RBCs.
   b. Under normal circumstances plasma of *(Rh positive blood? Rh negative blood? both Rh groups? neither Rh group?)* contains anti-Rh antibodies.
   c. Rh *(positive? negative?)* persons can develop these antibodies when they are exposed to Rh *(positive? negative?)* blood.
   d. An example of this occurs in fetal-maternal incompatibility when a mother who is Rh *(positive? negative?)* has a baby who is Rh *(positive? negative?)* and some of the baby's blood enters the mother's bloodstream. The mother develops anti-Rh antibodies which may cross the placenta in future pregnancies and hemolyze the RBCs of Rh *(positive? negative?)* babies. Such a condition is known as _____.
8. Explain how administration of Rh antibodies to a mother just after the birth of an Rh positive baby can prevent Rh problems in future pregnancies.

I. Interstitial fluid and lymph

(page 456)

1. Contrast these two body fluids in the table.

33

	Interstitial Fluid and Lymph	Plasma
a. Amount of protein (more or less)		
b. Formed elements present		Red blood cells, white blood cells, platelets
c. Location	Between cells and in lymph vessels	

J. Disorders, medical terminology

1. Define *anemia.*

34 2. Match names of types of anemia with descriptions below.

_____ a. Condition resulting from inadequate diet, such as deficiency of
iron, vitamin B_{12}, or amino acids

_____ b. Condition in which intrinsic factor is not produced, so absorption
of vitamin B_{12} is inadequate

_____ c. Inherited condition in which hemoglobin forms stiff rodlike structures causing erythrocytes to assume sickle shape and rupture,
reducing oxygen supply to tissues

_____ d. Rupture of red blood cell membranes due to variety of causes,
such as parasites, toxins, or antibodies

_____ e. Condition due to excessive bleeding, as from wounds, gastric ulcers, heavy menstrual flow

_____ f. Inadequate erythropoiesis as a result of destruction or inhibition of
red bone marrow

A. Aplastic
Hl. Hemolytic
Hr. Hemorrhagic
N. Nutritional
P. Pernicious
S. Sickle cell

3. Persons with one gene for sickle cell anemia are said to have *(sickle cell trait?
sickle cell anemia?).* These individuals have *(greater? less?)* resistance to
malaria. Explain why.

35 4. *A clinical challenge.* Mary (19 years old) has a red blood count of 7 mil-
lion/mm^3. She has the condition known as *(anemia? polycythemia?).* Mary
has a hematocrit done. It is more likely to be *(under 32? over 45?).* Her blood
is *(more? less?)* viscous than normal, which is likely to cause *(high? low?)*
blood pressure.

36 5. Infectious mononucleosis is caused by the Epstein-Barr *(bacteria? virus?).* It
affects mostly *(teenagers? elderly?).* The name of this infection is based upon
the fact that B lymphocytes become abnormal in appearance, resem-

bling _____. A differential count shows a great increase

in _____-cytes. Permanent damage usually *(does? does
not?)* result.

6. Leukemia is a form of cancer involving abnormally high production of ⬜37⬜ *(erythrocytes? leucocytyes?).* Briefly explain why the following symptoms are likely to occur.
   a. Anemia

   b. Hemorrhage

   c. Infection

7. Define and explain clinical significance of each of these terms.
   a. Gamma globulin

   b. Thrombocytopenia

   c. Septicemia

416   d. Indirect (mediate) transfusion

Answers to Numbered Questions in Learning Activities

1. (a) 8, 12. (b) 5–6, 2, 8–10.

2. (a) 14–16.5. (b) 12–15. (c) 14–20.

3. (a) Globin, hemes. (b) Iron (Fe), oxygen, 4. (c) Globin, carbamino-, 23, in plasma, bicarbonate.

4. 5 million.

5. (a) Lower, lower. (b) 5 billion (= 5 million × 1,000 mm/cc), about 6,000, about 2 million, 30 trillion (5 billion/ml × 6,000 ml).

6. C (rubricyte), A, B (reticulocyte), D.

7. (a) H. (b) REF. (c) E. (d) Fe. (e) B_{12}. (f) IF.

8. (a) 100. (b) 42. (c) 47. (d) 65. (e) 25.

9. Leucocytes are colorless since they lack hemoglobin (*leuko* = white). They are nucleated. Most are larger than erythrocytes. Some contain granules.

10. Defense, as by phagocytosis and antibody production.

11. (a) N. (b) L. (c) L. (d) B. (e) E. (f) M. (g) M, N. (h) B, E, N.

12. (a) Stimulate production of antibodies; protein; are not, pollen, toxin released by bacteria, structural parts of microorganisms. (b) B-lymphocyte, antibodies, proteins, inactivate. (c) T, killer.

13. (a) Cancer cells invade bone and cause its destruction; bone cancer is especially painful since bone is rigid and cannot expand, so tumors exert pressure from within bone, compressing nerves. (b) Breakdown of bone releases calcium; extra calcium in blood may deposit in soft tissues like kidneys, forming stones, blocking urine pathways, and leading to kidney failure. (c) Cancer cells use up marrow nutrients and prevent normal production of RBCs, WBCs, and platelets. (d) Decrease in WBCs limits antibody production; anemia can also alter health of tissues, predisposing them to infection.

14. (a) 5 million. (b) 5,000–9,000. (c) 700 (4,900,000 RBC/7,000 WBC).

15. (a) In, 15,000, leucocytosis. (b) Differential, WBC. (c) Neutrophils (or polymorphs or "polys"), 76, monocytes.

16. (a) Donor killer T-cells may attack recipient tissues, such as skin and gastrointestinal organs. (b) Before marrow is introduced into the recipient, donor killer T-cells can be removed by effects of chemicals like lectin or by monoclonal antibodies.

17. (a) T. (b) L. (c) M. (d) B, N. (e) E.

18 (a) W. (b) HE. (c) E. (d) A. (e) F. (f) G. (g) N. (h) GAF. **417**

19 (a) Vascular spasm (vasoconstriction of vessel wall). (b) Platelet plug formation. (c) Coagulation (clotting or clot formation).

20 Plasma, platelets, damaged tissues.

21 (a) Thrombin. (b) Fibrin.

22 (a) Extrinsic. (b) Intrinsic.

23 (a) III. (b) II. (c) I. (d) IV.

24 Vitamin K is necessary for synthesis of prothrombin. If fat absorption is reduced, this fat-soluble vitamin will not be absorbed properly.

25 Some clotting factor.

26 (a) They are enzymes that dissolve blood clots. (b) Tissue plasminogen activator causes conversion of the inactive enzyme plasminogen to the active form, plasmin. This dissolves fibrin of the clot. (c) t-PA is specific for clots, and so it can be administered directly into a vein and travel through blood to the clot site; urokinase and streptokinase must be sent by catheter directly to the clot site or else they would degrade other tissues en route. (d) They do not. They dissolve clots that have already formed.

27 (a) EDTA. (b) H. (c) W.

28 Chronic alcoholism results in destruction of normal liver tissue. Plasma proteins, normally made in liver, include factor II (prothrombin) which is necessary for normal prothrombin time. Factors VIII and IX (also made in liver) are required for normal clotting time. Other plasma coagulation factors made in liver include I (fibrinogen), V, VII, X, and XI.

29 (a) Clotting, can occur in absence of. (b) Antibody, does, antigens.

30 (a) Antigens, on surface of red blood cells. (b) A. (c) b, agglutinin (or isoantibody). (d) Anti-B (or b), B.

31 Agglutinogens or isoantigens, AB, their plasma lacks both agglutinin a (anti-A) and b (anti-B).

32 (a) Positive, positive. (b) Neither Rh group. (c) Negative, positive. (d) Negative, positive, positive, erythroblastosis fetalis.

33 (a) Less, more. (b) Contain some leucocytes, but no red blood cells or platelets, -. (c) -, within blood vessels.

34 (a) N. (b) P. (c) S. (d) Hl. (e) Hr. (f) A.

35 Polycythemia, over 45, more, high.

36 Virus, teenagers, monocytes, lympho, does not.

37 Abnormal bone marrow (with focus on production of only immature leucocytes) prevents production of (a) red blood cells, (b) platelets, and (c) mature leucocytes.

Questions 1-2: Arrange the answers in correct sequence.

_____ _____ _____ 1. Events in the coagulation process:

        A. Retraction or tightening of fibrin clot

        B. Fibrinolysis or clot dissolution by plasma

        C. Clot formation

_____ _____ _____ 2. Stages in the clotting process:

        A. Formation of thromboplastin

        B. Conversion of prothrombin to thrombin

        C. Conversion of fibrinogen to fibrin

Questions 3-11: Choose the one best answer to each question.

_____ 3. The normal red blood cell count in healthy males is _____ RBCs/mm^3.

    A. 5.4 million   B. 2 million   C. 0.5 million

    D. 250,000   E. 8,000

_____ 4. All of the following types of formed elements are produced in bone marrow EXCEPT:

    A. Neutrophils   B. Basophils   C. Platelets

    D. Erythrocytes   E. Lymphocytes

_____ 5. Normal prothrombin time is about:

    A. 12 sec   B. 1 to 4 min   C. 10 to 15 min

    D. 3 hr

_____ 6. Megakaryocytes are involved in formation of:

    A. Red blood cells   B. Basophils   C. Lymphocytes

    D. Platelets   E. Neutrophils

_____ 7. Choose the FALSE statement about Joyce, who has type O blood.

    A. Joyce has neither A nor B agglutinogens on her red blood cells.

    B. She has both anti-A and anti-B agglutinins in her plasma.

    C. She is called the universal donor.

    D. She can receive blood safely from both type O and type AB persons.

_____ 8. Choose the FALSE statement about blood.

    A. Blood is thicker than water.

    B. It normally has a pH of 7.0.

    C. The human body normally contains about 5 to 6 liters of it.

    D. It normally consists of more plasma than cells.

_____ 9. Choose the FALSE statement about factors related to erythropoiesis.

    A. Erythropoietin stimulates RBC formation.

    B. Oxygen deficiency in tissues serves as a stimulus for erythropoiesis.

    C. Intrinsic factor, which is necessary for RBC formation, is produced in the kidneys.

    D. A reticulocyte count of over 1.5 percent of circulating RBCs indicates that erythropoiesis is occurring rapidly.

_____ 10. Choose the FALSE statement about plasma.

    A. It is red in color.

    B. It is composed mainly of water.

    C. Its concentration of protein is almost four times that of interstitial fluid.

    D. It contains plasma proteins, primarily albumin.

_____ 11. Choose the FALSE statement about neutrophils.

    A. They are actively phagocytic.

    B. They are the most abundant type of leucocyte.

    C. Neutrophil count decreases during most infections.

    D. An increase in their number would be a form of leucocytosis.

Questions 12-20: Circle T (true) or F (false). If the statement is false, change the underlined word or phrase so that the statement is correct.

T F 12. As cells become more specialized, they become <u>more</u> dependent on other cells and fluids in the body.

T F 13. Plasmin is an enzyme that facilitates clot <u>formation.</u>

T F 14. <u>All types of white blood cells, as well as red blood cells and platelets,</u> originate from hemocytoblasts.

T F 15. Heparin serves as an anticoagulant by interfering with <u>stage 1</u> of clot formation.

T F 16. Erythroblastosis fetalis is most likely to occur with an Rh <u>positive mother and her Rh negative babies.</u>

T F 17. Thromboplastin is released from tissues in <u>intrinsic</u> clotting pathways, but thromboplastin formation depends on platelet coagulation factors in <u>extrinsic</u> pathways.

T F 18. Interstitial fluid has a <u>higher</u> concentration of protein than plasma has.

T F 19. In both black and white populations in the U.S., type O is <u>most</u> common and type AB is <u>least</u> common of the ABO blood groups.

T F 20. In 100 ml of blood there are usually about <u>15 ml</u> of formed elements, most of which are red blood cells.

Questions 21–25: fill-ins. Answer questions or complete sentences with the word or phrase that best fits.

_____ 21. Name five types of substances transported by blood.

_____ 22. Name three types of plasma proteins.

_____ 23. Name the type of leucocyte that is transformed into plasma cells which then produce anti-bodies.

_____ 24. Write a value for a normal leucocyte count: ____/mm^3.

_____ 25. All blood cells and platelets are derived from ancestor cells called ____.

The Cardiovascular System: The Heart

You have learned about blood, the principal circulating fluid in the body. Now you will consider the organ that pumps the blood through virtually all regions of the body; this is the heart. You will examine the position, coverings, and structure of the heart (Objectives 1–5). You will look at the blood supply of the heart (6), and while considering its electrical conduction system, you will learn about the electrocardiogram (ECG) (7–8). You will study the cardiac cycle (9–11) and examine factors that control heart rate and cardiac output (12–16). You will explore the problems of circulatory shock (17) and some advances in cardiology that may help cardiac patients (18–20). You will learn about heart development (21). And finally you will consider a number of disorders associated with the heart (22–27).

Topics Summary

A. Location, size, shape
B. Pericardium, walls and chambers, vessels, valves
C. Blood supply
D. Conduction system, electrocardiogram
E. Cardiac cycle
F. Cardiac output, regulation of heart rate
G. Circulatory shock and homeostasis
H. Advances in cardiology
I. Developmental anatomy
J. Disorders, medical terminology

Objectives

1. Describe the location of the heart in the mediastinum and identify the borders of the heart.
2. Describe the structure of the pericardium.
3. Contrast the structure and location of the epicardium, myocardium, and endocardium of the heart wall.
4. Identify and describe the chambers, great vessels, and valves of the heart.
5. Describe the surface anatomy features of the heart.
6. Discuss the route of blood in coronary (cardiac) circulation.
7. Explain the structural and functional features of the conduction system of the heart.
8. Describe an electrocardiogram (ECG) and explain its significance.
9. Explain the pressure changes associated with blood flow through the heart.
10. Describe the principal events of a cardiac cycle.
11. Contrast the sounds of the heart and their clinical significance.
12. Define cardiac output (CO) and explain what determines it.
13. Define Starling's law of the heart.
14. Contrast the effects of sympathetic and parasympathetic stimulation of the heart.
15. Define the role of baroreceptors in reflex pathways in controlling heart rate.
16. Explain how chemicals, temperature, emotions, sex, and age affect heart rate.
17. Define circulatory shock and explain the homeostatic mechanisms that compensate for it.
18. Describe the structure and function of an artificial heart.
19. Explain how heart disorders are diagnosed via cardiac catheterization and echocardiograph.
20. Describe the operation of a heart–lung machine.
21. Describe the development of the heart.
22. List the risk factors involved in heart disease.
23. Describe how atherosclerosis and coronary artery spasm contribute to coronary artery disease (CAD).
24. Contrast coarctation of the aorta, patent ductus arteriosus, septal defects, valvular stenosis, and tetralogy of Fallot as congenital heart defects.
25. Define atrioventricular (AV) block, atrial flutter, atrial fibrillation, and ventricular fibrillation as abnormalities of the conduction system of the heart (arrhythmias).
26. Define congestive heart failure (CHF) and cor pulmonale (CP).
27. Define medical terminology associated with the heart.

Learning Activities

(pages 462–463) **A. Location, size, shape**

1. 1. What is the main function of the heart?

2. 2. Closely examine Figure 20-1, page 462, in your text. Consider the location, size, and shape of your heart as you do this exercise. Trace its outline on your body.

    a. Your heart lies in the _____ portion of your thorax, between your two _____. About *(one-third? one-half? two-thirds?)* of the mass of your heart lies to the left of the midline of your body.

    b. Your heart is about the size and shape of your _____.

    c. The pointed part of your heart, called the _____, lies in the _____ intercostal space, about _____ cm (_____ inches) from the midline of your body.

    d. The base of your heart is located just inferior to your _____ _____ ribs.

    e. The regions of the heart which lie directly on top of the diaphragm are the right and left *(atria? ventricles?)*.

3. 3. What structures protect the heart?
    a. Anteriorly

    b. Posteriorly

    c. Inferiorly

(pages 463–468) **B. Pericardium, walls and chambers, vessels, valves**

1. List in order (from most superficial to deepest) the coverings over the heart. Briefly describe the structure and functions of each.

For extra review of the pericardium, look again at Chapter 4, Learning Activity 20 , page LG 77.

2. What is the function of *pericardial fluid?*

3. Inflammation of the pericardium is known as _____ 4

_____. This condition may lead to cardiac tamponade

which means _____.

4. Contrast according to structure and location:
   a. Epicardium

   b. Myocardium

   c. Endocardium

5. Refer to Figure LG 20-1 as you do this exercise about heart structure. 5

   a. The heart is divided into _____ spaces or chambers.

   b. The superior ones are known as _____. Each of these
   has an appendage, or _____. Most of the lining
   of the atria is *(smooth? ridged?).* The tissue separating the two atria
   is named the _____.

   c. The two inferior chambers of the heart are the _____.
   Their walls are *(thicker? thinner?)* than those of the atria. The two
   ventricles are separated by the _____

   _____. Externally, a groove, known as the
   _____, marks the location of the septum. (See Figure
   20-3, page 464, in the text.)

   d. The atrial myocardium is separated from that of the ventricles by the
   "cardiac skeleton" composed of _____, which also

   forms _____ there.

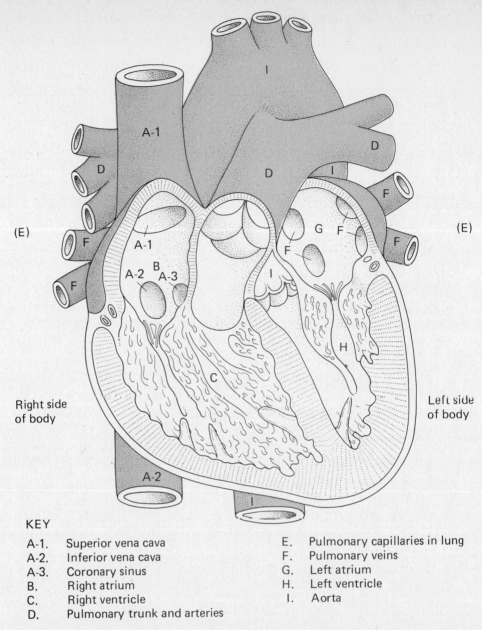

KEY

A-1. Superior vena cava
A-2. Inferior vena cava
A-3. Coronary sinus
B. Right atrium
C. Right ventricle
D. Pulmonary trunk and arteries

E. Pulmonary capillaries in lung
F. Pulmonary veins
G. Left atrium
H. Left ventricle
I. Aorta

Figure LG 20-1 Structure of the heart, anterior internal view. Letters follow the path of blood through the heart. Complete as directed.

e. Blood from all parts of the body except the lungs flows into the chamber named the _____. Blood from superior body parts enters the heart via the vein called the _____. The inferior vena cava returns blood from _____ _____. Blood from vessels supplying heart tissue returns to the right atrium via the vessel named the _____ _____. Identify openings from each of these vessels in the right atrium on Figure LG 20-1.

f. Blood in the right atrium passes into the chamber named the
_____. From here it is squeezed into a vessel named
the pulmonary _____ which leads to the pulmonary
(arteries? vein?) and smaller vessels in the _____.

g. Oxygenated blood returns to the heart by means of the pulmonary
_____. There are _____ of these, two from
each lung. They empty into the _____ of the heart.

h. Blood in the left atrium flows into the _____. This
chamber is the *(thickest? thinnest?)* chamber in the heart since it must
squeeze blood into the _____, which gives branches
leading to all body parts (except the lungs).

6. On Figure LG 20-1:
  a. Draw arrows indicating direction of blood flow.
  b. Color red the areas of the heart containing highly oxygenated blood; color
     blue the regions in which blood is low in oxygen.
  c. Label the four valves that control blood flow through the heart.

7. State the general function of valves. Describe the structure of valves.

8. Check your understanding of the valves by doing this matching exercise.
   More than one answer may be required for each description.

| 6 |

_____ a. Also called the mitral valve
_____ b. Prevents backflow of blood from
        right ventricle to right atrium
_____ c. Prevents backflow from pulmo-
        nary trunk to right ventricle
_____ d. Prevents backflow of blood into
        left atrium
_____ e. Have half-moon-shaped leaflets or
        cusps (2 answers)
_____ f. Also called atrioventricular (AV)
        valves (2 answers)

A. Aortic semilunar
B. Bicuspid
P. Pulmonary semilunar
T. Tricuspid

9. Describe each of the following according to structure and function. Also
   label them on Figure LG 20-1.
  a. Chordae tendineae

  b. Papillary muscles

426

|7|

The chordae tendineae and papillary muscles help to control the _____ valves.

10. Using Figure 20-5 (page 468) of your text, locate the sites on the surface of your own body where heart sounds related to each valve are best heard. Remember that these sounds are created by turbulence of blood when valves close; so these sites *(do? do not?)* necessarily lie directly over the valve itself.

(pages 468–469) **C. Blood supply**

1. Defend or dispute this statement: "The myocardium receives all of the oxygen and nutrients it needs from blood which is passing through its four chambers."

|8|

2. The arteries that supply heart tissue are the _____ arteries. They are branches from the *(ascending? arch? descending?)* portion of the aorta.

|9|

3. Match the coronary artery branches with descriptions given. More than one answer may be required.

_____ a. Supplies the left ventricle

_____ b. Supplies both the left atrium and left ventricle

_____ c. Major branch of the right coronary artery

_____ d. Located on the anterior surface of the heart in a groove between the ventricles

A. Anterior interventricular
C. Circumflex
M. Marginal
P. Posterior interventricular

4. The coronary sinus functions as *(an artery? a vein?)*. It collects blood that has passed through coronary arteries and capillaries into _____ veins. The coronary sinus finally empties this blood into the *(right? left?) (atrium? ventricle?)*.

|10|

5. Define *anastomosis* and explain how anastomoses of coronary blood vessels help the heart.

6. Define *ischemia* and explain how it is related to *angina pectoris*.

   List several common causes of angina.

7. A "heart attack" is the common name for a myocardial _____
   _____(MI). Answer the following questions about this
   problem.
   a. What does the term *infarction* mean?

   b. How might a thrombus or an embolus lead to an MI?

   c. Measurement of CPK level of blood is one test used to diagnose heart
      attacks. Write abbreviations for two other enzyme tests: _____
      _____ and _____. Explain why
      elevation of these enzyme levels may indicate that a myocardial infarction
      has occurred.

D. Conduction system, electrocardiogram   (pages 469–472)

1. Summarize the conduction system of the heart by doing this learning activity.   11
   a. The *(autonomic? somatic?)* nervous system innervates the heart. These
      nerves function to *(initiate? alter rate of?)* contraction of the heart. In
      other words, the heart *(can? cannot?)* contract in the absence of ANS
      stimulation. (More about ANS control in Chapter 21.)

   b. The heart's intrinsic regulating system is known as its _____
      _____ system. Cells forming this system are devel-
      oped embryologically from cardiac muscle cells; however, their specialty
      is *(contraction? impulse transmission?)*.

c. The first structure in the conduction pathway is the _____ _____ node (_____ node). It is also called the _____ since its rate of firing impulses alters the heart rate. State two factors that affect the rate of impulses of the pacemaker.

d. The SA node is located in the wall of the *(right? left?)* atrium. Its impulses spread over *(atria? ventricles?)* and cause them to contract.

e. What structural feature of the heart makes the AV node and the AV bundle necessary for conduction to the entire heart?

f. Arrange the remaining conductive structures of the heart in order of the transmission of impulses to the ventricular myocardium. ____ ____ ____

A. AV bundle
B. Conduction myofibers (Purkinje fibers)
C. Bundle branches

2. Contrast *electrocardiogram (ECG* or *EKG)* with *electrocardiograph.*

3. Draw a typical ECG recording, and label the following parts:

P. P wave   S-T. S-T segment
P-R. P-R interval   T. T wave
QRS. QRS wave (complex)

12 4. Now match parts of the ECG recording listed above with descriptions below.

_____ a. Deflection wave recorded during atrial depolarization, leading to contraction of atria

_____ b. Normally less than 0.2 sec

_____ c. Represents ventricular depolarization; masks the atrial repolarization that occurs during this time

_____ d. Time from the end of ventricular depolarization to the start of ventricular repolarization

_____ e. Associated with ventricular repolarization

5. *For extra review.* Explain why each of the following disorders would alter the ECG as indicated.

    a. Enlarged P wave as result of mitral stenosis

    b. Elongated P-R interval associated with atherosclerotic heart disease and rheumatic fever

    c. Enlarged R wave if ventricles are enlarged

6. Explain why a pacemaker may be used.

    Describe these parts of the pacemaker: *pulse generator, lead, electrode.*

E. Cardiac cycle

(pages 472-474)

1. Answer these questions about the movement of blood through the heart. $\boxed{13}$

    a. The movement of blood is controlled by two phenomena: _____ of the cardiac muscle and _____ _____ of the valves.

    b. The phase of contraction is called the _____; the phase of relaxation is the _____. The myocardium

contracts in response to stimulation by the _____
system.

c. Opening and closing of valves is controlled by _____

_____ in each heart chamber.

14 2. Refer to Figure LG 20-2 and summarize major aspects of the cardiac cycle in this learning activity.

a. The duration of one average cycle is ____ sec; *(54? 60? 75?)* complete cardiac cycles (or heart beats) occur per minute (if pulse rate is 75).

b. Label these parts of the ECG on Figure LG 20-2c: *P wave, QRS wave, T wave.*

c. As the P wave of the ECG occurs, nerve impulses spread across the

_____. During this initial *(0.1? 0.3?)* sec of the cycle, the atria begin their contraction; they are said to be in *(systole? diastole?).* As a result, pressure within the atria *(increases? decreases?),* as shown at point A in Figure LG 20-2b.

d. The _____ wave then signals contraction (or

_____) of the ventricles. Note that although the QRS wave itself happens over a very brief period, the ventricular contraction that follows occurs over a period of *(0.3? 0.5? 0.7?)* sec.

e. Trace a pencil lightly along the curve that shows changes in pressure within the ventricles following the QRS wave. Notice that pressure there *(increases? decreases?) (slightly? dramatically?).* In fact it quickly sur-

passes intraatrial pressure (at point _____ in the figure). As a result, AV valves are forced *(open? closed?).*

f. A brief time later ventricular pressure becomes so great that it even surpasses pressure in the great arteries (pulmonary artery and aorta). This

pressure forces blood against the undersurface of the _____

_____ valves, *(opening? closing?)* them. This occurs

at point _____ on the figure.

g. Continued ventricular systole ejects blood from the heart into the great vessels. In the aorta this creates a systolic blood pressure of about *(15? 80? 120?)* mm Hg.

h. What causes ventricular pressure to begin to drop? The cessation of

_____-polarization (and contraction) followed by repolarization

of these chambers (after the _____ wave) causes the ventricles to go into *(systole? diastole?).*

i. When the pressure within ventricles drops just below that in the great

arteries, blood in these vessels fills the _____ valves

and closes them. This occurs at point _____ in the figure. Ventricular pressure now drops at a rapid rate.

j. When intraventricular pressure becomes lower than that in atria, the force of blood within the atria causes the AV valves to *(open? close?).* This

happens at point _____ in the figure.

k. Note that the AV valves now remain open all the way to point *B* in the next cycle. This permits adequate filling of ventricles before their next contraction. Trace a pencil lightly along the curve on Figure LG 20-2e

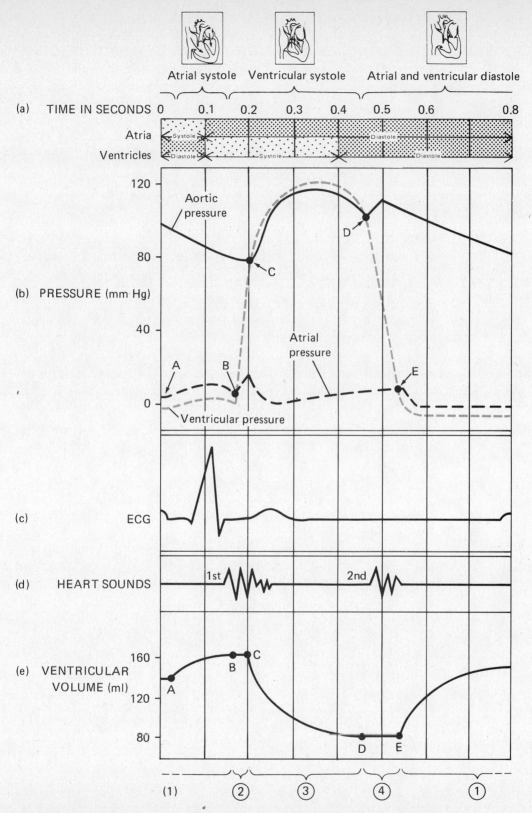

Figure LG 20-2 Cardiac cycle. (a) Systole and diastole of atria and ventricles related to time. (b) Pressures in aorta, atria, and ventricles. Points labeled *A* to *E* refer to Learning Activity 14 . (c) ECG related to the cardiac cycle. (d) Heart sounds related to the cardiac cycle. (e) Volume of blood in ventricles. Points *A* to *E* refer to Learning Activity 14 .

starting at point *E*. Note that the curve is steepest *(just after E? just before A of the next cycle?)*. In other words, the first part of the curve from *E* to *A* of the next cycle (ventricular diastole) is the time when ventricles are filling *(rapidly? slowly?)*, while the latter part is the period of *(rapid filling? diastasis?)*. In fact, _____ percent of all of the blood that will enter ventricles does so during the first two thirds of ventricular diastole (period *E–A*).

l. The remaining _____ percent enters ventricles between points *A* and *B,* that is, as a result of atrial *(systole? diastole?)*.

m. Summarize changes in ventricular volume by labeling regions 1–4 below Figure LG 20-2e. In part 1, the ventricles are relaxed and they are in a state of *(filling? ejection?)*. In part 2, ventricles are contracting but the semilunar valves have not yet opened. So blood cannot leave the ventricles yet. Thus this is the brief period named isovolumetric *(contraction? relaxation?)*. Once the semilunar valves open, the ventricular volume _____-creases. This period (3) is ventricular _____. In the short period 4, the ventricles have stopped contracting so pressure is rapidly dropping, but AV valves have not yet opened. So ventricles cannot fill yet. This is the period of _____

_____. Note that *iso-* means *same*.

n. The first heart sound *(lubb? dup?)* is created by turbulence of blood at the *(opening? closing?)* of the _____ valves. What causes the second sound?

Write *first* and *second* next to the parts of Figure LG 20-2d showing each of these heart sounds.

3. *For extra review,* label points *B, C, D,* and *E* to indicate which valves open or close at those points.

| 15 |

4. Explain what causes heart murmurs.

5. Define *auscultation*.

1. Define *cardiac output.* Name the two main factors that determine cardiac output.

2. Determine the average cardiac output in a resting adult.
   Cardiac output = stroke volume × heart rate

   $$= \underline{\hspace{2cm}} \text{ ml/stroke} \times \underline{\hspace{2cm}} \text{ strokes/min}$$

   $$= \underline{\hspace{2cm}} \text{ ml/min} (\underline{\hspace{2cm}} \text{ liter/min})$$

3. Do this exercise about volumes of blood in the heart at different times during the cardiac cycle. |16|
   a. During diastole ventricles *(eject? fill with?)* blood. At the end of their diastole the ventricular volume is about _____ ml of blood. This is known as the EDV, or _____ volume.

   b. As ventricles contract, they eject about _____ ml, known as their _____ volume (SV).

   c. So at the end of systole ventricles contain _____ ml, called their _____ volume (or _____).

   d. In summary, which statement is true? (Circle one.)
      A. ESV − SV = EDV
      B. ESV − EDV = SV
      C. EDV − SV = ESV

   e. *A clinical challenge.* Mrs. G. has an EDV of 120. But a myocardial infarction (MI) has weakened her ventricles so that they can eject a stroke volume (SV) of only 40 ml. Calculate Mrs. G's ESV.

   f. Another *clinical challenge.* Gene has tachycardia (rapid heartbeat, such as 140 beats/min). Consequently, each of his cardiac cycles is *(longer? shorter?)*, especially between points *E* and the next *A* on Figure LG 20-2. As a result, EDV will be *(higher? lower?)* than normal, leading to a lowered stroke volume also.

   g. Another factor that alters stroke volume is arterial pressure. High blood pressure creates a *(greater? smaller?)* resistance against which the ventricles must contract. So high blood pressure tends to decrease effectiveness of ventricular systole and ___-crease stroke volume.

4. Describe factors affecting stroke volume by doing this exercise. |17|
   a. During exercise skeletal muscles surrounding blood vessels squeeze *(more? less?)* blood back toward the heart. So venous return *(increases? decreases?)*, causing heart muscle fibers to stretch *(more? less?)*.

   b. A stretched muscle contracts with *(greater? less?)* force than a muscle which is only slightly stretched. This is a statement of _____ law. To get an idea of this, blow up a balloon slightly and let it go. Then blow it up quite full of air (much like the heart when stretched out by large venous return) and now let the balloon go.

In which stage do the walls of the balloon/heart compress the air/blood with greater force?

    c. A heart that is overstretched too often is much like a balloon that has been greatly expanded 5,000 times. Now the heart fibers are beyond the optimal length according to Starling's law. So now the force of contraction of the balloon/heart is ____-creased. This heart is now capable of a(n) ____-creased stroke volume. State two reasons why blood backs up in heart chambers, causing such overdilation of the heart.

    d. As a result of Starling's law, during exercise the ventricles of the normal heart contract *(more? less?)* forcefully, and stroke volume *(increases? decreases?).*

    e. State several reasons why venous return might be decreased, leading to reduced cardiac output.

18 5. At rest Dave has a cardiac output of 5 liters per minute. During a strenuous cross country run, Dave's maximal cardiac output is 25 liters per minute.

In other words, his heart has the reserve to pump out ____ liters per minute more during maximal exercise than at rest. Write your calculations here:

Cardiac reserve = maximal CO − resting CO

Dave's cardiac reserve is ____ percent of his resting cardiac output.

Note that cardiac reserve is usually *(higher? lower?)* in trained athletes than in sedentary persons.

6. Complete this exercise describing control of heart rate and blood pressure (BP). ⑲

a. Recall the two factors that determine cardiac output: _____ and _____. If one of these is altered from normal, the other one may compensate. For example, if SV is decreased (as by an MI), then the heart rate will normally ____-crease in an attempt to maintain adequate cardiac output.

b. Cardiac output is directly related to blood pressure (BP). This means that when the heart increases its output of blood, more blood is present in vessels, exerting *(increased? decreased?)* BP. Cardiac output must be kept at an appropriate level, however, or else BP will become too low or too high.

c. Cardiac and blood pressure control centers are located in the _____ of the brain. Sensory nerves must constantly inform the medulla of the need for adjustments. Cranial nerves IX and X carry these messages from receptors sensitive to BP. These baroreceptors are located in the walls of the _____ and the _____ .

d. When baroreceptors indicate that BP is increased slightly, BP must then be *(decreased slightly? increased even more?)* in order to maintain homeostasis. This can be accomplished by ____-creasing heart rate.

e. Consider the cardiac center as comparable to the control center of an automobile. The center has an accelerator portion (like a gas pedal) and an inhibitor portion (like the _____ of a car).

f. When BP is reported to be too high (as if the car is running too fast), the cardioacceleratory center must be *(stimulated? inhibited?)*; in the car analogy, you would *(step on the gas? take your foot off the gas?)*. At the same time the cardioinhibitory center (brakes) would be *(stimulated? inhibited?)*.

g. Dual control is carried out by autonomic nerve fibers. When BP is too high, sympathetic (accelerator) nerve messages would *(increase? decrease?)*, while parasympathetic nerve impulses along the _____ nerve to the heart would *(increase? decrease?)*.

h. As a result of these nerve messages heart rate would *(increase? decrease?)* and thereby *(raise? lower?)* BP in response to each occasion of slightly increased BP. This is a statement of _____ law. The mechanism is an example of *(positive? negative?)* feedback.

7. The Bainbridge reflex works by a *(positive? negative?)* feedback mechanism. ⑳ Describe this reflex.

21

8. Draw arrows to indicate whether each of the following factors increases (↑) or decreases (↓) heart rate.

   a. Epinephrine _____ heart rate.

   b. Increased levels of potassium (K⁺) _____ heart rate.

   c. Increased temperature, such as in fever, _____ heart rate.

   d. Fear and anger _____ heart rate.

(pages 477–478) **G. Circulatory shock and homeostasis**

1. *Shock* is said to occur when tissues receive inadequate _____ _____ supply to meet their demands for oxygen, nutrients, and waste removal. When the cause of shock is reduction of either cardiac output or blood volume, then this is said to be _____ _____ *shock.*

22

2. Complete the following exercise about compensatory mechanisms during mild circulatory shock.

   a. Blood vessels in nonessential areas are constricted so that blood is shunted into major vessels and BP is maintained. This occurs by means of powerful vasoconstricters such as _____, released from the adrenal medulla, and angiotensin II, formed when _____ is secreted from the kidneys. Angiotensin II also stimulates release of the hormone _____.

   b. Both aldosterone and _____ are hormones which cause water *(retention? elimination?).* Increase in fluid level of the body will *(increase? decrease?)* blood volume, and so will increase BP.

3. List six symptoms of circulatory shock.

23

4. *A clinical challenge.* Explain which compensatory mechanisms account for these symptoms of shock:

   a. Rapid pulse

b. Cool, pale, clammy skin

H. Advances in cardiology (pages 478-479)

1. Implantation of the first permanent artificial heart in a human took place in December, 19____. In this surgery patient Dr. Barney Clark's own atria were *(removed? left in place?)*. Briefly describe the structure of the implanted heart and its power source.

2. Briefly describe the procedure of cardiac catheterization.

3. In each case indicate whether a catheter should be placed into an artery (A), such as the femoral or brachial artery, or into a vein (V), such as the femoral or brachial vein.

   24

   _____ a. To record pressure in the right ventricle

   _____ b. To record pressure in the left ventricle

   _____ c. To determine oxygen content in a pulmonary artery

   _____ d. To insert dye into coronary arteries in order to diagnose narrowing of these vessels

4. Discuss *echocardiography*. Tell how the procedure is performed and for what purposes it may be used.

438

5. Explain how each of the following techniques aids open-heart surgery.
   a. Hypothermia

   b. Heart-lung bypass

(page 480) **I. Developmental anatomy**

1. The heart is derived from _____-derm. The heart begins to develop during the *(third? fifth? seventh?)* week. Its initial formation consists of two endothelial tubes which unite to form the _____

   25

   _____ tube.

   26
2. Match the regions of the primitive heart at right with the related parts of a mature heart listed below.

   _____ a. Superior and inferior vena cava

   _____ b. Right and left ventricles

   _____ c. Parts of the fetal heart connected by foramen ovale

   _____ d. Aorta and pulmonary trunk

A. Atrium
BCTA. Bulbus cordis and truncus arteriosus
SV. Sinus venosus
V. Ventricle

J. Disorders, medical terminology

1. List five risk factors related to heart attacks. Explain why each factor increases the likelihood of an attack.

2. Outline the sequence of changes in atherosclerosis in this learning activity. [27]

   a. Atherosclerosis involves damage to the tunica _____ _____ of arteries. State several factors that can lead to damage of this lining.

   b. Next _____-cytes stick to the damaged lining. These cells soon become _____ which are capable of ingesting cholesterol. Name one other type of cell that takes in cholesterol: _____.

   c. The resulting fatty lesion in the arterial lining is known as atherosclerotic _____. The rough surface of plaque may snag platelets. These may then cause _____ formation at the site, possibly leading to unwanted emboli. Platelets may also release a chemical called _____.

   d. Name two other types of cells that release PDGF: _____ _____ and _____. What is the effect of this chemical?

3. Distinguish classes of chemicals related to atherosclerosis by doing this exercise. [28]

   a. HDL and LDL both consist of complexes of _____ combined with _____. HDL refers to (high? low?) density lipoprotein, while LDL means _____.

b. *(HDL? LDL?)* are considered culprits in causing atherosclerosis since they may carry cholesterol and deposit this lipid in walls of arteries. So a *(high? low?)* blood level of LDL is desirable.

c. High levels of LDL may occur in blood of persons who have too *(many? few?)* LDL receptors on their cells. Consequently, LDL remains in blood and may travel to blood vessel walls and take up residence there.

d. A high level of HDL tends to *(lead to? prevent?)* atherosclerosis. List two ways that HDLs may be increased in your own blood.

4. Match the diagnostic tests listed at right with the descriptive clues below.

_____ a. Involves treadmill or bicycle

_____ b. Radioisotope is injected into vein

_____ c. Dye is injected into artery; procedure indicates level of blood supply to heart

CA. Coronary angiogram
SE. Stress ECG
TS. Thallium scan

5. Explain the role of each of the following in a coronary bypass surgery.
a. Saphenous vein

b. Cold saline and potassium

6. List several factors that may lead to a coronary artery spasm.

7. Match each congenital heart disease at right with the related description below.

[29]

_____ a. Connection between aorta and pulmonary artery is retained after birth, allowing backflow of blood to right ventricle.

_____ b. Foramen ovale fails to close.

_____ c. Septum between ventricles does not develop properly.

_____ d. Valve is narrowed, limiting forward flow and causing increase in workload of heart.

_____ e. Aorta is abnormally narrow.

C. Coarctation of aorta
IASD. Interatrial septal defect
IVSD. Interventricular septal defect
PDA. Patent ductus arteriosus
VS. Valvular stenosis

8. Which of the conditions in Learning Activity 29 result in left-to-right shunts?

Explain why this is so.

9. In tetralogy of Fallot four defects are present. An opening is present in the septum between the *(atria? ventricles?)*. The aorta emerges from _____

_____

The _____ artery is stenosed. Because of the restricted circulation through the stenosed artery and pulmonary semilunar valve, the right ventricle has to work harder and therefore becomes *(smaller? enlarged?)*. As a result, a right-to-left shunt occurs. What does this mean?

10. Do this exercise on arrhythmias.
    a. *(All? Not all?)* arrhythmias are serious.
    b. Conduction failure across the AV node results in an arrhythmia known

       as _____ .
    c. Which is more serious? *(Atrial? Ventricular?)* fibrillation. Explain why.

    d. Excitation of a part of the heart other than the normal pacemaker (SA

       node) is known as a(n) _____ focus. Such excitations, as by caffeine or lack of sleep, or by more serious problems such as ischemia, may lead to an extra early ventricular systole known as

       _____ .

11. Briefly describe each of these cardiac conditions.
    a. Congestive heart failure (CHF)

    b. Cor pulmonale

    c. Cardiomegaly

| 1 | It is a pump, pumping about 1,000 gal/day. |

| 2 | (a) Mediastinal, lungs, two-thirds. (b) Fist. (c) Apex, fifth, 8, 3. (d) Second. (e) Ventricles. |

| 3 | (a) Sternum, costal cartilages. (b) Bodies of vertebrae, ribs. (c) Diaphragm. |

| 4 | Pericarditis; compression of heart due to accumulation of fluid in the pericardial space. |

| 5 | (a) 4. (b) Atria, auricle, smooth, interatrial septum. (c) Ventricles, thicker, interventricular septum, coronary sulcus. (d) Connective tissue, valves. (e) Right atrium, superior vena cava, parts of the body inferior to the heart, coronary sinus. (f) Right ventricle, trunk, arteries, lungs. (g) Veins, 4, left atrium. (h) Left ventricle, thickest, aorta. |

| 6 | (a) B. (b) T. (c) P. (d) B. (e) A, P. (f) B, T. |

| 7 | Bicuspid and tricuspid (or AV). |

| 8 | Coronary, ascending. |

| 9 | (a) A, C, P. (b) C. (c) M, P. (d) A. |

| 10 | Two or more arteries supplying blood to an area; if one branch of a coronary is obstructed, blood can travel through an alternative (collateral) route to reach the heart tissue. |

| 11 | (a) Autonomic, alter rate of, can. (b) Conduction, impulse transmission. (c) Sinoatrial (SA); pacemaker, ANS nerve impulses and chemicals such as hormones. (d) Right, atria. (e) The fibrous cardiac skeleton (including AV valves) completely separates atrial myocardium from ventricular myocardium. (f) A C B. |

| 12 | (a) P. (b) P-R. (c) QRS. (d) S-T. (e) T. |

| 13 | (a) Contraction and relaxation, opening and closing. (b) Systole, diastole, conduction. (c) Pressure changes. |

| 14 | (a) 0.8, 75. (b) See Figure 20-7b (page 471) in your text. (c) Atria, 0.1, systole, increases. (d) QRS, systole, 0.3. (e) Increases, dramatically, *B*, closed. (f) Semilunar, opening, *C*. (g) 120. (h) De-, T, diastole. (i) Semilunar, *D*. (j) Open, *E*. (k) Just after *E*, rapidly, diastasis, 70. (l) 30, systole. (m) Filling, contraction, de, ejection, isovolumetric relaxation. (n) Lubb, closing, AV; closing of the semilunar valves; see Figure 20-8d (page 473) of text. |

| 15 | AV valves close, *B*; SL valves open, *C*; SL valves close, *D*; AV valves open, *E*. |

| 16 | (a) Fill with, 120–130, end-diastolic. (b) 70, stroke. (c) 50, end-systolic, ESV. (d) C. (e) 120 − 40 = 80. (f) Shorter, lower. (g) Greater, de. |

| 17 | (a) More, increases, more. (b) Greater, Starling's, the more expanded balloon/heart exerts greater force on air/blood. (c) De, de, examples: hypertension of the arteries of the body causes backflow into left ventricle, dilating it; mitral valve malfunction backlogs blood into the left atrium, dilating it; pulmonary artery thickening (stenosis) leads to dila- |

tion of the right ventricle. (d) More, increases. (e) Lack of exercise, MI, loss of blood (hemorrhage).

18 $20 = 25 - 5$; $400 (= 20/5 \times 100\%)$, higher.

19 (a) Stroke volume and heart rate, in. (b) Increased. (c) Medulla, carotid sinus, aorta. (d) Decreased slightly, de. (e) Brakes. (f) Inhibited, take your foot off the gas, stimulated. (g) Decrease, vagus (X), increase. (h) Decrease, lower, Marey's, negative.

20 Positive: increase in venous return to right atrium (as in exercise) increases heart rate, allowing heart to more quickly pump the blood, for example, to skeletal muscles.

21 (a) ↑. (b) ↓. (c) ↑. (d) ↑.

22 (a) Epinephrine, renin, aldosterone. (b) ADH, retention, increase.

23 (a) Effects of epinephrine and sympathetic nerves. Recall that if BP is decreased (as in shock), according to Marey's law, heart rate (pulse) will increase. (b) Sympathetic nerves cause vasoconstriction of skin vessels (so less blood in skin makes skin cool and pale); sweating is also a sympathetic response, so the cool skin is moist (clammy).

24 (a) V (since all veins except pulmonary lead back to the right side of the heart). (b) A (catheter moves retrograde, that is, backward, through the aorta to the left side of the heart). (c) V (since pulmonary artery can be reached via right side of heart). (d) A (since coronary arteries branch off of the aorta).

25 Meso, third, primitive heart.

26 (a) SV. (b) V. (c) A. (d) BCTA.

27 (a) Interna; carbon monoxide from smoking, diabetes mellitus, diet high in cholesterol and certain triglycerides. (b) Mono; macrophages; smooth muscle cells. (c) Plaque; clot; platelet-derived growth factor (PDGF). (d) Macrophages and endothelial cells; promote growth of smooth muscle cells, thus narrowing arterial lumen.

28 (a) Lipids (like cholesterol and triglycerides), proteins, high, low density lipoproteins. (b) LDL, low. (c) Few. (d) Prevent, exercise, diets low in animal fats and not smoking.

29 (a) PDA. (b) IASD. (c) IVSD. (d) VS. (e) C.

30 IASD, IVSD, and PDA. In all three cases, blood is forced from high pressure area in left side of heart or aorta back to lower pressured right side of heart.

31 (a) Not all. (b) Heart block. (c) Ventricular; atrial fibrillation reduces effectiveness of the heart by only about 30 percent, but ventricular fibrillation causes the heart to fail as a pump. (d) Ectopic, premature ventricular contractions (PVCs).

Questions 1–4: Arrange the answers in correct sequence.

_____ _____ _____ 1. Pathway of the conduction system of the heart:

    A. AV node
    B. AV bundle and bundle branches
    C. SA node
    D. Purkinje fibers

_____ _____ _____ _____ _____ 2. From most superficial to deepest:

    A. Epicardium (visceral pericardium)
    B. Myocardium
    C. Parietal pericardium, fibrous layer
    D. Pericardial space containing pericardial fluid
    E. Parietal pericardium, serous layer

_____ _____ _____ _____ _____ 3. Route of a red blood cell supplying oxygen to myocardium of left atrium:

    A. Arteriole, capillary, and venule within myocardium
    B. Branch of great cardiac vein
    C. Coronary sinus leading to right atrium
    D. Left coronary artery
    E. Circumflex artery

_____ _____ _____ _____ _____ 4. Route of a red blood cell now in the right atrium:

    A. Left atrium
    B. Left ventricle
    C. Right ventricle
    D. Pulmonary artery
    E. Pulmonary vein

_____ 5. Which of the following factors will tend to decrease heart rate?
  A. Stimulation by cardiac nerves
  B. Release of the transmitter substance norepinephrine in the heart
  C. Activation of neurons in the cardioaccelerator center
  D. Increase of vagal nerve impulses
  E. Increase of sympathetic nerve impulses

_____ 6. The average cardiac output for a resting adult is about ____/min.
  A. 1 qt   B. 5 pt   C. 1.25 gal (5 liters)
  D. 0.5 liter   E. 2,000 ml

_____ 7. When intraventricular pressure exceeds intraatrial pressure, what event occurs?
  A. AV valves open   B. AV valves close
  C. Semilunar valves open   D. Semilunar valves close

_____ 8. The second heart sound is due to turbulence of blood flow as a result of what event?
  A. AV valves opening   B. AV valves closing
  C. Semilunar valves opening   D. Semilunar valves closing

_____ 9. Choose the FALSE statement about the circumflex artery.
  A. It is a branch of the left coronary artery.
  B. It provides the major blood supply to the right ventricle.
  C. It lies in a groove between the left atrium and left ventricle.
  D. Damage to this vessel would leave the left atrium with virtually no blood supply.

_____ 10. Choose the FALSE statement about heart structure.
  A. The heart chamber with the thickest wall is the left ventricle.
  B. The apex of the heart is more superior in location than the base.
  C. The heart has four chambers.
  D. The left ventricle forms the apex and most of the left border of the heart.

_____ 11. All of the following are defects involved in Tetralogy of Fallot EXCEPT:
  A. Ventricular septal defect   B. Enlarged right ventricle
  C. Stenosed mitral valve
  D. Aorta emerging from both ventricles

_____ 12. Choose the TRUE statement.
  A. The T wave is associated with atrial depolarization.
  B. The normal P-R interval is about 0.4 sec.
  C. Myocardial infarction means strengthening of heart muscle.
  D. Myocardial infarction is commonly known as a "heart attack" or a "coronary."

_____ 13. Choose the FALSE statement.
  A. Pressure within the atria is known as intraarterial pressure.
  B. During most of ventricular diastole the semilunar valves are closed.
  C. During most of ventricular systole the semilunar valves are open.
  D. Diastole is another name for relaxation of heart muscle.

_____ 14. Which of the following structures are located in ventricles?

    A. Trabeculae carneae   B. Fossa ovalis

    C. Ligamentum arteriosum   D. Musculi pectinati

Questions 15–20: Circle T (true) or F (false). If the statement is false, change the underlined word or phrase so that the statement is correct.

T   F   15. The blood in the left chambers of the heart contains <u>higher</u> oxygen content than blood in the right chambers.

T   F   16. Most ventricular filling occurs during atrial <u>systole.</u>

T   F   17. At the point when intraarterial pressure surpasses intraventricular pressure, semilunar valves <u>open.</u>

T   F   18. The pulmonary <u>artery carries</u> blood from the lungs to the left atrium.

T   F   19. The normal cardiac cycle <u>does not</u> require direct stimulation by the autonomic nervous system.

T   F   20. During <u>about half</u> of the cardiac cycle, atria and ventricles are contracting simultaneously.

Questions 21–25: fill-ins. Complete each sentence with the word or phrase that best fits.

_____ 21. Cyanosis is likely to result from congenital ____ to ____ shunt since blood does not pass through lungs.

_____ 22. ____ is a cardiovascular disorder that is the leading cause of death in the U.S.

_____ 23. An ECG is a recording of ____ of the heart.

_____ 24. ____ is a term which means abnormality or irregularity of heart rhythm.

_____ 25. According to Marey's law of the heart, if blood pressure against baroreceptors decreases (as in shock), heart rate ____-creases.

The Cardiovascular System: Vessels and Routes

21

In the preceding two chapters you learned about blood and the heart. In this chapter you will study mechanisms of circulation and pathways taken by blood. You will contrast the five principal types of blood vessels (Objectives 1–2). You will discuss factors that control circulation and blood pressure at rest and in exercise (3–9, 18). You will study clinical methods for determining pulse rate and blood pressure and the clinical significance of these measurements (10–12). Next you will examine circulatory routes in different regions of adult and fetal bodies (13–17). You will consider development of the cardiovascular system and effects of age upon it (19–20). Finally you will learn about disorders and medical terminology related to blood vessels and blood pressure (21–22).

Topics Summary

A. Blood vessels
B. Physiology of circulation
C. Checking circulation
D. Circulatory routes: systemic arteries
E. Circulatory routes: systemic veins
F. Circulatory routes: hepatic portal, pulmonary, and fetal circulation
G. Exercise and the cardiovascular system
H. Aging and developmental anatomy
I. Disorders, medical terminology, drugs

Objectives

1. Contrast the structure and function of arteries, arterioles, capillaries, venules, and veins.
2. Define a blood reservoir and explain its importance.
3. Relate the importance of cardiac output (CO), blood volume, and peripheral resistance to blood pressure.
4. Explain the role of the vasomotor center in controlling blood pressure.
5. Contrast the roles of baroreceptors and chemoreceptors in regulating blood pressure.
6. Describe the effects of epinephrine, norepinephrine (NE), antidiuretic hormone (ADH), angiotensin II, histamine, and kinins on blood pressure.
7. Define autoregulation and explain its importance.
8. Discuss the various pressures involved in the movement of fluids between capillaries and interstitial spaces.
9. Explain how skeletal muscle contractions, valves in veins, and breathing assist in the return of venous blood to the heart.
10. Define pulse and identify the arteries where pulse may be felt.
11. Define blood pressure (BP) and explain one clinical method for recording systolic and diastolic pressure.
12. Contrast the clinical significance of systolic, diastolic, and pulse pressures.
13. Identify the principal arteries and veins of systemic circulation.
14. Identify the major blood vessels of pulmonary circulation.
15. Trace the route of blood involved in hepatic portal circulation and explain its importance.
16. Contrast fetal and adult circulation.
17. Explain the fate of fetal circulation structures once postnatal circulation is established.
18. Explain the effects of exercise on the cardiovascular system.
19. Describe the effects of aging on the cardiovascular system.
20. Describe the development of blood vessels and blood.
21. List the causes and symptoms of aneurysms, coronary artery disease (CAD), hypertension, and deep-venous thrombosis (DVT).
22. Define medical terminology associated with the cardiovascular system.

447

Learning Activities

(pages 488–492) **A.** Blood vessels

|1| 1. Trace the route of a drop of blood.

Aorta → artery → _____ →

_____ → _____ →

_____ → heart

2. Write a brief description of the structure and function of each of the following parts of a blood vessel. (Refer to Figure 21–1, page 488, in your text for help.)
a. Tunica interna (intima)

b. Tunica media

c. Tunica externa (adventitia)

d. Lumen

e. Valve

|2| 3. State the two major properties of arteries.

Explain what structural features account for these properties.

|3| 4. Complete this exercise about alterations in the diameter of blood vessels.
a. Decrease in the diameter of vessels is known as *(vasoconstriction? vasodilation?)*. The smooth muscle fibers which cause change in vessel size are located in the tunica_____.
b. Stimulation of *(sympathetic? parasympathetic?)* nerves is responsible for vasoconstriction. In most cases vasodilation is caused by *(parasympathetic? lack of sympathetic?)* stimulation.

|4| 5. Most parts of the body receive blood from *(only one? more than one?)* artery. State the significance of the following in maintaining or disrupting homeostasis.
a. Anastomoses and collateral circulation

6. Explain how *digital subtraction angiography (DSA)* is used as a diagnostic tool for determining certain vascular disorders.

7. Describe the gradual changes in structure of an arteriole in its course between an artery and a capillary.

8. Explain how the distribution of capillaries in the body correlates to the activity of the tissue. Name several areas where you would expect capillary supply to be:
   a. Extensive

   b. Limited

   c. Absent

5 | 9. Match the structure at right with the correct description below.

_____ a. Composed of one continuous layer of endothelial cells which permit exchange of nutrients and wastes between blood and cells

_____ b. Porous vessels forming choroid plexuses in ventricles of brain and ciliary processes of eye

_____ c. Vessels that are wider and more winding than capillaries; lined with phagocytes, so they serve protective function in organs such as liver and spleen

_____ d. Ring of smooth muscle that controls blood flow into true capillaries

_____ e. Regulate blood flow from arteries through capillaries (three answers)

_____ f. Aorta, for example

_____ g. Brachial and femoral arteries, for example

A. Arteriole
A(e,c). Artery: elastic or conducting
A(m,d). Artery: muscular or distributing
CC. Continuous capillaries
FC. Fenestrated capillaries
M. Metarterioles
PS. Precapillary sphincter
S. Sinusoids

6 | 10. Complete this exercise about structure of veins.

a. Veins have *(thicker? thinner?)* walls than arteries. This structural feature relates to the fact that the pressure in veins is *(more? less?)* than in arteries. The pressure difference is demonstrated when a vein is cut; blood leaves a cut vein in *(rapid spurts? an even flow?)*.

b. Gravity exerts back pressure on blood in veins located inferior to the heart. To counteract this, veins contain_____.

c. When valves weaken, veins become enlarged and twisted. This condition is called_____. This occurs more often in *(superficial? deep?)* veins. Why?

11. Name two parts of the body where vascular (venous) sinuses are located.

7 | 12. Answer these questions about the blood reservoirs of the body.

a. At any given moment, about 60% of the blood of your body is contained within *(arteries and arterioles? capillaries? veins, venules, and venous sinuses?)*. State two locations where many such reservoir vessels are located.

b. When blood pressure drops suddenly, *(sympathetic? parasympathetic?)* nerves stimulate smooth muscle in walls of veins. Consequently, veins are

(dilated? constricted) causing_____-creased return of venous blood to the heart. According to Starling's law, the heart will then contract with *(greater? less?)* force, increasing blood pressure.

B. Physiology of circulation

(pages 492–496)

1. State the one main factor that explains why blood flows from one vessel to the next.

```
8
```

Two additional factors that aid blood flow are the *(large? small?)* diameter of veins, which offers *(more? less?)* resistance to capillary blood, and also

the contraction of_____, which squeeze venous blood back towards the heart.

2. Refer to Figure 21-5 (page 493) of the text, and complete this exercise.

```
9
```

   a. Blood pressure in the aorta and large arteries such as the brachial *(does? does not?)* vary during the cardiac cycle. In capillaries and veins blood pressure *(does? does not?)* change noticeably during the cardiac cycle.

   b. In the aorta and brachial artery blood pressure (BP) is about 120 mm Hg immediately following ventricular contraction. This is called *(systolic? diastolic?)* BP. As ventricles relax (or go into _____),

   blood is no longer ejected into these arteries. However, the normally *(elastic? rigid?)* walls of these vessels recoil against blood, pressing it onward with a diastolic BP of_____mm Hg.

   c. *For extra review.* Refer again to Figure LG 20–2a, page LG 431 and identify BP values in the aorta during the cardiac cycle. Note that the slope of that curve is *(more? less?)* steep since the time frame is more expanded than in Figure 21–5, page 493 of the text. (The latter shows BP changes during a number of cardiac cycles.)

   d. Write the normal ranges of BP values for the following types of vessels:

   arterioles, _____mm Hg; capillaries, _____mm Hg; venules and veins,

   _____mm Hg; venae cavae,_____mm Hg.

   e. The pressure in the right atrium is normally about_____mm Hg. If the right side of the heart is weakened and cannot adequately pump blood into the lungs, or if pulmonary vessels are sclerosed, offering resistance to blood flow, blood will tend to backlog in the right atrium causing a (an) *(increase? decrease?)* in right atrial pressure.

3. Define *arterial blood pressure.*

4. Complete Figure LG 21-1 which summarizes factors determining arterial blood pressure (BP). Indicate directions of arrows and add missing words.

5. If directions of arrows in Figure LG 21-1 are reversed, blood pressure can be lowered. Check your understanding of factors that affect arterial blood pressure by circling those which will tend to *decrease* it.

   A. Increase in cardiac output, as by increased heart rate or stroke volume.

   B. Increase in (vagal) impulses from cardioinhibitory center.

   C. Decrease in blood volume, as following hemorrhage

   D. Increase in blood volume, as by excess salt intake and water retention

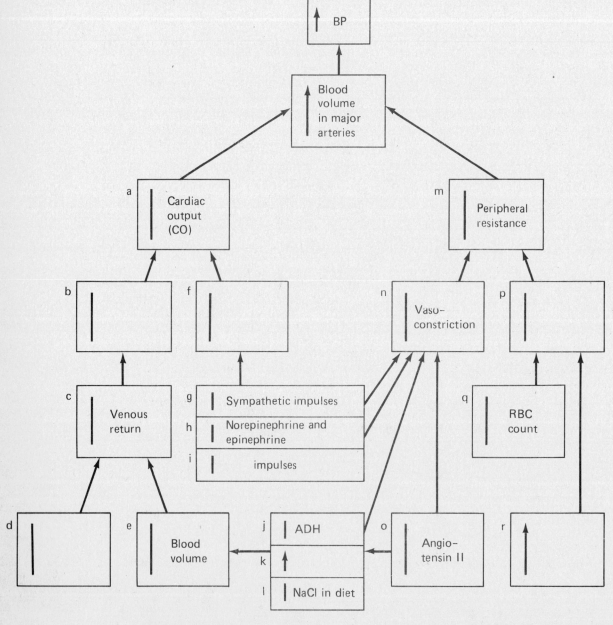

Figure LG 21-1 Summary of factors controlling blood pressure (BP). Arrows indicate increase (↑) or decrease (↓). Complete as directed in Learning Activity 10 .

E. Increased peripheral resistance due to vasoconstriction of arterioles (blood prevented from leaving large arteries and entering small vessels)

F. Decrease in sympathetic impulses to the smooth muscle of arterioles

G. Decreased viscosity of blood via loss of blood protein or red blood cells, as by hemorrhage

H. Use of medications called dilators since they dilate arterioles, especially in areas such as abdomen and skin.

6. Review the blood pressure control mechanisms involving afferent pathways from pressoreceptors to the cardiac center and efferent responses by the heart (Chapter 20, Activity 19, page LG 434).

7. Additional control of blood pressure is provided by the vasomotor center. Describe its function in this exercise. 12

   a. Like the cardiac center, the vasomotor center is located in the _____. Its function is to control the diameter of _____ and thereby to alter BP.

   b. The vasomotor center "is told" to alter BP by pressure sensitive receptors located in the _____ and _____. When these pressoreceptors sense that BP is low, they send afferent impulses to both the cardiac and vasomotor centers. The cardiac response is to *(increase? decrease?)* heart rate.

   c. The vasomotor center responds to low BP by *(increasing? decreasing?)* sympathetic nerve impulses to arterioles, causing them to *(vasoconstrict? vasodilate?)*. Since blood is thus prevented from entering these narrowed arterioles, blood stays in major arteries, *(increasing? decreasing?)* arterial BP.

   d. Chemoreceptors sensitive to _____ and _____ can also stimulate the vasomotor center. Low oxygen or high carbon dioxide will tend to cause *(vasoconstriction? vasodilation?)* and consequently an *(increase? decrease?)* in BP. As a result, blood can move more rapidly to the lungs to pick up oxygen and give up carbon dioxide.

8. Tell whether each of the following factors tends to increase or decrease BP. Then briefly describe how it causes vasoconstriction or vasodilation. Answer specific questions also.

   a. Anger _____ -creases BP.

   Explain why a person might faint in extreme anger.

   b. Angiotensin II and aldosterone _____ -crease BP. (For help, review Chapter 20, Activity 22, page LG 436.)

c. Histamine_____-creases BP.

[13] 9. Describe the process of autoregulation in this activity.

    a. Autoregulation is a mechanism for regulating blood flow through specific tissue areas. Unlike cardiac and vasomotor mechanisms, autoregulation occurs by *(local? autonomic nervous system?)* control.

    b. In other words, when a tissue (such as an active muscle) is hypoxic, that is, has a *(high? low?)* oxygen level, the cells of that tissue release *(vasoconstrictor? vasodilator?)* substances. These may include _____

_____acid, built up by active muscle, or

_____, which is a product of ATP catabolism. Other products of metabolism that serve as dilator substances include the gas

_____and the_____ion.

    c. The effect of these vasodilators is to cause precapillary sphincters to *(contract? relax?),* resulting in *(increased? decreased?)* blood flow into the active tissue.

10. Recall that pressure is the principal factor causing blood to move through the vessels of the body. Pressure is also the major factor determining movement of substances across capillary membranes. Refer to Figure LG

[14]   21-2 and do this learning activity.

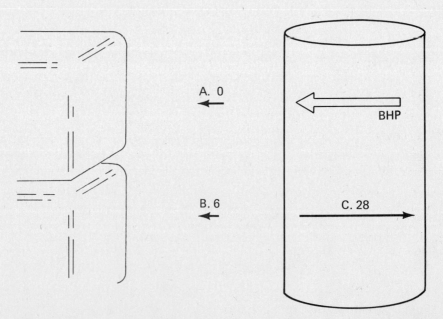

Figure LG 21-2 Practice diagram showing forces controlling movement across membranes and for determining P_{eff} at arterial end of capillary. Numbers are pressure values in mm Hg.

Complete as directed in Learning Activity [14] .

a. At the arterial end of a capillary, blood pressure exerts a pressure known as blood_____pressure (or BHP). Write a typical value for this pressure on the arrow marked BHP on the figure. (For help, refer to Figure 21-5, page 493, in the text.)

b. Now label arrows *A* and *B* in the figure, showing sources of two other pressures that effect movement of substances out of blood.

c. One force counteracts the three pressures described above. This force is blood_____pressure (BOP). Write this label at arrow *C* on the figure. BOP is due mainly to the presence of large molecules of_____in all normal blood plasma.

d. Imagining these four forces as in an unbalanced tug-of-war game, calculate the *effective filtration pressure* (P_{eff}) which determines the amount and direction of movement across the membrane.

Write this value on the figure with an arrow showing direction of movement.

e. At venous ends of capillaries, one of these four pressures is substantially different. It is *(BHP? IFHP? BOP? IFOP?)*. The value of this pressure is about_____mm Hg.

f. Calculate P_{eff} at venous ends of capillaries.

g. You discovered that *(none? most?)* of the fluid that is pushed out at arterial ends of capillaries moves back into blood at venous ends. In Chapter 22 you will hear how this fluid is removed from intercellular spaces by the_____system.

11. Do this exercise about factors that influence blood flow back to the heart and therefore affect blood pressure. [15]

a. Any factor that increases venous return to the heart will increase cardiac output (due to_____law of the heart). This will *(increase? decrease?)* blood pressure. *(For extra review* of Starling's law of the heart, return to Chapter 20, Activity [17], page LG 433.)

b. Figure 21-5 (page 493) of your text shows that velocity of blood flow is greatest in *(arteries? capillaries? veins?)*. Flow is slowest in _____. As a result, ample time is available for exchange between capillary blood and tissues.

c. As blood moves from capillaries into venules and veins, its velocity *(increases? decreases?)*, enhancing venous return.

d. Although each individual capillary has a cross-sectional area which is microscopic, the body contains so many capillaries that their total cross-sectional area is enormous. (For comparison, look at the eraser end on one pencil; you are seeing the cross-sectional area of that one pencil. If you bind together 100 pencils with a strong rubber band and observe their eraser ends as a group, you see that the individual pencils contribute to a relatively large total cross-sectional area.) All of the capillaries in the body have a total cross-sectional area of about____cm².

e. *(A direct? An indirect?)* relationship exists between cross-sectional area and velocity of blood flow. This is evident in the aorta also. This vessel has a cross-section of about _____ cm² and has a very *(high? low?)* velocity.

f. The increased respiratory rate during active exercise also enhances venous return to the heart. During inspiration the size of the thorax enlarges causing *(increased? decreased?)* intrathoracic pressure, while intraabdominal pressure *(increases? decreases?)*. A gradient is therefore established which pushes blood upward to the heart.

12. How do skeletal muscle contractions and venous valves increase venous return?

(pages 496–498) **C. Checking circulation**

1. Define *pulse*.

2. In general, where may pulse best be felt? List six places where pulse can readily be palpated. Try to locate pulse at these points on yourself or on a friend.

3. Normal pulse rate is about____beats/min. *Tachycardia* means *(rapid? slow?)* pulse rate; _____ means slow pulse rate.

4. Describe the method commonly used for taking blood pressure by answering these questions about the procedure.

   a. Name and briefly describe the instrument used to check BP.

   b. The cuff is placed over the_____artery. How can you tell that this artery is compressed after the bulb of the sphygomanometer is squeezed?

   c. As the cuff is deflated, the first sound heard indicates_____

   _____ pressure, since reduction of pressure from the cuff below systolic pressure permits blood to begin spurting through the artery.

   d. Diastolic pressure is indicated by the point at which the sounds _____. At this time blood can once again flow freely through the artery.

5. *A clinical challenge.* Mrs. Kinney has a systolic pressure of 128 mm Hg and diastolic of 83 mm Hg. Write her blood pressure in the form in which BP is usually expressed:_____. Determine her pulse pressure.

Would this be considered a normal pulse pressure?

D. Circulatory routes: systemic arteries

1. Study Figure 21-11 (page 499) in your textbook. Then describe the general plan of systemic circulation. Remember that this is a "closed circuit," that is, all blood vessels are continuous from one side of the heart back to the other side of the heart.

2. In Chapter 20 you studied vessels that transfer blood between heart and lungs and provide for gas exchange in lungs. This is the *(systemic? pulmonary?)* circulation. *(For extra review* study Chapter 20, Activity 5f,g , page LG 425.)

17 3. Use Figure LG 21-3 to do this learning activity about systemic arteries.
   a. The major artery from which all systemic arteries branch is the

   _____. It exits from the chamber of the heart known as the *(right? left?)* ventricle. The aorta can be divided into three

   portions. They are the_____(N1 on the figure), the

   _____(N2), and the_____(N3).
   b. The first arteries to branch off of the aorta are the

   _____arteries.Labelthese(at O) on the figure. These

   vessels supply blood to the_____.
   c. Three major arteries branch from the aortic arch. Write the names of these (at *K, L,* and *M*) on the figure, beginning with the vessel closest to the heart.
   d. Locate the two branches of the brachiocephalic artery. The right

   subclavian artery (letter ____ supplies blood to _____

   _____.

   Vessel *E* is named the *(right? left?)* _____

   _____artery. It supplies the right side of the head and neck.
   e. As the subclavian artery continues into the axilla and arm, it bears different names (much as a road does when it passes into new towns). Label *G* and *H*. Note that *H* is the vessel you studied as a common site for

   measurement of_____.
   f. Label vessels *I* and *J*, and then continue the drawing of these vessels into the forearm and hand. Notice that they anastomose in the hand. Vessel *(I? J?)* is often used for checking pulse in the wrist.

Right side
of body

Left side
of body

A

B

C

D

E

F

K

L

M

G

N

2

1

3

O

H

P

Q

R

S

T

I

J

U

V

W

X

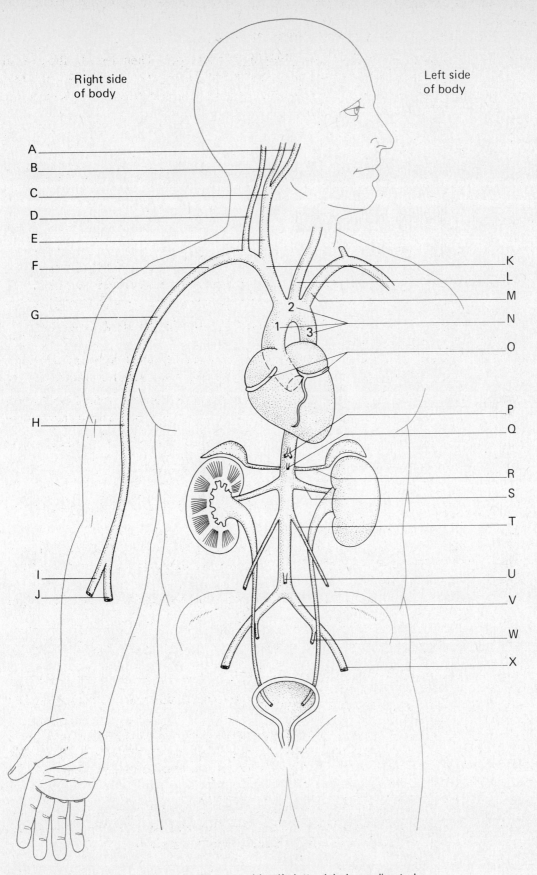

Figure LG 21-3 Major arteries from the aorta. Identify letter labels as directed.

460

g. Besides supplying the arms, subclavian arteries each send a branch that ascends the neck through foramina in cervical vertebrae (see Figure 21-14b, page 505, in the text). This is the _____ artery. The right and left vertebral arteries join at the base of the brain to form the _____ artery. (Refer to *C* and *D* on Figure LG 21-4.)

18 4. The right and left common carotid arteries ascend the neck in positions considerably *(anterior? posterior?)* to the vertebral arteries. Find a pulse in one of your own carotids. These vessels each bifurcate (divide into two vessels) at about the level of your mandible. Identify the two branches (*A* and *B*) on Figure LG 21-3. Which supplies the brain and eye? *(Internal? External?)* What structures are supplied by the external carotids? (See Exhibit 21-4, page 504, in the text for help.)

19 5. Refer to Figure LG 21-4 and answer these questions about the vital blood supply to the brain.
a. An anastomosis, called the cerebral arterial circle of _____ _____ is located just inferior to the brain. This circle closely surrounds the pituitary gland.

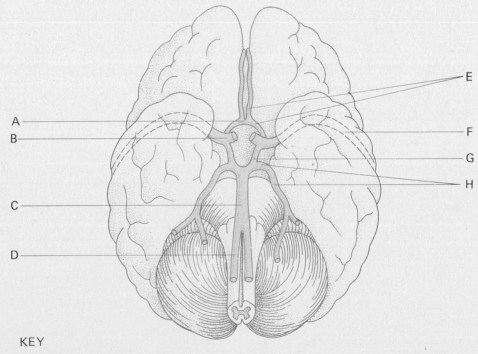

KEY

A. Anterior communicating cerebral artery (1)
B. Internal carotid arteries (cut) (2)
C. Basilar artery (1)
D. Vertebral arteries (2)

E. Anterior cerebral arteries (2)
F. Middle cerebral arteries (2)
G. Posterior communicating cerebral arteries (2)
H. Posterior cerebral arteries (2)

Figure LG 21-4 Arterial supply to the brain including vessels of the cerebral arterial circle of Willis. View is from the undersurface of the brain. Numbers in parentheses indicate whether arteries are single (1) or paired (2). Refer to Learning Activity **19** .

b. The circle is supplied with blood by two pairs of vessels: anteriorly, the *(internal? external?)* carotid arteries, and, posteriorly, the _____ arteries (joined to form the basilar artery).

c. The circle supplies blood to the brain by three pairs of cerebral vessels, identified on Figure LG 21-4 by letter labels _____, _____, and _____. Vessels *A* and *G* complete the circle; they are known as anterior and posterior _____ arteries. Now cover the KEY and correctly label each of the lettered vessels.

6. *A clinical challenge.* Relate symptoms to the interruption of normal blood supply to the brain as described in each case. Explain your reasons. [20]

   a. A patient's basilar artery is sclerosed. Which part of the brain is most likely to be deprived of normal blood supply? *(Frontal lobes of cerebrum? Occipital lobes of cerebrum, as well as cerebellum and pons?)*

   b. Paralysis and loss of some sensation are noted on a patient's *right* side. Angiography points to sclerosis in a common carotid artery. Which one is more likely to be sclerosed? *(Right? Left?)* Explain.

   c. The middle cerebral artery which passes between temporal and frontal lobe (see broken lines on Figure LG 21-4) is blocked. Which sense is more likely to be affected *(Hearing? Vision?)* Explain.

   d. Anterior cerebral arteries pass anteriorly from the circle of Willis and then arch backward along the medial aspect of each cerebral hemisphere. Disruption of blood flow in the right anterior cerebral artery will be most likely to cause effects in the *(right? left?)* *(hand? foot?)*. *(Hint:* The affected brain area would include region *Z* in Figure LG 15-1 and/or region *R* in Figure LG 15-2, pages LG 306 and 308. Note that these regions are on the medial aspect of the right hemisphere.)

7. Contrast the *visceral branches* of aorta with the *parietal branches* of aorta. Then list several examples of each.

Visceral   Parietal

8. Complete the table describing visceral branches of the abdominal aorta. Note that arteries are listed in the order in which they emerge from the aorta (from superior to inferior). Label these on Figure LG 21-3 (vessels *P* to *U*).

21

Artery	Structures Supplied
a. Celiac (three major branches) 1. 2. 3.	1. Liver 2. 3.
b.	Small intestine and part of large intestine
c.	Adrenal (suprarenal) glands
d. Renal	
e.	Testes (or ovaries)
f. Inferior mesenteric	

22

9. The aorta ends at about level_____ vertebra by dividing into right and left _____arteries. Label these at *V* on Figure LG 21-3.

10. Label the two branches (*W* and *X*) of each common iliac artery. Name structures supplied by each. (See Exhibit 21-7, page 508 in the text.)
   a. Internal iliac artery

b. External iliac artery

11. Trace one possible route of a drop of blood from the common iliac artery to the little toe of the right foot. (Refer to Figure 21-16a–c, pages 509–510, of the text for help.) The pathway is started for you:
Right common iliac artery → right external iliac artery →

12. *For extra review.* Suppose that the right femoral artery were completely occluded. Describe an alternate (or collateral) route that would enable blood to pass from the right common iliac artery to the little toe of the right foot. 23

Right common iliac artery →

E. Circulatory routes: systemic veins

(pages 510–519)

1. Name the three main vessels that empty venous blood into the right atrium of the heart. 24

_____

_____

_____

Next to each, write the major regions of the body drained by the vessel.

2. Review coronary blood flow. (See the Mastery Test for Chapter 20, question 3, page LG 444.)

25   3. Do this exercise about venous return from the head.

a. Venous blood from the brain drains into vessels known as_____

_____ . These are (*structurally? functionally?*) like veins. Describe their structure.

b. Name the sinus that lies directly beneath the sagittal suture:

_____ .

c. Blood from all of the cranial vascular sinuses eventually drains into the

_____ veins which descend in the neck. These veins are positioned close to the_____ arteries.

d. Where are the jugular arteries and carotid veins located?

26   4. Imagine yourself as a red blood cell currently located in the *left brachial vein.* Answer these questions.

a. In order for you to flow along in the bloodstream to the *left brachial artery,* which of these must you pass through? *(A brachial capillary? The heart and a lung?)*

b. In order for you to flow along to the *right arm,* must you pass through the heart? *(Yes? No?)* Both sides of the heart, that is, right and left? *(Yes? No?)* One lung? *(Yes? No?)*

5. Once you are familiar with the pathways of systemic arteries, you know much about the veins which accompany these arteries. Demonstrate this by tracing routes of a drop of blood along the following deep vein pathways.

27   a. From the thumb side of the left forearm to the right atrium

b. From the medial aspect of the left knee to the right atrium

6. Contrast veins with arteries in this exercise.
   a. Veins have *(higher? lower?)* blood pressure than arteries and a *(faster? slower?)* rate of blood flow.
   b. Veins are therefore *(more? less?)* numerous, in order to compensate for slower blood flow in each vein. The "extra" veins consist of vessels located just under skin (in subcutaneous fascia); these are known as

   _____ veins.

7. Now trace a drop of blood through these superficial pathways. [29]
   a. From the skin on the thumb side of the left forearm to the right atrium

   b. From the skin on the medial aspect of the left knee to the right atrium

8. On text Figures 21-19 (page 514) and 21-21 (page 518), locate the following veins. Identify the location at which each empties into a deep vein. [30]
   a. Basilic

   b. Median cubital

   c. Great saphenous

   d. Small saphenous

9. Veins connecting the superior and inferior venae cavae are named _____ veins. Identify the azygos system of veins on Figure 21-20 (page 517) in the text. What parts of the thorax do they drain? Explain how these veins may help to drain blood from the lower part of the body if the inferior vena cava is obstructed.

10. Name the visceral veins that empty into the abdominal portion of the inferior vena cava. Then color them blue on Figure LG 26-1 (page LG 599). Contrast accompanying systemic arteries by coloring them red on that figure. Note on Figure LG 26-1 that the inferior vena cava normally lies on the (*right? left?*) side of the aorta. One way to remember this is that the inferior vena cava is returning blood to the (*right? left?*) side of the heart.

| 31 |

(pages 520-524) **F. Circulatory routes: hepatic portal, pulmonary, and fetal circulation**

1. Note on Figure LG 26-1 (page LG 599) that blood from digestive organs (such as blood in superior and inferior mesenteric veins) does *not* drain directly into the inferior vena cava. Complete this exercise about this blood route.

| 32 |

   a. Virtually all blood from digestive organs, namely,_____

   _____

   as well as blood from the spleen, empties into veins which lead to the single _____ vein. This vessel enters the undersurface of the liver.

   b. Look at Figure 21-23 (page 521) in the text and name the major veins that drain into this large hepatic portal vein.

   c. Note the uniqueness of this vein: it carries (*port* = carry) blood from a set of organs to another organ. Other veins of the body empty into

   _____

   _____.

d. You have studied another vessel which enters the liver. This is the _____ artery, which carries oxygenated blood. (See Figure LG 24-4, page LG 553.)

e. In the liver blood from both the hepatic artery and the portal vein mixes (see Figure 24-17, page 613, in the text) as it travels through tortuous capillarylike structures named _____.

f. Blood leaves the liver via the right and left _____, which exit from the top of the liver and empty into the inferior vena cava.

g. What functions are served by the special hepatic portal pathway?

2. What aspects of fetal life require the fetus to possess special cardiovascular structures not needed after birth?

3. Complete the following table about fetal structures. (Note that the structures are arranged in order of blood flow from fetal aorta back to fetal aorta.) | 33 |

Fetal Structure	Structures Connected	Function	Fate of Structure After Birth
a.		Carries fetal blood low in oxygen and nutrients and high in wastes	
b. Placenta	(Omit)		
o.	Placenta to livor and ductus venosus		
d.		Branch of umbilical vein, bypasses liver	

Fetal Structure	Structures Connected	Function	Fate of Structure After Birth
e. Foramen ovale			
f.			Becomes ligamentum arteriosum

34

4. For each of the following pairs of fetal vessels, circle the vessel with higher oxygen content.
   a. Umbilical artery/umbilical vein
   b. Ductus arteriosus/ductus venosus
   c. Femoral artery/femoral vein
   d. Pulmonary artery/pulmonary vein
   e. Aorta/thoracic portion of inferior vena cava

(page 524) **G. Exercise and the cardiovascular system**

1. List several examples of aerobic exercises.

2. Describe changes that occur in the following aspects of body function as a result of physical training.
   a. Cardiac output

   b. Oxygen requirement of skeletal muscles

c. Blood pressure

d. HDL level

3. Identify one factor that may be responsible for the psychological "high" experienced during regular workouts, for example, by runners. $\boxed{35}$

H. Aging and developmental anatomy **(pages 524-525)**

1. Draw arrows next to each factor listed below to indicate whether it increases (↑) or decreases (↓) with normal aging. $\boxed{36}$

    _____ a. Size and strength of cardiac muscle cells

    _____ b. Cardiac output (CO)

    _____ c. Deposits of cholesterol in blood vessels supplying brain

    _____ d. Blood pressure

2. Write a paragraph describing embryonic development of blood vessels. Include the following key terms: *15–16 days, mesoderm, mesenchyme, blood islands, spaces, endothelial, tunics.*

470

3. List four sites of blood formation in the embryo and fetus.

(pages 525–526) **I. Disorders, medical terminology**

1. Define *aneurysm*.

List four possible causes of aneurysms.

2. Define *hypertension*. What clinical measurement determines the diagnosis of the condition?

37 3. What blood pressure values are generally regarded as threshold level of hypertension?_____

4. Contrast *primary hypertension* and *secondary hypertension*.

5. Match each organ at right with the related disorder that may cause secondary hypertension.

_____ a. Production of renin which catalyzes formation of Angiotensin II,
a powerful vasoconstrictor

38

_____ b. Pheochromocytoma, a tumor that
releases large amounts of norepinephrine and epinephrine

_____ c. Release of excessive amounts of
aldosterone which promotes salt
and water retention.

AM. Adrenal medulla
AC. Adrenal cortex
K. Kidney

6. Describe effects of hypertension upon the following body structures:
   a. Heart

   b. Cerebral blood vessels

   c. Kidneys

7. Explain how each of the following therapeutic measures can help control hypertension.
   a. Sodium restriction

   b. Cessation of smoking

   c. Exercise

   d. Diuretics

   e. Vasodilators

8. *A clinical challenge.* Explain how a deep vein thrombosis (DVT) of the left
femoral vein can lead to a pulmonary embolism.

Answers to Numbered Questions in Learning Activities

1 Arteriole, capillary, venule, vein.

2 Elasticity, contractility.

3 (a) Vasoconstriction, media. (b) Sympathetic, lack of sympathetic.

4 More than one.

5 (a) CC (b) FC. (c) S. (d) PS. (e) A, M, PS. (f) A(e,c). (g) A(m,d).

6 (a) Thinner, less, an even flow. (b) Valves. (c) Varicose veins, superficial, skeletal muscles around deep veins prevent overstretching.

7 (a) Veins, venules and venous sinuses; abdominal organs and skin. (b) Sympathetic, constricted, in, greater.

8 Continuous drop in blood pressure from one vessel to the next; large, less, skeletal muscles.

9 (a) Does, does not. (b) Systolic, diastole, elastic, 70 to 80. (c) Less. (d) 40 to 25, 25 to 12, 12 to 5, 2. (e) 0, increase.

10 (a) ↑ CO. (b) ↑ Stroke volume. (c) ↑ Venous return. (d) ↑ exercise. (e) ↑ Blood volume. (f) ↑ Heart rate. (g) ↑ Sympathetic impulses. (h) ↑ Norepinephrine and epinephrine. (i) ↓ Vagal impulses. (j) ↑ ADH. (k) ↑ Aldosterone. (l) ↑ NaCl in diet. (m) ↑ Peripheral resistance. (n) ↑ Vasoconstriction. (o) ↑ Angiotensin II. (p) ↑ Viscosity of blood. (q) ↑ RBC count. (r) ↑ Plasma proteins.

11 B, C, F, G, H.

12 (a) Medulla, arterioles. (b) Aorta (arch), carotid sinus, increase. (c) Increasing, vasoconstrict, increasing. (d) Oxygen, carbon dioxide, vasoconstriction, increase.

13 (a) Local. (b) Low, vasodilator, lactic, adenosine, carbon dioxide, hydrogen. (c) Relax, increased.

14 (a) Hydrostatic, 30. (b) A, IFHP; B, IFOP. (c) Osmotic, protein. (d) + 8 mm Hg (or ← 8). (e) BHP, +15 (or ← 15). (f) −7 (or → 7) mm Hg. (g) Most, lymphatic.

15 (a) Starling's, increase. (b) Arteries, capillaries. (c) Increases. (d) 2,500. (e) An indirect, 2.5, high. (f) Decreased, increases.

16 128/83, 45 (128-83); yes, it is in the normal range.

17 (a) Aorta; left; ascending aorta, arch of the aorta, descending aorta. (b) Coronary, heart. (c) *K,* brachiocephalic; *L,* left common carotid; *M,* left subclavian. (d) F; in general, the right extremity and right side of thorax, neck, and head; right common carotid. (e) *G,* Axillary; *H,* brachial; blood pressure. (f) *I,* radial; *J,* ulnar; *I.* (g) Vertebral, basilar.

18 Anterior; internal; external portions of head (face, tongue, ear, scalp) and neck (throat, thyroid).

19 (a) Willis. (b) Internal, vertebral, (c) *E, F, H,* communicating.

20 (a) Occipital lobes of cerebrum, as well as cerebellum and pons. (b) Left, since this artery supplies the left side of the brain. (Remember that neurons in tracts cross to the opposite side of the body, but arteries do not.) (c) Hearing, since it is perceived in the temporal lobe. (Visual areas are in occipital lobes). (d) Left foot.

21 Letters *P–U* refer to Figure LG 21-3. (a) Celiac, *P:* hepatic to liver, 1; gastric to stomach, 2; splenic to spleen, 3. (b) Superior mesenteric, *Q.* (c) Suprarenal (adrenal), *R.* (d) Renal, *S,* to kidney. (e) Gonadal, *T.* (f) Inferior mesenteric, *U,* to left side of large intestine and colon.

22 L-4, common iliac.

23 Right external iliac artery → right lateral circumflex artery (right descending branch) → right anterior tibial artery → right dorsalis pedis artery → to arch and digital arteries in foot.

24 Superior vena cava, inferior vena cava, coronary sinus.

25 (a) Intracranial vascular (venous) sinuses, functionally, endothelium with no smooth muscle but supported by dura mater. (b) Superior sagittal. (c) Internal jugular, common carotid. (d) There are no such vessels.

26 (a) The heart and a lung. (b) Yes, Yes, Yes.

27 (a) Left radial vein → left brachial vein → left axillary vein → left subclavian vein → left brachiocephalic vein → superior vena cava. (Note that there are left and right brachiocephalic veins, but only one artery of that name.) (b) Small veins in medial aspect of left knee → left popliteal vein → left femoral vein → left external iliac vein → left common iliac vein → inferior vena cava. (c) Right renal vein → inferior vena cava.

28 (a) Lower, slower. (b) More, superficial.

29 (a) Left cephalic vein → left axillary vein (a deep vein) → pathway continues in answer 27a . (b) Left great saphenous vein → left femoral vein → pathway continues as in answer 27b .

30 (a) Axillary. (b) Axillary. (c) Femoral. (d) Popliteal.

31 Right, right.

32 (a) Stomach, intestines, pancreas, gallbladder; hepatic portal. (b) Superior and inferior mesenteric, splenic, gastric, cystic. (c) Larger veins which lead to the right atrium. (d) Hepatic. (e) Sinusoids. (f) Hepatic veins. (g) Blood containing ingested substances can be "inspected and processed." While in liver sinusoids, blood is cleaned, modified, detoxified. Nutrients and other substances are metabolized and stored.

33

Fetal Structure	Structures Connected	Function	Fate of Structure After Birth
a. Umbilical arteries (2)	Fetal internal iliac arteries to placenta	Carries fetal blood low in oxygen and nutrients and high in wastes	Lateral umbilical ligaments
b. Placenta	(Omit)	Site where maternal and fetal blood exchange gases, nutrients and wastes	Delivered as "afterbirth"
c. Umbilical vein (1)	Placenta to liver and ductus venosus	Carries blood high in oxygen and nutrients, low in wastes	Round ligament of the liver
d. Ductus venosus	Umbilical vein to inferior vena cava	Branch of umbilical vein, bypasses liver	Ligamentum venosum
e. Foramen ovale	Right and left atria	Bypasses lungs	Fossa ovalis
f. Ductus arteriosus	Pulmonary artery to aorta	Bypasses lungs	Becomes ligamentum arteriosum

34 (a) Umbilical vein. (b) Ductus venosus. (c) Femoral artery. (d) Pulmonary artery (no oxygenation occurs in fetal lungs). (e) Thoracic portion of inferior vena cava (since umbilical vein blood enters it via ductus venosus).

35 Endorphins.

36 (a) ↓. (b) ↓. (c) ↑. (d) ↑.

37 140/90

38 (a) K. (b) AM. (c) AC.

39 Embolus (a "clot-on-the-run") travels through femoral vein to external and common iliac veins to inferior vena cava to right side of heart. It lodges in pulmonary arterial branches, blocking further blood flow into the pulmonary vessels.

MASTERY TEST: Chapter 21

Questions 1–2: Choose ALL correct answers to each question.

_____ 1. In the most direct route from the left leg to the left arm of an adult, blood must pass through all of these structures:

      A. Inferior vena cava   B. Brachiocephalic artery
      C. Capillaries in lung   D. Hepatic portal vein
      E. Left subclavian artery   F. Right ventricle of heart
      G. Left external iliac vein

_____ 2. In the most direct route from the fetal right ventricle to the fetal left leg, blood must pass through all of these structures: **475**

    A. Aorta   B. Umbilical artery
    C. Lung   D. Ductus arteriosus
    E. Ductus venosus   F. Left ventricle
    G. Left common iliac artery

Questions 3–9: Choose one answer to each question. Note whether the underlined word or phrase makes the statement true or false.

_____ 3. Choose the FALSE statement.

    A. In order for blood to pass from a vein to an artery, it must pass through chambers of the <u>heart.</u>
    B. In its passage from an artery to a vein a red blood cell must ordinarily travel through a <u>capillary.</u>
    C. The wall of the femoral artery is <u>thicker</u> than the wall of the femoral vein.
    D. Most of the smooth muscle in arteries is in the <u>tunica interna.</u>

_____ 4. Choose the FALSE statement.

    A. Arteries <u>contain valves, but veins do not.</u>
    B. Decrease in the size of the lumen of a blood vessel by contraction of a smooth muscle is called <u>vasoconstriction.</u>
    C. <u>End-arteries</u> are vessels that do not anastomose.
    D. Sinusoids are <u>wider and more tortuous (winding)</u> than capillaries.

_____ 5. Choose the FALSE statement.

    A. Exercise tends to <u>increase</u> HDL levels.
    B. Nicotine is a <u>vasodilator which helps control</u> hypertension.
    C. <u>Cor pulmonale</u> refers to heart disease such as right-sided heart failure resulting from resistance to blood flow through lungs.
    D. Phlebitis is an <u>inflammation of a vein.</u>

_____ 6. Choose the FALSE statement.

    A. The vessels which act as the major regulators of blood pressure are <u>arterioles.</u>
    B. The <u>only</u> blood vessels which carry out exchange of nutrients, oxygen, and wastes are <u>capillaries.</u>
    C. The velocity of blood flow is <u>faster</u> in capillaries than in veins.
    D. At any given moment more than 50 percent of the blood in the body is in <u>veins.</u>

_____ 7. Choose the TRUE statement.

    A. The basilic vein is located <u>lateral</u> to the cephalic vein.
    B. The great saphenous vein runs along the <u>lateral</u> aspect of the leg and thigh.
    C. The vein in the body most likely to become varicosed is the <u>femoral.</u>
    D. The hemiazygos and accessory hemiazygos veins lie to the <u>left</u> of the azygos.

_____ 8. Choose the TRUE statement.

    A. A normal blood pressure for the average adult is <u>160/100.</u>
    B. During exercise, blood pressure will tend to <u>increase.</u>
    C. Sympathetic impulses to the heart and to arterioles tend to <u>decrease</u> blood pressure.
    D. Decreased cardiac output causes <u>increased</u> blood pressure.

_____ 9. Choose the TRUE statement.

    A. Cranial venous sinuses are composed of the <u>typical three layers</u> found in walls of all veins.
    B. Cranial venous sinuses all eventually empty into the <u>external jugular veins.</u>
    C. Internal and external jugular veins <u>do</u> unite to form common jugular veins.
    D. Most parts of the body supplied by the internal cartoid artery are ultimately drained by the <u>internal jugular vein.</u>

Questions 10–14: Choose the one best answer to each question.

_____ 10. Hepatic portal circulation carries blood from:

    A. Kidneys to liver
    B. Liver to heart
    C. Stomach, intestine, and spleen to liver
    D. Stomach, intestine, and spleen to kidneys
    E. Heart to lungs
    F. Pulmonary artery to aorta

_____ 11. Which of the following is a parietal (rather than a visceral) branch of the aorta?

    A. Superior mesenteric artery   B. Bronchial artery
    C. Celiac artery   D. Lumbar artery
    E. Esophageal artery

_____ 12. All of these vessels are in the leg or foot EXCEPT:

    A. Saphenous vein   B. Azygos vein
    C. Peroneal artery   D. Dorsalis pedis artery
    E. Popliteal artery

_____ 13. All of these are superficial veins EXCEPT:

    A. Median cubital   B. Basilic   C. Cephalic
    D. Great saphenous   E. Brachial

_____ 14. Which fetal structure contains more highly oxygenated blood?

    A. Umbilical artery   B. Umbilical vein

Questions 15–20: Arrange the answers in correct sequence.

_____ _____ _____ _____ 15. Route of a drop of blood from the right side of the heart to the left side of the heart:

    A. Pulmonary artery
    B. Arterioles
    C. Capillaries
    D. Venules and veins

_____ _____ _____ _____ 16. Blood pressure in vessels, from highest to lowest:

        A. Aorta and other arteries

        B. Arterioles

        C. Capillaries

        D. Venules and veins

_____ _____ _____ _____ 17. Total cross-sectional areas of vessels, from highest to lowest:

        A. Aorta and other arteries

        B. Arterioles

        C. Capillaries

        D. Venules and veins

_____ _____ _____ _____ 18. Velocity of blood in vessels, from highest to lowest:

        A. Aorta and other arteries

        B. Arterioles

        C. Capillaries

        D. Venules and veins

_____ _____ _____ _____ _____ 19. Route of a drop of blood from small intestine to heart:

        A. Superior mesenteric vein

        B. Hepatic portal vein

        C. Small vessels within the liver

        D. Hepatic vein

        E. Inferior vena cava

_____ _____ _____ 20. Layers of blood vessels, from most superficial to deepest:

        A. Tunica interna

        B. Tunica externa

        C. Tunica media

Questions 21–25: fill-ins. Complete each sentence or answer the question with the word(s) or phrase that best fits.

_____ 21. A weakened section of a blood vessel forming a balloonlike sac is known as a(n) _____.

_____ 22. The large superficial vein along the anteromedial side of the leg and thigh is the _____ vein.

_____ 23. In congestive heart failure (CHF), increase in right atrial pressure may be indicated by distension of the _____ veins in the neck.

_____ 24. In taking a blood pressure, the cuff is first inflated over an artery. As the cuff is then slowly deflated, the first sounds heard indicate the level of ____ blood pressure.

_____ 25. Most arteries are paired (one on the right side of the body, one on the left). Name five vessels that are unpaired.

22

The Lymphatic System and Immunity

The lymphatic system works side by side with the cardiovascular system in transporting fluids. It also provides vital defense mechanisms. In this chapter you will examine components, structure, and functions of the lymphatic system and follow circulatory pathways of lymph (Objectives 1–5). You will consider the variety of bodily defense factors that provide immunity (6–7, 11). You will study antigens and antibodies (8–9) and contrast B cells and T cells in immunity (10). You will discuss roles of the immune system in causes and treatments of cancer and other disorders (12–13, 15). You will look at the development of the lymphatic system (14). And you will also learn about medical terminology associated with the lymphatic system and immunity (16).

Topics Summary

A. Lymphatic organization
B. Lymphatic tissue
C. Lymph circulation
D. Nonspecific resistance to disease
E. Immunity (specific resistance to disease)
F. Monoclonal antibodies; immunology and cancer
G. Developmental anatomy
H. Disorders and medical terminology

Objectives

1. Describe the components of the lymphatic system and list its functions.
2. Describe the structure and origin of lymphatics and contrast them with veins.
3. Describe the histological aspects of lymph nodes, tonsils, spleen, and thymus gland and explain their functions.
4. Trace the general plan of lymph circulation from lymphatics into the thoracic duct or right lymphatic duct.
5. Discuss how edema develops.
6. Explain the difference between nonspecific and specific resistance.
7. Discuss the roles of the skin and mucous membranes, antimicrobial substances, phagocytosis, inflammation, and fever as components of nonspecific resistance.
8. Define immunity. explain the relationship between an antigen and an antibody.
9. Contrast the main characteristics of antigens and antibodies.
10. Contrast the role of T cells in cellular immunity and the role of B cells in humoral immunity.
11. Explain the role of the skin in immunity.
12. Define a monoclonal antibody and explain its importance.
13. Discuss the relationship of immunology to cancer.
14. Describe the development of the lymphatic system.
15. Describe the clinical symptoms of the following disorders: hypersensitivity (allergy), tissue rejection, autoimmune diseases, acquired immune deficiency syndrome (AIDS), severe combined immunodeficiency (SCID), and Hodgkin's disease.
16. Define medical terminology associated with the lymphatic system.

(pages 530–531) **A. Lymphatic organization**

1   1. List the components of the lymphatic system.

2. Discuss the organization of the lymphatic system in the following categories:
   diffuse lymphatic tissue, solitary nodules, aggregates of nodules, and *organs.*

2   3. Describe the functions of the lymphatic system in this exercise.

   a. Lymphatic vessels drain from tissue spaces_____
   which are too large to readily pass into blood capillaries. They also trans-

   port_____from the gastrointestinal (GI) tract to the
   bloodstream.

   b. The defensive functions of the lymphatic system are carried out largely

   by white blood cells named_____-cytes. *(B? T?)* lymphocytes destroy foreign substances directly, while B lymphocytes lead to

   the production of_____.

4. Contrast in terms of structure and location:
   a. Blood capillaries/lymph capillaries

   b. Veins/lymphatics

5. What is a *lymphangiogram?* Explain how it is used clinically.

B. Lymphatic tissue

1. Do this exercise about functions of lymph nodes. Refer to lettered structures in Figure LG 22-1. [3]
    a. Lymph is conveyed into a node at several points as at *(A)* by *(afferent? efferent?)* lymphatic vessels.
    b. Lymph enters channel *B* and then channel *C* in the node. Label those parts of the figure. These channels are lined with cells that carry out the process of _____.
    c. Finally lymph exits from the node by means of an *(afferent? efferent?)* lymphatic vessel (letter____on the figure). This is located in a depression of the node known as a _____.
    d. Lymphocytes are formed and added to lymph in areas ____ and ____ in the figure. The germinal center of the nodule is labeled with letter _____, while area *F* is known as a _____cord.

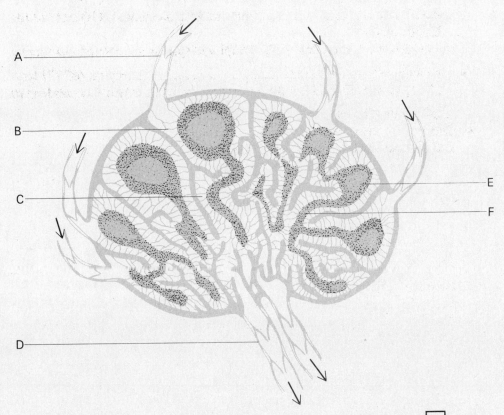

Figure LG 22-1 Structure of a lymph node. Letters refer to Learning Activity [3]

2. Describe how lymph is "processed" as it circulates through lymph nodes.

How does this function of lymph nodes explain the fact that nodes ("glands") become enlarged and tender during infections?

3. Explain how the lymphatic system plays a role in metastasis of cancer.

4. Name three lymphatic organs.

5. *A clinical challenge.* Which tonsils are most commonly removed in a tonsillectomy?_____State one disadvantage of loss of tonsillar tissue.

4

5

6. The spleen is located in the *(upper right? upper left? lower right? lower left?)* quadrant of the abdomen, immediately inferior to the

_____.

7. Contrast *white pulp* with *red pulp* of the spleen.

8. List three functions of the spleen. 6 **483**

For extra review. Name other parts of the body that could take over each of these functions if the spleen were removed.

9. The thymus gland is located in the *(neck? mediastinum?)*. Its position is 7
   just posterior to the_____bone. Its size is relatively *(large? small?)* in children, *(increasing? decreasing?)* in size in adulthood.
10. Summarize the role of the thymus gland in immunity.

C. Lymph circulation

(pages 536–537)

1. Describe the general pathway of lymph circulation. Refer to plasma, interstitial fluid, and cells.

   For extra review. Refer to Figure LG 1-2, page LG 11. Draw arrows on the figure to show movement of fluids leading to lymph formation.
2. Now explain how lymph finally returns to blood. Be sure to include *lymph nodes* and *lymph trunks* in your description.

3. Answer the following questions about the largest lymphatics. 8
   a. In general, the thoracic duct drains *(three-fourths? one-half? one-fourth?)* of the body's lymph. This vessel starts in the lumbar region as a
      dilation known as the_____. The cisterna chyli receives lymph from which areas of the body?

484

b. In the neck the thoracic duct receives lymph from three trunks. Name these.

c. One-fourth of the lymph of the body drains into the _____ duct. The regions of the body from which this vessel collects lymph are:

d. The thoracic duct empties into the junction of the blood vessels named left _____ and left_____ veins. The right lymphatic duct empties into veins of the same name on the right side.

e. In summary, lymph fluid starts out originally in plasma or cells and circulates as_____ fluid. Then as lymph it undergoes extensive cleaning as it passes through nodes. Finally fluid and other substances in lymph are returned to_____ in veins of the neck.

4. The lymphatic system does not have a separate heart for pumping lymph. Describe two principal factors which are responsible for return of lymph from the entire body to major blood vessels in the neck. (Note that these same factors also facilitate venous return.)

9

5. *A clinical challenge.* Explain why a person with a broken left clavicle might exhibit these two symptoms:

10

a. Edema, especially of the left arm and both legs

b. Excess fat in feces

(pages 537–542) **D. Nonspecific resistance to disease**

1. Contrast *resistance* and *susceptibility*.

2. Identify the types of resistance described in each case as specific (S) or nonspecific (N). [11]

      _____ a. Resistance against a single pathogen, such as the bacterium causing tuberculosis

      _____ b. Specific antibody against a specific antigen (such as a type of bacterium or its toxin)

      _____ c. Mechanical barriers such as skin and mucous membranes or chemicals such as acid stomach secretions

      _____ d. Phagocytosis

      _____ e. The inflammatory process

3. List three structural features of skin which make it a protective barrier against microbes.

4. Mucous membranes are *(highly? not very?)* susceptible to penetration by bacteria. Why?

5. Explain how each of the following provides the body with protection against microorganisms.
a. Lacrimal fluid (tears)

b. Saliva

c. "Ciliary escalator" of the respiratory tract

6. *A clinical challenge.* Explain why a decreased rate of urine production may lead to urinary tract infection (UTI).

[12]

7. Which organs produce the chemicals listed below? Explain how each chemical factor guards against microbial invasion.
   a. Gastric juice

   b. Lysozyme

8. In Activity [3], page LG 485 you described structural features of skin which offer protection. Now state two chemical aspects of skin that provide resistance against microbes.

[13]

9. You have considered a number of mechanical and chemical factors which offer protection to the body. Now list three other antimicrobial substances made by the body which provide nonspecific resistance to disease.

[14]

10. Complete this exercise about interferon.
    a. Interferon is produced by *(viruses? cells infected with virus?)*. Name two types of human cells that have been shown to produce interferon.

    b. In what way does interferon offer protection to the body?

c. Name several types of viruses against which interferon offers protection.

d. List several types of cancer that seem to respond to interferon.

e. State two mild and two serious side effects that are linked to therapeutic use of interferon.

11. Describe *complement* in this exercise.
    a. Complement refers to a group of *(lipids? proteins?)* in blood *(cells? serum?)*. Its name indicates that it helps (or complements) the action of the body's *(antibodies? antigens?)*.
    b. Once an antibody recognizes an antigen (such as a microbe) and helps complement attach to the microbe, complement destroys the microbe in several ways. One method is disruption (or _____

    _____) of the microbial membrane. Another is by chemically   coating   the   microbe

    (the process of _____). The chemical attraction of a

    phagocyte to an antigen is known as _____.
    c. In addition, complement helps to cause release of _____

    _____ from certain cells (such as mast cells).
12. Define *properdin* and discuss functions of properdin.

13. Contrast these classes of phagocytic cells, stating two examples of each.
    a. Microphages

    b. Macrophages

14. Cite several examples of mechanical or chemical trauma that could cause inflammatory responses.

15. Defend or dispute this statement: "Inflammation is a process that can be both helpful and harmful."

16. List the four fundamental symptoms of inflammation and a fifth symptom that can occur.

17. Explain how these symptoms of inflammation occur by completing the following description of the stages of inflammation.
    a. *Vasodilation* refers to an increase in _____ _____ This process occurs as a result of release of the chemical named _____ from cells damaged by the trauma. Four types of cells which are responsible for release of histamine are _____

    _____.

    b. *Increased permeability* of blood vessels means that substances normally

    retained in blood are permitted to _____

    _____.

    c. Three other types of chemicals that play a role in both vasodilatation and increased permeability are _____.

d. One effect of vasodilation and increased permeability is that large numbers of _____ cells reach the site of the inflammation. How may such cells be helpful?

e. Clotting elements of the blood can also arrive at the scene of infection. A soluble protein called _____ is converted into the insoluble protein fibrin. How does the formation of a clot help to control infection?

f. Distressing symptoms associated with inflammation occur simultaneously and are directly related to the helpful activities described above. Symptoms of redness and heat are both due to _____ _____. Swelling (edema) is the result of _____ _____. List four factors associated with inflammation that cause pain.

g. *Phagocytic migration,* due to increased blood flow and increased production (initiated by _____-promoting factor), brings large numbers of blood cells to the inflamed region. White blood cells, especially those named _____, adhere to the insides (endothelium) of vessel walls, an occurrence called _____ _____ .

h. Some neutrophils squeeze through walls of blood vessels to reach the damaged area. This process is known as _____ (literally, "to leap through"). What do neutrophils do once they reach the "scene of the crime"?

i. Diapedesis occurs as a result of the fact that neutrophils are attracted by chemicals present in the damaged area. Such chemical attraction is known as _____.

j. Whereas _____ are the predominant phagocytic cells in early stages of inflammation, later phagocytic migration involves two other kinds of cells. Monocytes are white blood cells which become transformed into large cells called _____. Additional macrophages migrate to the scene from other parts of the body. What are functions of macrophages?

18. Explain the relationship between the terms in each pair.
    a. Pus/abscess

19. *ACTive learning.* Role play in a study group, lab activity, or classroom dramatization the nonspecific defense against microbial invasion of the body. Determine which wall of the room is the skin (or mucosa) and which door/window serves as a break in the skin (or mucosa). Individuals assume roles in the following categories: (a) Invading microbes (determine whether viral or bacterial). (b) Chemicals produced by skin and mucosa to attempt to thwart entrance of the microbe. (c) Other nonspecific antimicrobial chemicals: demonstrate how interferon will assist if the invader is viral; several persons portraying proteins of the complement system and properdin show specific activities that help to eliminate the invader. (d) Finally two different types of phagocytic cells/persons dramatize adherence and ingestion of the microbe. Demonstrate whether any microbes survive phagocytosis. Determine the outcome of the infection: who wins—invaders or host?

(pages 542–548)   **E. Immunity (specific resistance to disease)**

18 1. Do this exercise on antigens.

a. An antigen is defined as "any chemical substance which, when introduced into the body, causes_____

_____." In general, antigens are *(parts of the body? foreign substances?).*

b. Antigens have two characteristics:_____, the ability to stimulate specific antibodies, and *reactivity,* the ability to_____

_____ _____.

c. An antigen with both of these characteristics is called a_____

_____ antigen.

d. Chemically, antigens:
   A. Are all protein
   B. May be proteins, proteins combined with other organic compounds, or polysaccharides

e. Can an entire microbe serve as an antigen? *(Yes? No?)* List the parts of microbes which may be antigenic.

f. If you are allergic to pollen in the spring or fall or to certain foods, the pollen or foods serve as *(antigens? antibodies?)* to you.

g. Antibodies form against *(the entire? only a specific region of the?)* antigen. This region is known as the_____. Most antigens have *(only one? only two? a number of?)* antigenic determinant sites, and so are called_____-valent.

h. In general, antigens have a molecular weight greater than *(10,000? 10 million?)*. Each determinant site may have a molecular weight of only about_____.

i. An individual determinant site has the characteristic of *(reactivity? immunogenicity? both?)* as a result of its low molecular weight. A determinant site with reactivity but not immunogenicity is called a

_____.

j. A hapten can form a complete antigen by _____

_____. For example, in individuals who are allergic to penicillin, penicillin serves as a *(complete antigen? hapten?)* which combines with a larger protein in the body to form a complete antigen.

2. Describe antibodies in this exercise.
   a. Contrast antigen with antibody.

   b. Most antibodies are *(multivalent? bivalent?)*. `19`

3. Chemically, all antibodies are composed of proteins named _____

_____. Since they are involved with immunity, these are called immunoglobulins, abbreviated _____.

Write the name of the related class of Ig next to each description. `20`

_____ a. The only type of Ig to cross the placenta, it provides specific resistance to newborns. Also significantly enhances phagocytosis and neutralization of toxins in persons of all ages since it is the most abundant type of antibody.

_____ b. Found in secretions such as mucus, saliva, and tears, so protects against oral, vaginal, and respiratory infections.

_____ c. Involved in allergic reactions, for example, to certain foods, pollen, or bee venom.

_____ d. Antibodies that can cause agglutination and lysis, and in this way can destroy invading microbes.

_____ e. Antibodies which act as receptors for antigens on surface of B-lymphocytes, and in this way can lead to production of antibodies against the antigen.

4. Refer to Figure LG 22-2 and do this exercise. `21`
   a. This diagram shows that an antibody contains *(two? four? six?)* chains of

_____. Two chains are heavy, containing_____

_____amino acids each. Color these red. Two are light. Color these blue.

492

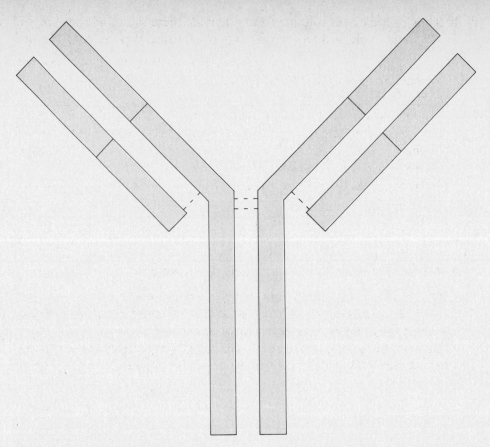

Figure LG 22-2 Diagrammatic representation of an antibody molecule. Complete as directed in Learning Activity [21].

b. These chains are held together by _____ bonds. Label these.

c. All antibodies of a specific class, for example, IgG, contain the same *constant portions*. Label these regions *C*. *Variable portions* differ for each specific antibody. Label these *V*. The site that is uniquely designed to bind with a specific antigen is *(C? V?)*.

d. The antibody in this figure is in the *(T? Y?)* shape, as it would be in the *(combined? uncombined?)* state. In the margin, draw a small diagram of this antibody in the uncombined shape. Complement can fix to the antibody in the *(T, uncombined? Y, combined?)* state.

[22] 5. Contrast two types of immunity by writing C before descriptions of cellular immunity and H before those describing humoral immunity.

_____ a. Final products are sensitized T lymphocytes

_____ b. Involves formation of antibodies which circulate in the bloodstream

_____ c. Especially effective against bacterial and viral infections

_____ d. Effective against fungi, parasites, certain viruses, cancer, and tissue transplants

_____ e. Involves B cells that develop into plasma cells

[23] 6. Complete this exercise about lymphocytes.

a. Name the two main types of lymphocytes: ____ cells and ____ cells. Both of these originate in _____ tissue.

b. Some of these cells migrate to the thymus; there these cells become *(B? T?)* cells. This involves processing of T cells, that is, they develop

_____. Thus, lymphocytes have the ability to differentiate into cells that perform specific immune functions against specific antigens.

c. Immunological competence is conferred *(shortly before and after birth? throughout a lifetime? mostly later in life?)*. Therefore removal of the thymus gland *(shortly after birth? at age 30?)* is likely to have more deleterious effects upon immunity.

d. What is the basis for the name B cells?

e. After processing, both T and B cells then move to_____

_____ where they reside until needed.

7. Refer to Figure LG 22-3 and do this Learning Activity about the sequence of events that occurs when one type of antigen enters the body. 24

a. How many different types of T cells are in the body? *(One? Twelve? Thousands?)* How do T cells differ from one another?

At any given time *(most? few?)* T cells are active. When one type of antigen enters the body, one type of T cell will be activated.

b. First the invading antigens must be processed by cells such as

_____ *(A* in Figure LG 22-3). The macrophages initiate a series of events that may lead to the destruction of the invader. In the first of these, the macrophages present the antigens to T cells (which we will consider in more detail in *C* and *E).*

c. However, in order for T cells to respond to the presented antigens and help eliminate them, T cells must be stimulated by chemicals known as

_____. Name two cells that secrete interleukin I.

_____ *(A)* and_____ *(P).*

d. One function of interleukin 1 is stimulation of *(helper? killer?)* T cells *(C).* These cells have several functions as shown on the figure. One is secretion of proteins that enhance the phagocytic effects of macrophages *(A).* Note that action on the figure.

e. A second job description for helper T cells is production of a chemical

named_____ *(D),* which stimulates cloning of T cells. (See *E.*) What is a clone?

494

O ← Antigens enter → Q

A

secrete

process and present antigen

P

B. Interleuken 1

process and present antigen

activate

activate

stimulate cloning

attract and activate

C

cooperate with

I

secrete

may become

inhibit

D

N. Suppressor T cells

L

stimulate cloning

differentiate and clone to form

later becomes

inhibit

E

J

release

secrete

H

recruit and stimulate

secrete

decrease viral replication

secrete

F

may become

G

directly kill antigen

complex with antigens

M. T Memory cells

K

Figure LG 22-3 Diagram of sequence of events in immune response (specific resistance to disease). Ovals contain names of cells; rectangles contain names of chemicals. Solid lines indicate enhancing effects; broken lines show inhibitory relationships. Complete as directed in Activities 24 , 26 , and 27 .

KEY

A. Macrophages
B. Interleukin 1
C. Helper T cells (lymphocytes)
D. Interleukin 2
E. Killer T cells (lymphocytes)
F. Lymphokines
G. Lymphotoxins
H. Interferon

I. B cells (lymphocytes)
J. Plasma Cells
K. Antibodies
L. B memory cells (lymphocytes)
M. T memory cells (lymphocytes)
N. Suppressor T cells (lymphocytes)
O. Langerhans cells
P. Keratinocytes
Q. Granstein cells

f. As a result of presentation of the antigens (by the macrophages) and stimulation by interleukin 2, a large number of killer T cells are now said to be _____ toward the specific invader antigens and ready to combat them.

g. Now consider the array of effects of these killer T cells. They secrete chemicals called _____ *(F)*. What is the overall effect of lymphokines? (More specifics about these in Activity 25 .)

h. Killer T cells also release chemicals named _____ *(G)*. By what means do these directly destroy the antigens?

i. One more function of killer T cells is secretion of interferon *(H)*. This chemical decreases replication of invaders that are *(viral? bacterial?)*, and also exhibits a *(positive? negative?)* feedback effect upon killer T cells.

j. In summary, macrophages *(A)* recognize a type of antigen. With the help of interleukins *(B and D)*, T cells are called to action *(C and E)*. Both types of T cells release lymphokines (as at *F)* that enhance the phagocytic activities of macrophages *(A)*. Killer T cells *(E)* also directly kill or decrease replication of antigens *(G and H)*. All of these actiities involve either direct or indirect actions of cells (T cells and macrophages). So the defenses just described are providing *(humoral? cell-mediated?)* immunity.

8. In Activity 24g you considered overall effects of lymphokines. Now match the names of chemicals in this class with their specific actions listed below.
   25

____ a. Attracts additional macrophages to the scene of the infection

____ b. Activates the macrophages

____ c. Prevents macrophages from moving away from the scene of the infection

____ d. Stimulates mitosis among lymphocytes so that numbers increase

MAF. Macrophage activating factor

MCF. Macrophage chemotactic factor

MF. Mitogenic factor

MIF. Migration inhibitory factor

9. Contrast natural killer (NK) with killer T cells.

10. Now that you have considered cell-mediated immune responses, check your understanding of humoral responses in this exercise. Again refer to Figure LG 22-3.
    26

a. In humoral immunity, the final products (antibodies) are made by (*T? B?*) lymphocytes. In order for these cells to be activated to perform their roles, B cells *(must be? do not necessarily need to be?)* presented with a specific type of antigen that fits that type of B cell. Name a cell that does such presenting._____*(A)* on the figure. This activity is enhanced by release of the chemical _____ *(B)*. These initial steps of humoral immunity are quite *(different from? similar to?)* early stages of cell-mediated immunity.

b. B cells can then differentiate and divide to form a clone of _____ cells *(J)*. Plasma cells then *(stay in lymphoid tissue? circulate in blood?)*. There they secrete _____*(K)* that can complex with the specifically related antigen. This complex can fix the nonspecific proteins called _____to further enhance the destruction of the antigens. *(For extra review* of complement, refer back to Learning Activity 15 .)

c. Describe a typical rate of antibody procution: _____ molecules/sec of antibody may be secreted for a period of several *(minutes? days? years?)* until the plasma cell dies.

d. Some plasma cells do not differentiate into plasma cells right away. Instead they remain as_____cells *(L)*. Once a person is sensitized against a specific antigen, either by infection or by a(n) *(initial? booster dose?)* immunization, the person produces some memory cells. These will provide a *(more? less?)* intense response upon subsequent exposure to the antigen, as during another infection or by a _____dose.

e. T cells *(do? do not?)* produce memory cells. (See *M* on the figure.) Both the B and T memory cells evoke the intense secondary response (as to a booster dose). This is also called an_____response to an antigen.

f. Note one other factor that enhances humoral responses. *(Helper? Killer?)* T cells influence B cell differentiation to plasma cells. Identify that relationship on the figure. Now find the type of lymphocyte that can inhibit both plasma cell formation and killer T functions. This is a _____ cell (____ on the figure).

11. What is the meaning of the term antibody titer?

12. Earlier in this chapter you studied nonspecific resistance offered by skin. Now examine specific immunity provided by skin. Refer to Figure LG 22-3.

27

a. Antigens entering skin may bind to epidermal cells known as _____cells *(O in the figure)*. These cells than present the antigens to_____cells *(C)*. Draw an arrow showing that relationship on the figure.

b. Other skin cells called keratinocytes *(P)* secrete_____

_____ *(B)*. State its two effects.

c. Note that antigens may enter_____cells *(Q)* in skin.

These cells then interact with suppressor T cells *(N)*, _____

_____-creasing immune responses, therefore provid-
ing some regulation of immunity in skin. Draw an arrow showing that
relationship on the figure.

F. Monoclonal antibodies; immunology and cancer   (page 549)

1. Define *monoclonal antibodies.*

These antibodies are produced by hybridoma cells. Explain how such a cell
can be produced.

2. Describe three clinical uses of monoclonal antibodies.

3. Answer these questions about the relationship between immunology and
cancer.
   a. What are tumor-specific antigens? How are they related to immunologic
   surveillance?

28

b. Most researchers believe that tumors are destroyed by *(cellular? humoral?)* immune responses. Therefore *(T? B?)* cells sensitized to the cancer antigens play the major role.

c. State two ways that cancer cells may carry out *immunological escape.*

(pages 549–550) **G. Developmental anatomy of the lymphatic system**

29

1. The lymphatic system derives from _____-derm, beginning about the *(fifth? seventh? tenth?)* week of gestation.

30

2. Lymphatic vessels arise from lymph sacs that form from early *(arteries? veins?).* Name the embryonic lymph sac that develops vessels to each of the following regions of the body.

   a. Pelvic area and lower extremities: _____ lymph sac

   b. Upper extremities, thorax, head, and neck: _____ lymph sac.

(pages 550–552) **H. Disorders, medical terminology**

1. Name four examples of allergens.
2. Another term for allergic is _____.

31

3. Based on what you know about antigens and antibodies, complete this exercise about how one type of allergic reaction—anaphylaxis—occurs.

   a. In the first stage B cells respond to an allergen (antigen) by producing

   Ig_____ antibodies. These attach to certain types of body cells, namely,

   _____ cells and _____. These

   cells are then said to be _____ to the antigen.

   b. At a later time exposure to the same antigen causes the IgEs to bind together, leading to release of chemicals from mast cells and basophils.

   Among these chemicals is one named _____.

   c. Some effects of these chemicals which are associated with anaphylaxis are: *(increased? decreased?)* capillary permeability, leading to extra fluid in

   tissue spaces (the condition called _____) and also to lowered blood pressure; *(increased? decreased?)* smooth muscle contraction of bronchial tubes, making breathing *(easier? more difficult?); (increased? decreased?)* mucous production, causing a "runny nose."

4. Transplantation is likely to lead to tissue rejection because the transplanted organ or tissue serves as an *(antigen? antibody?).* The body tries to reject this foreign tissue by producing _____ _____. | 32 |

5. Match types of transplants with descriptions below. | 33 |

_____ a. Between animals of different species   A. Allograft

_____ b. The most successful type of transplant   I. Isograft

                                                           X. Xenograft

_____ c. Between individuals of same species, but of different genetic backgrounds, such as a mother's kidney donated to her daughter or a blood transfusion

6. Describe how immunosuppressant therapy is used in transplant situations.

Name one drug often used for transplant patients since it is selectively immunosuppressant.

7. Do this exercise on autoimmune disease. | 34 |
   a. Normally the body *(does? does not?)* produce antibodies against its own tissues. Such self-recognition is called immunological_____ _____. Possible explanations for immunological tolerance are based on actions of_____cells *(N* in Figure LG 22-3). These cells normally *(stimulate? inhibit?)* production of antibodies against "self."

   b. Such inhibition is lost in_____diseases, so the body starts attacking itself.

8. Match the condition at right with the description below. | 35 |

_____ a. A curable malignancy usually arising in lymph nodes   A. Autoimmune disease

_____ b. Multiple sclerosis (MS), rheumatoid arthritis (RA), and systemic lupus erythematosus (SLE) are examples   H. Hodgkin's disease

                                                    L. Lymphangioma

_____ c. Benign tumor of lymph vessels   S. Severe combined immunodeficiency (SCID)

_____ d. Characterized by lack of both B and T cells

9. Complete this exercise about the condition known as AIDS.

    a. AIDS is an acronym which stands for _____

    _____ AIDS victims have too few *(B? T?)* cells.

    b. Mortality rate of AIDS is about_____percent. Describe the two main causes of death of AIDS victims.

    c. The causative microbe of AIDS is a *(virus? bacterium?)*. Name it.

Answers to Numbered Questions in Learning Activities

1   Lymph, vessels (lymphatics), lymph nodules, nodes and other lymph organs.

2   (a) Proteins, lipids. (b) Lympho, T, antibodies.

3   (a) Afferent. (b) *B*, cortical sinus; *C*, medullary sinus; phagocytosis. (c) Efferent, *D*, hilus. (d) *E, F, E*, medullary.

4   Palatine: as lymphatic tissue does become inflamed at times, it does protect against foreign substances.

5   Upper left, diaphragm.

6   Lymphocyte and antibody production; blood reservoir; phagocytosis of microbes, red blood cells, and platelets.

7   Mediastinum (note that the *thyroid* is in the neck), sternum, large, decreasing.

8   (a) Three-fourths, cisterna chyli; digestive and other abdominal organs and both lower extremities. (b) Left jugular, left subclavian, and left bronchomediastinal. (c) Right lymphatic; right upper extremity and right side of thorax, neck, and head. (d) Internal jugular, subclavian. (e) Interstitial, plasma.

9   Skeletal muscle contraction squeezing lymphatics which contain valves directing flow of lymph. Respiratory movements. (See Chapter 21, Activity **15f**, page LG 456.)

10 Fracture of this bone might block the thoracic duct which enters veins close to the left clavicle. Lymph therefore backs up with these results: (a) Extra fluid remains in tissue spaces normally drained by vessels leading to the thoracic duct. (b) Lymph capillaries in the intestine normally absorb fat from foods. Slow lymph flow can decrease such absorption leaving fat in digestive wastes.

11 (a) S. (b) S. (c) N. (d) N. (e) N.

12 Adequate flow of urine normally flushes the urinary passageways, preventing microbial colonization there.

13 Acidic pH of skin and unsaturated fatty acids in sebum.

14 Interferon, complement, properdin.

15 (a) Proteins, serum, antibodies. (b) Lysis, opsonization, chemotaxis. (c) Histamine. (d) Dilates and increases permeability of capillaries.

16 Redness, heat, swelling, pain; loss of function to injured area.

17 (a) Diameter of blood vessels; histamine; mast cells, basophils, platelets, and neutrophils. (b) Pass out of blood vessels to damaged area. (c) Kinins, prostaglandins (especially E series), and serotonin (5-HT). (d) White blood; they can destroy invading microorganisms and remove toxic products, dead cells, and other debris. (e) Fibrinogen; clot prevents hemorrhage and traps invading organisms and toxins to prevent their spread. (f) Increased blood flow to the area (due to increase in heart rate, metabolism, vasodilation, and permeability); accumulation of fluid in tissue (due to increased blood flow and permeability); injury to nerve fibers, irritation of nerves by toxic chemicals, pressure due to swelling, stimulation by kinins and prostaglandins. (g) Leucocytosis, neutrophils, margination. (h) Diapedesis, see answers to (d) above. (i) Chemotaxis. (j) Neutrophils, macrophages; similar to those of white blood cells, described in (d) above, except more virulent since macrophages are larger than neutrophils.

18 (a) The body to produce a specific antibody, foreign substances. (b) Immunogenicity, react with the specific antibody. (c) Complete. (d) B. (e) Yes; flagella, capsules, cell walls, and toxins produced by microbes. (f) Antigens. (g) Only a specific region of the, antigenic determinant, a number of, multi-. (h) 10,000, 200 to 1,000. (i) Reactivity, hapten (or partial antigen). (j) Combining with another molecule, hapten.

19 Bivalent.

20 Globulins, Ig. (a) IgG. (b) IgA. (c) IgE. (d) IgM. (e) IgD.

21 (a) Four, polypeptide, over 400. (b) Disulfide. (c) V. (d) Y; combined; Y, combined.

22 (a) C. (b) H. (c) H. (d) C. (e) H.

23 (a) B, T, bone marrow (myeloid). (b) T, immunological competence. (c) Shortly before and after birth, shortly after birth. (d) Similar cells in birds were known to be processed in a bursa; the processing site of these cells in humans may be the bone marrow. (e) Lymphoid tissue.

24 (a) Thousands; have specific sites on them which can react with only a certain type of antigen; few. (b) Macrophages. (c) Interleukins, macrophages, keratinocytes. (d) Helper. (e) Interleukin 2; a proliferation of cells derived from one cell, in this case, from the one killer T cell specific for this antigen. (f) Sensitized. (g) Lymphokines, recruit and further enhance activity of both macrophages and killer T cells. (h) Lymphotoxins, lyse invading cells by producing holes in their plasma membranes. (i) Viral, positive. (j) Cell-mediated.

25 (a) MCF. (b) MAF. (c) MIF. (d) MF.

26 (a) B, must be, macrophages (or Langerhans: see Learning Activity 27), interleukin(*I*), similar to. (b) Plasma, stay in lymphoid tissue,-antibodies, complement. (c) 2,000, days. (d) Memory, initial, more, booster. (e) Do, anamnestic. (f) Helper, suppressor T *(N)*.

27 (a) Langerhans, helper T. (b) Interleukin 1; stimulates helper T cells to secrete interleukin 2 *(D)*, and also activates B cells to convert to clones of plasma cells. (c) Granstein, de.

28 Cellular, T.

29 Meso, fifth.

30 Veins. (a) Posterior. (b) Jugular.

31 (a) E, mast, basophils, sensitized. (b) Histamine. (c) Increased, edema, increased, more difficult, increased.

32 Antigens; antibodies and also T lymphocytes against the tissue.

33 (a) X. (b) I. (c) A.

34 (a) Does not, tolerance, suppressor T, inhibit. (b) Autoimmune.

35 (a) H. (b) A. (c) L. (d) S.

MASTERY TEST: Chapter 22

Questions 1–3: Arrange the answers in correct sequence.

_____ _____ _____ _____ 1. Flow of lymph through a lymph node:

        A. Afferent lymphatic vessel
        B. Medullary sinus
        C. Cortical sinus
        D. Efferent lymphatic vessel

_____ _____ _____ _____ 2. From first to last in cell-mediated immunity:

        A. Lymphokines and lymphotoxins are released
        B. Interleukin 2 stimulates killer T cells
        C. Helper T cells are activated by interleukin 1 and presentation of the specific antigen
        D. The invading antigen enters macrophage

_____ _____ _____ _____ _____ 3. Activities in humoral immunity, in chronological order

    A. B cells develop in bone marrow or other part of the body

    B. B cells differentiate and divide (clone), forming plasma cells

    C. B cells migrate to lymphoid tissue

    D. B cells are activated by specific antigen

    E. Antibodies are released and are specific against the antigen that activated the B cell

Questions 4–18: Choose the one best answer to each question. Take particular note of underlined terms.

_____ 4. Both the thoracic duct and the right lymphatic duct empty directly into:

    A. Axillary lymph nodes   B. Superior vena cava
    C. Cisterna chyli   D. Subclavian arteries
    E. Junction of internal jugular and subclavian veins

_____ 5. Which cells are the most dominant microphages?

    A. Histiocytes   B. Basophils   C. Lymphocytes
    D. Neutrophils   E. Eosinophils

_____ 6. All of these are examples of nonspecific defenses EXCEPT:

    A. Antigens and antibodies   B. Saliva   C. Complement
    D. Interferon   E. Skin   F. Phagocytes

_____ 7. Which of these is an acid fluid (about pH 2) that guards against bacterial invasion?

    A. Lysozyme   B. Gastric juice   C. Sebum
    D. Saliva   E. Lacrimal fluid   F. Mucus

_____ 8. All of these are major sites of lymphatic tissue EXCEPT:

    A. Tonsils   B. Thymus   C. Kidneys   D. Spleen
    E. Lymph nodes

_____ 9. Choose the FALSE statement about lymphatic vessels.

    A. Lymph capillaries are _more_ permeable than blood capillaries.
    B. Lymphatics have _thinner_ walls than veins.
    C. Like arteries, lymphatics contain _no_ valves.
    D. Lymph vessels are _blind-ended._

_____ 10. Choose the FALSE statement about nonspecific defenses.

    A. Complement functions protectively by _being converted into histamine._
    B. Histamine _increases_ permeability of capillaries so leukocytes can more readily reach the infection site.
    C. Complement and properdin are both _enzymes found in serum._
    D. Opsonization _enhances_ phagocytosis.

_____ 11. Choose the FALSE statement about T cells.

    A. Some are called "killer cells."

    B. They are called T cells because they are processed in the thymus.

    C. They are involved primarily in humoral immunity.

    D. Like B cells, they originate from stem cells in bone marrow.

_____ 12. Choose the FALSE statement about cisterna chyli.

    A. It receives lymph which originated in the lower extremities and in pelvic viscera.

    B. It drains lumbar and intestinal trunks.

    C. It is located in the posterior of the thorax.

    D. It empties lymph directly into the thoracic duct.

_____ 13. Choose the FALSE statement about lymph nodes.

    A. Lymphocytes are produced here.

    B. Lymph nodes are distributed evenly throughout the body, with equal numbers in all tissue.

    C. Lymph may pass through several lymph nodes in a number of regions before returning to blood.

    D. Lymph nodes are shaped roughly like kidney (or lima) beans.

_____ 14. Choose the FALSE statement about lymphatic organs.

    A. The palatine tonsils are the ones most often removed in a tonsillectomy.

    B. The spleen is the largest lymphatic organ in the body.

    C. The thymus reaches its maximum size at age 40.

    D. The spleen is located in the upper left quadrant of the abdomen.

_____ 15. Choose the FALSE statement about the lymphatic system.

    A. Skeletal muscle contraction aids lymph flow.

    B. Mucous membranes are less effective than skin in preventing entrance of microbes into the body.

    C. Interferon is produced by viruses.

    D. An allergen is an antigen, not an antibody.

_____ 16. Choose the TRUE statement about antigens.

    A. Antigens are usually made by the body.

    B. All antigens are proteins.

    C. Microbes serve as antigens, not antibodies.

    D. Most antigens are univalent.

_____ 17. Choose the TRUE statement about antibodies.

    A. Most antibodies are foreign substances, not made by the body.

    B. Antibodies are usually composed of one light and one heavy polypeptide chain.

    C. In the uncombined state antibodies are in the Y shape.

    D. All antibodies are protein.

_____ 18. All of the following are chemicals released by killer T cells EXCEPT:

    A. Interleukin 1   B. Lymphotoxins

    C. Interferon

    D. Macrophage activating factor (MAF)

Questions 19–20: Circle T (true) or F (false). If the statement is false, change the underlined word or phrase so that the statement is correct.

T F 19. Immunogenicity means the ability of an antigen to <u>react with</u> a specific antibody.

T F 20. A person with autoimmune disease produces <u>fewer than normal antibodies.</u>

Questions 21–25: fill-ins. Complete each sentence with the word or phrase that best fits.

_____ 21. ____ Lymphocytes provide humoral immunity, and ____ lymphocytes and macrophages offer cellular immunity.

_____ 22. ____ are a group of chemicals secreted by killer T cells that attract and activate macrophages and recruit even more killer T cells.

_____ 23. ____ cells inhibit both plasma cells and killer T cells.

_____ 24. The class of immunoglobins most associated with allergy are Ig ____.

_____ 25. Three examples of mechanical factors that provide nonspecific resistance to disease are ____.

The Respiratory System

In this chapter you will learn about the system that provides for gaseous exchange between the external environment and body cells. You first will study the passageways that lead to the lungs (Objectives 1–7) and then will consider the composition of the lungs themselves (8–10). You will discuss respiration, including the processes of ventilation, external respiration, and internal respiration (11–16). Next you will learn about transport of gases within the bloodstream (17–18). Then you will examine the control of respiration (20–21). You will also look at the changes in the respiratory system during development and aging (22–23). Finally you will study clinical aspects of the respiratory system, including disorders and medical terminology (19, 24–25).

Topics Summary

A. Upper respiratory passageways
B. Lower respiratory passageways, lungs
C. Respiration: ventilation
D. Respiration: exchange of respiratory gases
E. Transport of respiratory gases
F. Control of respiration, modified forms of respiration
G. Aging and development of the respiratory system
H. Disorders, medical terminology

Objectives

1. Identify the organs of the respiratory system.
2. Compare the structure and function of the external and internal nose.
3. Differentiate the three regions of the pharynx and describe their roles in respiration.
4. Describe the structure of the larynx and explain its function in respiration and voice production.
5. Explain the structure and function of the trachea.
6. Contrast tracheostomy and intubation as alternative methods for clearing obstructed air passageways.
7. Describe the location and structure of the tubes that form the bronchial tree.
8. Identify the coverings of the lungs and the division of the lungs into lobes.
9. Describe the composition of a lobule of the lung.
10. Explain the structure of the alveolar-capillary (respiratory) membrane and its function in the diffusion of respiratory gases.
11. List the events involved in inspiration and expiration.
12. Explain how compliance and airway resistance relate to breathing.
13. Define coughing, sneezing, sighing, yawning, sobbing, crying, laughing, and hiccuping as modified respiratory movements.
14. Compare the volumes and capacities of air exchanged during respiration.
15. Define Boyle's law, Dalton's law, and Henry's law.
16. Explain how external and internal respiration differ.
17. Describe how the oxygen-carrying capacity of the blood is affected by PO_2, and PCO_2, temperature, and DPG.
18. Explain how the respiratory gases are transported by the blood.
19. Distinguish the various types of hypoxia.
20. Explain how the respiratory center functions in establishing the basic rhythm of respiration.
21. Describe how various neural and chemical factors may modify the rate of respiration.
22. Describe the development of the respiratory system.
23. Describe the effects of aging on the respiratory system.
24. Define bronchogenic carcinoma (lung cancer), bronchial asthma, bronchitis, emphysema, pneumonia, tuberculosis, respiratory distress syndrome (RDS) of the newborn, respiratory failure, sudden infant death syndrome (SIDS), coryza (common cold), influenza (flu), pulmonary embolism (PE), pulmonary edema, as disorders of the respiratory system.
25. Define medical terminology associated with the respiratory system.

Learning Activities

(pages 555-560) **A.** Upper respiratory passageways

1. Explain how the respiratory and cardiovascular systems work together to accomplish gaseous exchange among the atmosphere, blood, and cells.

2. Name the structures of the nose that are designed to carry out each of the following functions.
   a. Warm, moisten, and filter air

   b. Sense smell

   c. Assist in speech

3. On Figure LG 23-1, cover the KEY and identify structures associated with the nose and palate: *A, B, C, M, N,* and *R.*

  1 4. State the functions of the pharynx.

5. On Figure LG 23-1, identify the three portions of the pharynx at letters *Q, T,* and *U.* Now write here which of these structures are located in each pharyngeal region: adenoids, palatine tonsils, lingual tonsils, openings (fauces) from oral cavity, openings into auditory tubes, and passageways into both larynx and esophagus.

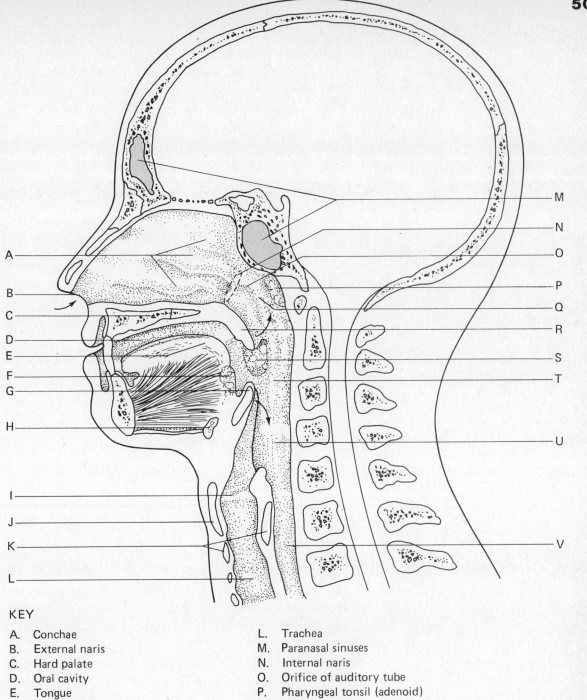

KEY

A. Conchae
B. External naris
C. Hard palate
D. Oral cavity
E. Tongue
F. Lingual tonsil
G. Epiglottis
H. Hyoid bone
I. Vocal folds
(true vocal cords)
J. Thyroid cartilage
K. Cricoid cartilage

L. Trachea
M. Paranasal sinuses
N. Internal naris
O. Orifice of auditory tube
P. Pharyngeal tonsil (adenoid)
Q. Nasopharynx
R. Soft palate
S. Palatine tonsil
T. Oropharynx
U. Laryngopharynx
V. Esophagus

Figure LG 23-1 Sagittal section of the right side of the head with the nasal septum removed.

a. Nasopharynx

b. Oropharynx

c. Laryngopharynx

6. Describe the structure and function of each of these cartilages of the larynx, and identify these on Figure LG 23-1.
   a. Epiglottis

   b. Thyroid cartilage

   c. Cricoid cartilage

7. Explain how the larynx prevents food from entering the trachea.

8. The _____ is a space between the true vocal cords. The [2]
_____ cartilages are pyramid-shaped cartilages of the larynx attached to the vocal cords.

9. Tell how the larynx produces sound. Explain how pitch is controlled and what causes male pitch usually to be lower than female pitch.

10. After a larynx is removed (for example, due to cancer), what other struc- [3] tures help the laryngectomee to speak?

11. *For extra review* of the upper airways, match structures at right with descriptions below. [4]

_____ a. Turbinates and meati are located here.

_____ b. Enlarged adenoids here may block openings into auditory tubes, causing middle ear infections.

_____ c. Palatine tonsils are located here.

_____ d. This structure leads directly into the esophagus.

_____ e. Cricoid, epiglottis, and thyroid cartilages are here.

_____ f. Vocal cords here enable voice production.

_____ g. Internal nares (choanae) are located between these two struc-tures (2 answers).

_____ h. Fauces are located between these two structures (2 answers).

L. Larynx
LP. Laryngopharynx
M. Mouth
N. Nose
NP. Nasopharynx
OP. Oropharynx

512 (pages 561–
565) **B. Lower respiratory passageways, lungs**

    1. What is the function of each of these parts of the trachea?
      a. Ciliated columnar cells

      b. Goblet cells

      c. C-shaped cartilage rings

    5 2. What is the *carina*?

      What is the clinical significance of the carina?

    3. Define these terms and give a clinical application for each procedure.
      a. Bronchoscopy

      b. Tracheostomy

      c. Intubation

      d. Bronchogram

4. *A clinical challenge.* Anna has aspirated a small piece of candy. Dr. Lennon expects to find it in the right bronchus rather than in the left. Why? 6

5. The lower respiratory passageways are known as the bronchial tree. Refer to Figure LG 23-2 and do this exercise. 7

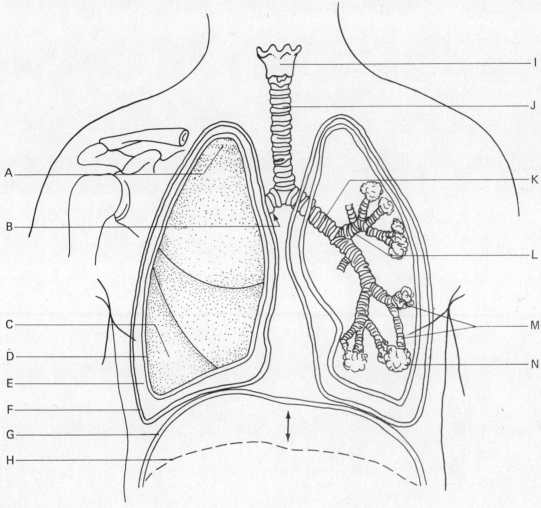

KEY

A. Apex
B. Hilus
C. Base
D. Visceral pleura
E. Pleural cavity
F. Parietal pleura
G. Diaphragm (relaxed)
H. Diaphragm (contracted)

I. Larynx
J. Trachea
K. Primary bronchus
L. Secondary (lobar) bronchus
M. Tertiary (segmental) bronchi and bronchioles
N. Alveoli

Figure LG 23-2 Diagram of lungs with pleural coverings and bronchial tree. Refer to Learning Activities 7 and 9 .

513

a. In what way do the passageways resemble a tree?

b. Identify structures *I–N* on this figure.

6. Describe the histological changes in cartilage, smooth muscle, and epithelium from primary bronchi to terminal bronchioles.

7. Explain how the absence of cartilage rings in bronchioles is significant during an asthma attack.

8. On Figure LG 23-2 color the *visceral pleura* red and the *parietal pleura* blue. Locate the *pleural cavity*.

9. What is the function of serous fluid secreted by the pleura?

8
10. Write the correct term for each of these conditions.

   a. Inflammation of the pleura:_____

   b. Air in the pleural cavity:_____

   c. Blood in the pleural cavity:_____

9
11. Answer these questions about the lungs. (As you do the exercise, locate the parts of the lung on Figure LG 23-2.)

   a. The broad, inferior portion of the lung which sits on the diaphragm is

   called the_____. The upper narrow apex of each lung

   extends just superior to the_____. The costal surfaces

   lie against the_____.

b. Along the mediastinal surface is the_____ where the root of the lung is located. This root consists of_____

_____.

c. Answer these questions with *right* or *left*. Which lung is thicker and broader?_____ In which lung is the cardiac notch located?_____ Which lung has just two lobes and so only two lobar bronchi?_____ Which lung has a horizontal fissure?_____ Write *S* (superior), *M* (middle), and *I* (inferior) on the three lobes of the right lung.

12. Contrast a *lobe* with a *segment* of a lung.

13. The lungs contain an estimated 130,000 lobules including a total of perhaps 300 million alveoli. Describe the structure of a lobule in this exercise. |10|

a. Arrange in order the structures through which air passes as it enters a lobule en route to alveoli. ____ ____ ____

A. Alveolar ducts
B. Respiratory bronchiole
T. Terminal bronchiole

b. Each lobule contains a vessel named a(n)_____ which brings in pulmonary blood *(high? low?)* in oxygen. The arteriole leads to a network of_____ which surround the alveoli. After blood is oxygenated, it enters a pulmonary_____; this carries blood out of the lobule. (Another type of vessel, a _____, also supplies each lobule.)

c. In order for air to pass from alveoli to blood in pulmonary capillaries, it must pass through the_____ (respiratory) membrane. Identify structures in this pathway: *A–E* on Figure LG 23-3.

d. *For extra review,* now label layers *(1–6)* of the alveolar-capillary membrane in the insert in Figure LG 23-3.

e. *A clinical challenge.* Normally the respiratory membrane is extremely *(thick? thin?).* Suppose pulmonary capillary blood pressure rises dramatically, pushing extra fluid out of blood. The presence of excess fluid in interstitial areas and ultimately in alveoli is known as pulmonary

_____. Diffusion occurs *(more? less?)* readily through the fluid-filled membrane.

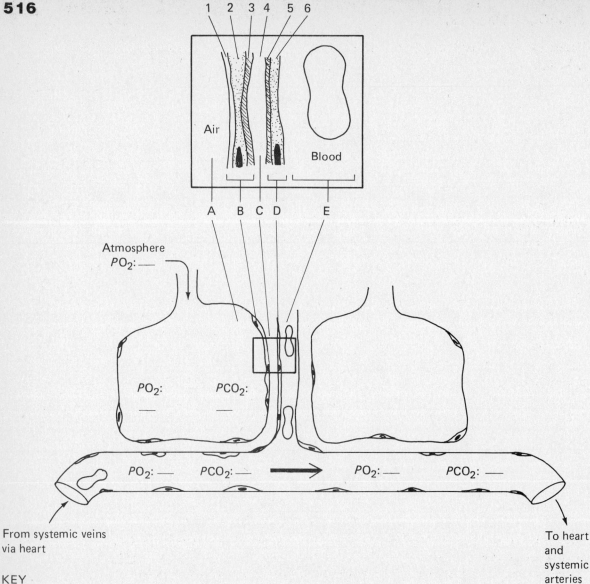

KEY

A. Alveolus
B. Alveolar wall
C. Interstitial space
D. Capillary wall
E. Blood plasma and blood cells

Figure LG 23-3 Diagram of alveoli and pulmonary capillary. Insert shows detailed alveolar-capillary membrane structure. Numbers refer to Learning Activity $\boxed{10d}$. Complete partial pressures (in mm Hg) as directed in Learning Activity $\boxed{19}$.

f. The alveolar epithelial layer of the respiratory membrane contains three types of cells. Forming most of the layer are flat *(squamous? septal? macrophage?)* cells which are ideal for diffusion of gases. Septal cells produce an important phospholipid substance called_____ _____which reduces tendency for collapse of alveoli due to surface tension. Macrophage cells help to remove_____ _____from lungs.

C. Respiration: ventilation

1. Define the three basic processes of respiration.
   a. Ventilation

   b. External respiration

   c. Internal respiration

2. Refer to Figure 23-11 (page 568) in the text and describe the process of ventilation in this exercise.

   a. In diagram (a), just before the start of inspiration, pressure within the lungs (called_____) is_____mm Hg. This is *(more than? less than? the same as?)* atmospheric pressure.

   b. At the same time, pressure within the pleural cavity (called _____ pressure) is_____mm Hg. This is *(more than? less than? the same as?)* intrapulmonic and atmospheric pressure.

   c. The first step in inspiration occurs as the muscles in the floor and walls of the thorax contract. These are the_____ and _____ muscles. Note in diagram (b) (and also in Figure LG 23-3) that the size of the thorax *(increases? decreases?)*. Since the two layers of pleura tend to adhere to one another, the lungs will *(increase? decrease?)* in size also.

   d. Increase in volume of a closed space such as the pleural cavity causes the pressure there to *(increase? decrease?)* to_____mm Hg. Since the lungs also increase in size (due to pleural cohesion), intrapulmonic pressure also *(increases? decreases?)* to_____mm Hg. This inverse relationship between volume and pressure is a statement of_____'s law.

e. A pressure gradient is now established. Air flows from high pressure area *(alveoli? atmosphere?)* to low pressure area *(alveoli? atmosphere?).* So air flows *(into? out of?)* lungs. Thus *(inspiration? expiration?)* occurs. By the end of inspiration, sufficient air will have moved into the lungs to make

pressure there equal to atmospheric pressure, that is,_____mm Hg.

f. Diagram (c) shows how expiration occurs. As the diaphragm and inter-costal muscles *(contract? relax?),* the size of the thoracic cage *(increases? decreases?).* Elastic fibers in the walls of the alveoli and respiratory passageways recoil also, causing the volume of the lungs to *(increase? decrease?).*

g. As volume decreases, pressure *(increases? decreases?)* to_____mm Hg. Air then flows *(into? out of?)* lungs until pressure within the lungs equals atmospheric pressure. At the end of expiration, intrapulmonic pressure is

again _____ mm Hg as in diagram (a).

12 3. Answer these questions about compliance.

a. Imagine trying to blow up a new balloon. Initially, the balloon resists your efforts. Compliance is the ability of a substance to yield elastically to a force; in this case it is the ease with which a balloon can be inflated. So a new balloon has a *(high? low?)* level of compliance, whereas a balloon that has been inflated many times has *(high? low?)* compliance.

b. Similarly, alveoli that inflate easily have *(high? low?)* compliance. The

presence of a coating called_____lining the inside of alveoli prevents alveolar walls from sticking together during expiration, and so *(increases? decreases?)* compliance.

c. Surfactant production is especially developed during the final weeks before birth. A premature infant may lack adequate surfactant; this disorder is

known as_____

_____.

d. Collapse of all or part of a lung may occur as a result of lack of surfactant

or other factors. This lung collapse is known as _____

_____.

4. Refer to Exhibit 23-1 (page 570) in your text. Read carefully descriptions of modified respiratory movements such as sobbing, yawning, sighing, or laughing while demonstrating those actions yourself.

5. Each minute the average adult takes_____breaths (respirations). Check

your own respiratory rate and write it here:_____breaths per minute.

13 6. Of the total amount of air that enters the lungs with each breath, about *(99%? 70%? 30%? 5%?)* actually enters the alveoli. The remaining amount of air is much like the last portion of a crowd trying to rush into a store: it does not succeed in entering the alveoli during an inspiration, but just reaches airways and then is quickly ushered out during the next expiration.

Such air is known as_____volume and constitutes

about_____ml of a typical breath.

7. Match the lung volumes and capacities with the descriptions given. You may find it helpful to refer to Figure 23-13 (page 571) in the text. `14`

_____ a. The amount of air taken in with each inspiration during normal breathing is called _____.

_____ b. At the end of a normal expiration the volume of air left in the lungs is called _____. Emphysemics who have lost elasticity of their lungs cannot exhale adequately, so this volume will be large.

_____ c. Forced exhalation can remove some of the air in FRC. The maximum volume of air which can be expired beyond normal expiration is called _____. This volume will be small in emphysema patients.

_____ d. Even after the most strenuous expiratory effort, some air still remains in the lungs; this amount, which cannot be removed voluntarily, is called _____.

_____ e. The volume of air which represents a person's maximum breathing ability is called _____. This is the sum of ERV, TV, and IRV.

_____ f. Adding RV to VC gives _____.

_____ g. The excess air a person can take in after a normal inhalation is called _____.

_____ h. IRV + TV = _____.

ERV. Expiratory reserve volume

FRC. Functional residual capacity

IC. Inspiratory capacity

IRV. Inspiratory reserve volume

RV. Residual volume

TLC. Total lung capacity

TV. Tidal volume

VC. Vital capacity

8. Indicate normal volumes for each of the following. `15`

a. TV = _____ ml (about _____ quart)

b. TLC = ml (_____ liters)

c. VC = _____ ml

d. ERV = _____ ml

e. RV = _____ ml

9. Maureen is breathing at the rate of 15 breaths per minute. She has a tidal volume of 480 ml per breath. Her _minute volume of respiration_ is _____. `16`

D. **Respiration: exchange of respiratory gases**

1. Define each of these laws related to gases.
   a. Boyle's law

   b. Dalton's law

   c. Henry's law

17 2. Check your understanding of these laws by matching the correct law with the condition that it explains.

   _____ a. If a patient breathes air highly concentrated in oxygen (as in a hyperbaric chamber), a higher percentage of oxygen will dissolve in the blood and tissues.

   _____ b. The total atmospheric pressure (760 mm Hg) is due mostly to pressure caused by nitrogen, partly to PO_2, and slightly to PCO_2.

   _____ c. Under high PN_2 more nitrogen dissolves in blood. The *bends* occurs when pressure decreases and nitrogen forms bubbles in tissue as it comes out of solution.

   _____ d. As the size of the thorax increases, the pressure within it decreases.

   B. Boyle's law
   D. Dalton's law
   H. Henry's law

18 3. Refer to Figure LG 23-4 and do this exercise.
   a. The atmosphere contains enough gaseous molecules to exert pressure upon a column of mercury to make it rise about _____ mm. The atmosphere is said to have a pressure of 760 mm Hg; this is equivalent to ____ inches of Hg. (See Figure LG 23-4a.)

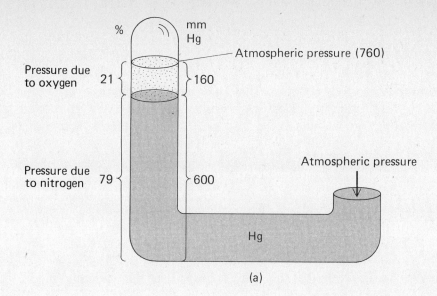

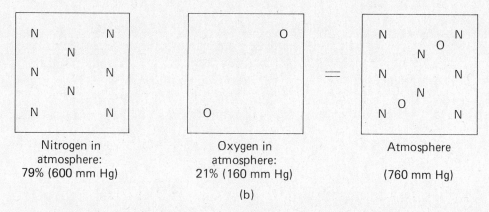

Figure LG 23-4 Atmospheric pressure. (a) Effect of gas molecules on column of mercury (Hg). (b) Partial pressures contributing to total atmospheric pressure. Refer to Learning Activity 18 .

b. The air is about _____ percent nitrogen and _____ percent oxygen. (See Figure LG 23-4b.) Only a small amount (_____ percent) of carbon dioxide is present in air.

c. Dalton's law explains that of the total 760 mm of atmospheric pressure, a certain amount is due to each type of gas. Determine the portion of the total pressure due to nitrogen molecules:

$$79\% \times 760 \text{ mm Hg} = \underline{\hspace{2cm}} \text{ mm Hg}$$

This is the partial pressure of nitrogen, or _____.

d. Calculate the partial pressure of oxygen.

e. *For extra review.* Calculate the PCO_2 of the atmosphere.

19 4. Answer these questions about external and internal respiration.
   a. External respiration is defined as the exchange of gases between *(atmospheric? alveolar?)* air and _____, that is, across the

   _____ membrane.
   b. A primary factor in the diffusion of gas across a membrane is the difference in concentration of the gas (reflected by _____ pressures) on the two sides of the membrane. On Figure LG 23-3 write values for PO_2 (in mm Hg) in each of the following areas. (Refer to text Figure 23-14, page 574, for help.)
      Atmospheric air (Recall this value from Figure LG 23-4.)
      Alveolar air (Note that this value is lower than that for atmospheric PO_2 since some alveolar O_2 enters blood)
      Blood entering lungs
   c. Calculate the PO_2 difference (gradient) between alveolar air and blood entering lungs.

   ____ mm Hg − ____ mm Hg = ____ mm Hg
   d. Three other factors that increase exchange of gases between alveoli and blood are: *(large? small?)* surface area of lungs; *(thick? thin?)* respiratory membrane; and *(increased? decreased?)* blood flow through lungs (as in exercise).
   e. By the time blood leaves lungs to return to heart and systemic arteries, its PO_2 is normally *(greater than? the same as? less than?)* PO_2 of alveoli. Write the correct value on Figure LG 23-3.
   f. Now fill in all three PCO_2 values on Figure LG 23-3.
   g. *A clinical challenge.* Erin has blood drawn from her brachial artery to determine her arterial blood gases. Her PO_2 is 56 and her PCO_2 is 48. Are these typical values for a healthy adult?

   h. Internal respiration is the exchange of gases between _____

   _____ and _____.
   i. What is the normal PCO_2 for blood leaving tissues (and also for blood entering lungs)? _____ mm Hg
20 5. With increasing altitude the air is "thinner," that is, gas molecules are farther apart, so atmospheric pressure is lower. Atop a 25,000 foot mountain, this pressure is only 282 mm Hg. Oxygen still accounts for 21 percent of the pressure. What is PO_2 at that level?

From this calculation you can see limitations of life (or modifications which must be made) at high altitudes. If atmospheric PO_2 is 59.2, neither alveolar nor blood PO_2 could surpass that level.

E. Transport of respiratory gases

(pages 575–579)

1. Answer these questions about oxygen transport. Refer to Figure 23-15 (page 576) in your text.

   21

   a. One hundred ml of blood contains about ____ ml of oxygen. Of this, over 19.5 ml is carried as _____. Only a small amount of oxygen is carried in the dissolved state since oxygen has a *(high? low?)* solubility in blood or water.

   b. Oxygen is attached to the _____ atoms in hemoglobin. The chemical formula for oxyhemoglobin is _____. When hemoglobin carries all of the oxygen it can hold, it is said to be fully _____ _____. High PO_2 in alveoli will tend to *(increase? decrease?)* oxygen saturation of hemoglobin.

   c. Refer to Figure 23-16 (page 577) in your text. Note that arterial blood, with a PO_2 of about 105 mm Hg, has its hemoglobin ____% saturated with oxygen. (This may be expressed as $SO_2 = $ ____%.)

   d. List four factors that will enhance the dissociation of oxygen from hemoglobin so that oxygen can enter tissues. *(Hint:* Think of conditions within active muscle tissue.)

   e. *For extra review,* demonstrate the effect of temperature on the oxygen–hemoglobin dissociation curve. Draw a vertical line on Figure 23-18 (page 577) of the text at $PO_2 = 40$ (the value for venous blood). Compare the percent saturation of hemoglobin (SO_2) at these two body temperatures:

      38°C (100.4°F) ____% SO_2

      43°C (109.4°F) ____% SO_2

   In other words, at a higher body temperature, *(more? less?)* oxygen will be attached to hemoglobin, while *(more? less?)* oxygen will enter tissues to fuel metabolism.

f. By the time blood enters veins to return to the heart, its oxygen saturation of hemoglobin *(SO₂)* is about ____%. Note on Figure 23-16 of your text that although *PO₂* drops from 105 in arterial blood to 40 in venous blood, oxygen saturation drops *(more? less?)* dramatically, that is, from 97% to 75%. Of what significance is this?

22 2. Carbon monoxide has about_____times the affinity that oxygen has for hemoglobin. State the significance of this fact.

3. List two examples of uses of 100% oxygen and discuss potential problems of such practice.

4. Define *hypoxia.*

Contrast anemic hypoxia with stagnant hypoxia.

5. Write the percentage of CO_2 carried in each of these forms. 23

a. Dissolved in plasma: _____

b. Bound to an amino group in the globin portion of hemoglobin: _____

c. As bicarbonate ion: _____

6. Complete this exercise. 24

a. Carbon dioxide (CO_2) produced in cells of your body diffuses into red blood cells. Water is present here also. Finish the chemical equation showing combination of these two common chemicals.

$$CO_2 + H_2O \rightarrow \underline{\qquad}$$

b. The enzyme that catalyzes this reaction is_____

c. Carbonic acid (H_2CO_3) tends to dissociate. Complete the equation giving the chemical formula and name of the products.

$$H_2CO_3 \rightarrow \underline{\qquad\qquad} + \underline{\qquad\qquad}.$$

d. It is important to remember that an increasing level of carbon dioxide in the body (as when respiratory rate is slow) will tend to cause a buildup of acid (hydrogen ion) in the body. Write the reactions involved in sequence.

$$CO_2 + \underline{\quad} \rightarrow \underline{\qquad} \rightarrow H^+ + \underline{\quad}.$$

e. The hydrogen ion produced in red blood cells in tissue capillaries is chemically bound to _____. (See Figure 23-19a, page 579, in the text.) Remember that increased acidity is one of the four factors which causes oxygen to dissociate from hemoglobin and therefore to be free to diffuse into tissue cells.

f. The bicarbonate ion (HCO_3^-) produced moves out of _____ _____ and into _____. What anion must diffuse into the red blood cell to replace the HCO_3^- (anion) that left? _____ This is known as the chloride _____.

7. Note that as the red blood cells reach lung capillaries, the same reactions you just studied occur, but in reverse. Study Figure 23-19b (page 000) in the text carefully. Then list the major steps that occur in the lungs so that CO_2 can be exhaled.

F. Control of respiration, modified forms of respiration

1. State the location and function of each of these respiratory control areas.
   a. Medullary rhythmicity area

   b. Apneustic area

   c. Pneumotaxic area

25 2. Answer these questions about respiratory control.
    a. The main chemical change that stimulates respiration is increase in blood
       level of _____ which is directly related to (*decrease in PO₂? increase in PCO₂?*) of blood.

b. Cells most sensitive to changes in blood CO_2 are located in the *(medulla? pons? aorta and carotid arteries?)*.

c. An increase in arterial blood PCO_2 is called_____.

Write an arterial PCO_2 value that is hypercapnic. ____ mm Hg *(Even slight? Only severe?)* hypercapnia will stimulate the respiratory system, leading to *(hyper? hypo?)*-ventilation.

d. State two locations of chemoreceptors sensitive to changes in PO_2.

(Even slight? Only severe?) decreases in PO_2 level of blood will stimulate these chemoreceptors and lead to hyperventilation. Give an example of a PO_2 low enough to evoke such a response.____ mm Hg

e. Increase in body temperature (as in fever), as well as stretching of the anal sphincter, will cause ____-crease in the respiratory rate.

G. Aging and development of the respiratory system

(pages 582–583)

1. Describe possible effects of these age-related changes. ⟦26⟧
   a. Pulmonary blood vessels become sclerosed, so are more rigid and resistant to blood flow.

   b. Chest wall becomes more rigid as bones and cartilage lose flexibility.

   c. Decreased macrophage and ciliary action of lining of respiratory tract.

2. Describe development of the respiratory system in this exercise. ⟦27⟧

   a. The laryngotracheal bud is derived from_____-derm. List structures formed from this bud.

   b. Identify portions of the respiratory system derived from mesoderm.

H. Disorders, medical terminology and drugs

1. A host of effects upon the respiratory system are associated with smoking. Describe these two.

   a. Bronchogenic carcinoma (Include the roles of basal cells and excessive mucus production.)

   b. Emphysema (Note that as walls of alveoli break down, surface area of the respiratory membrane ＿＿-creases, so amount of gas diffused ＿＿-creases also.)

28 2. Match the condition with the correct description.

   ＿＿＿＿ a. Permanent inflation of lungs due to loss of elasticity; rupture and merging of alveoli, followed by their replacement by fibrous tissue

   ＿＿＿＿ b. Inflammation of bronchi with excessive mucus production for at least three months per year for at least two consecutive years

   ＿＿＿＿ c. Acute infection or inflammation of alveoli which fill with fluid

   ＿＿＿＿ d. Spasms of small passageways with wheezing and dyspnea

   ＿＿＿＿ e. Caused by a species of Mycobacterium; lung tissue is destroyed and replaced with inelastic connective tissue

   ＿＿＿＿ f. Difficult, painful breathing

   ＿＿＿＿ g. Occurs with PO_2 drops below 50 mm Hg and PCO_2 rises above 50 mm Hg

   ＿＿＿＿ h. A group of conditions known as chronic obstructive pulmonary disease (COPD) (3 answers)

   BA. Bronchial asthma
   CB. Chronic bronchitis
   D. Dyspnea
   E. Emphysema
   P. Pneumonia
   RF. Respiratory failure
   TB. Tuberculosis

3. Discuss respiratory distress syndrome (RDS) in this exercise.
   a. Before birth fetal lungs are filled with *(air? fluid?)*. Inflation of lungs after birth depends largely on the presence of a chemical known as

   _____ . This chemical is a *(lipoprotein? phospholipid?)* produced by *(basal cells of bronchi? septal cells of alveoli?)*.
   b. Surfactant *(raises? lowers?)* surface tension so that alveolar walls are less likely to stick together. Thus surfactant *(facilitates? inhibits?)* lung inflation.
   c. RDS especially targets *(premature? full term?)* infants. Explain why.

4. To what does SIDS refer?

   What is the mortality rate of SIDS?

   Discuss possible causes of SIDS.

5. Contrast pulmonary embolism *(PE)* with pulmonary edema *(P edema)* by identifying the related descriptions.

   _____ a. Defined as abnormal accumulation of fluid in interstitial areas and/or alveoli of lungs
   _____ b. Likely to occur if infection such as pneumonia increases permeability of pulmonary capillaries
   _____ c. Likely to occur if left ventricle fails so blood backlogs and increases pressure in pulmonary capillaries

_____ d. Defined as presence of blood clot or other foreign material in pulmonary vessels, thus obstructing blood flow

_____ e. Likely to result from deep vein thrombosis (DVT), especially in bedridden persons, or from release of fat from marrow of fractured bones

6. Explain how the abdominal thrust (Heimlich maneuver) helps to remove food which might otherwise cause death by choking.

31 7. Write the *ABC*'s of CPR.

Answers to Numbered Questions in Learning Activities

1 Passageway for air and food, and resonating chamber for speech sounds.

2 Glottis, arytenoid.

3 Oral and nasal cavities, pharynx and paranasal sinuses continue to act as resonating chambers. Muscles of face, tongue, lips, and pharynx also help in forming words.

4 (a) N. (b) NP. (c) OP. (d) LP. (e) L. (f) L. (g) N, NP. (h) M, OP.

5 Point at which the trachea bifurcates into primary bronchi.

6 The right bronchus is more vertical and slightly wider than the left.

7 (a) They branch repeatedly much like an inverted tree (with the trachea as the main trunk). (b) See KEY to Figure LG 23-1.

8 (a) Pleurisy. (b) Pneumothorax. (c) Hemothorax.

9 (a) Base, clavicle, ribs. (b) Hilus, bronchi, pulmonary vessels and nerves. (c) Right, left, left, right.

10 (a) T R A. (b) Arteriole, low, capillaries, venule, lymphatic. (c) Alveolar-capillary. See KEY to Figure LG 23-3. (d) Surfactant, *1;* alveolar epithelium, *2;* epithelial basement membrane, *3;* interstitial space, *4;* capillary basement membrane, *5;* capillary endothelium, *6.* (e) Thin, edema, less. (f) Squamous, surfactant, debris.

11 (a) Intraalveolar or intrapulmonic, 760, the same as. (b) Intrathoracic or intrapleural, 756; less than. (c) Diaphragm, intercostal, increases, increase. (d) Decrease, 754, decreases, 758, Boyle. (e) Atmosphere, alveoli, into, inspiration, 760. (f) Relax, decreases, decrease. (g) Increases, 763, out of, 760.

12 (a) Low, high. (b) High, surfactant, increases. (c) Respiratory distress syndrome (or hyaline membrane disease). (d) Atelectasis.

13 70%, dead air (or dead space), 150.

14 (a) TV. (b) FRC. (c) ERV. (d) RV. (e) VC. (f) TLC. (g) IRV. (h) IC.

15 (a) 500, 0.5. (b) 6,000, 6. (c) 4,800. (d) 1,200. (e) 1,200.

16 7,200 ml/min ($=$ 15 breaths/min $\times$ 480 ml/breath).

17 (a) H. (b) D. (c) H. (d) B.

18 (a) 760, 30. (b) 79, 21, 0.04. (c) 600.4, PN_2. (d) 21/100 $\times$ 760 mm Hg $=$ about 160 mm Hg $= PO_2$. (e) 0.04/100 $\times$ 760 mm Hg $=$ 0.3 mm Hg $= PCO_2$.

19 (a) Alveolar, blood (in pulmonary capillaries), respiratory (alveolar-capillary). (b) Partial, 160, 105, 40. (c) 105 $-$ 40 $=$ 65. (d) Large, thin, increased. (e) The same as, 105. (f) Blood entering lungs: 45, alveoli and blood leaving lungs, both 40. (g) No. Typical brachial arterial values are same as for alveoli or blood leaving lungs: $PO_2 = 105$, $PCO_2 = 40$. These values indicate inadequate gas exchange. (h) Blood (in systemic capillaries), tissues (interstitial fluid). (i) 45.

20 282 $\times$ 21% $=$ 59.2 mm Hg.

21 (a) 20, oxyhemoglobin, low. (b) Iron, HbO_2, saturated, increase. (c) 97, 97. (d) Increase in temperature, PCO_2, acidity, and DPG. (e) 63, 36, less more. (f) 75, less. In the event that respiration is temporarily halted, even venous blood has much oxygen attached to hemoglobin and available to tissues.

22 200, oxygen-carrying capacity is drastically reduced.

23 (a) 7. (b) 23. (c) 70.

24 (a) H_2CO_3 (carbonic acid). (b) Carbonic anhydrase. (c) Bicarbonate ion ($HCO_3{}^-$), hydrogen ion (H^+). (d) H_2O, H_2CO_3, $HCO_3{}^-$. (e) Hemoglobin. (f) Blood cells, plasma, chloride (Cl), shift.

25 (a) H^+, increase in PCO_2. (b) Medulla. (c) Hypercapnia, any value higher than 40, Even slight, hyper. (d) Aortic and carotid bodies, Only severe, usually below 60. (e) In.

26 (a) Blood backs up into right ventricle, causing it to pump extra hard and hypertrophy. This is cor pulmonale. (b) Decreased vital capacity and so decreased arterial PO_2. (c) At risk for pneumonia.

27 (a) Endo; lining of larynx, trachea, bronchial tree, and alveoli. (b) Smooth muscle, cartilage, and other connective tissues of airways.

28 (a) E. (b) CB. (c) P. (d) BA. (e) TB. (f) D. (g) RF. (h) BA, CB, E.

29 (a) Fluid, surfactant, phospholipid, septal cells of alveoli. (b) Lowers, facilitates. (c) Premature. Septal cells do not produce adequate amounts of surfactant until between weeks 28–32 of the 39-week human gestational period. A baby born at 7 months (30 weeks), for example, would be at high risk for RDS.

30 (a) P edema. (b) P edema. (c) P edema. (d) PE. (e) PE.

31 Establish Airway, ventilate (Breathing), and reestablish Circulation.

MASTERY TEST: Chapter 23

Questions 1–5: Arrange the answers in correct sequence.

_____ _____ _____ 1. From first to last, the steps involved in inspiration:
  A. Diaphragm and intercostal muscles contract
  B. Thoracic cavity and lungs increase in size
  C. Intrapulmonic pressure decreases to 758 mm Hg

_____ _____ _____ 2. From most superficial to deepest:
  A. Parietal pleura
  B. Visceral pleura
  C. Pleural cavity

_____ _____ _____ _____ _____ 3. From superior to inferior:
  A. Bronchioles
  B. Bronchi
  C. Larynx
  D. Pharynx
  E. Trachea

_____ _____ _____ _____ _____ 4. Pathway of inspired air:
  A. External nares
  B. Internal nares
  C. Meati
  D. Nasopharynx
  E. Vestibule

_____ _____ _____ _____ _____ 5. Pathway of inspired air:
  A. Alveolar ducts
  B. Bronchioles
  C. Lobar bronchi
  D. Primary bronchi
  E. Segmental bronchi
  F. Alveoli

Questions 6–9: Choose the one best answer to each question.

_____ 6. Which of these values (in mm Hg) would be most likely for PO_2 of blood in the femoral artery?
  A. 40   B. 45   C. 100   D. 160   E. 760   F. 0

_____ 7. Pressure and volume in a closed space are inversely related, as described by _____ law.
  A. Boyle's   B. Starling's   C. Dalton's   D. Henry's

_____ 8. Choose the correct formula for carbonic acid:

    A. HCO_3^- B. H_3CO_2 C. H_2CO_3 D. HO_3C_2

    E. H_2C_3O

_____ 9. A procedure in which an incision is made in the trachea and a tube inserted into the trachea is known as a:

    A. Tracheostomy   B. Bronchogram   C. Intubation

    D. Pneumothorax

Questions 10–20: Circle T (true) or F (false). If the statement is false, change the underlined word or phrase so that the statement is correct.

T   F   10. In the chloride shift Cl^- moves into red blood cells in exchange for <u>H^+</u>.

T   F   11. When chemoreceptors sense <u>increase in PCO_2 or increase in acidity of blood (H^+)</u>, respiratory rate will be stimulated.

T   F   12. Both increased temperature and increased acid content tend to cause oxygen to <u>bind more tightly to</u> hemoglobin.

T   F   13. Under normal circumstances intrapleural pressure is <u>always negative.</u>

T   F   14. The pneumotaxic and apneustic areas controlling respiration are located in the <u>pons.</u>

T   F   15. Fetal hemoglobin has a <u>lower</u> affinity for oxygen than maternal hemoglobin does.

T   F   16. Most CO_2 is carried in the blood in the form of <u>bicarbonate.</u>

T   F   17. The alveolar wall <u>does</u> contain macrophages that remove debris from the area.

T   F   18. Intrapulmonic pressure means the same thing as <u>intrapleural</u> pressure.

T   F   19. Inspiratory reserve volume is normally <u>larger than</u> expiratory reserve volume.

T   F   20. The PO_2 and PCO_2 of blood leaving the lungs <u>are about the same</u> as PO_2 and PCO_2 of alveolar air.

Questions 21–25: fill-ins. Complete each sentence with the word or phrase that best fits.

_____ 21. The process of exchange of gases between alveolar air and blood in pulmonary capillaries is known as _____.

_____ 22. Take a normal breath and then let it out. The amount of air left in your lungs is the capacity called ____ and it usually measures about ____ ml.

_____ 23. The Bohr effect states that when more H^+ ions are bound to hemoglobin, less ____ can be carried by hemoglobin.

_____ 24. The epiglottis, thyroid, and cricoid cartilages are all parts of the ____.

_____ 25. ____ is a chemical that lowers surface tension and therefore increases inflatability (compliance) of lungs.

The Digestive System

Food is vital to homeostasis since food supplies the building blocks for all structures and the energy for various functions in the body. In this chapter you will study the system that changes complex foods into molecules which the body can utilize. You will define digestion and survey the general structure of digestive organs and their peritoneal coverings (Objectives 1–4). You will learn about chemical and mechanical processes that occur in each part of the system: mouth, pharynx, and esophagus (5–10), stomach (11–12), pancreas, liver, and gallbladder (13–17), small intestine (18–20), and large intestine (21–22). You will consider effects of aging and developmental anatomy (23–24). Finally you will study some common disorders, medical terminology, and drugs associated with the digestive system (25–26).

Topics Summary

A. Regulation of food intake, digestive processes, general organization
B. Mouth, pharynx, esophagus
C. Stomach
D. Accessory organs: pancreas, liver, gallbladder
E. Small intestine
F. Large intestine
G. Aging and developmental anatomy of the digestive system
H. Disorders, medical terminology

Objectives

1. Describe the mechanism that regulates food intake.
2. Define digestion and distinguish between the chemical and mechanical phases.
3. Identify the organs of the gastrointestinal (GI) tract and the accessory organs of digestion.
4. Discuss the structure of the wall of the gastrointestinal (GI) canal.
5. Explain the structure of the mouth and its role in mechanical digestion.
6. Describe the location, histology, and functions of the salivary glands.
7. Identify the mechanisms that regulate the secretion of saliva.
8. Identify the parts of a typical tooth and compare deciduous and permanent dentitions.
9. Discuss the stages of swallowing.
10. Describe the role of the esophagus in digestion.
11. Describe the anatomy and histology of the stomach and explain the relationship between its structural features and digestion.
12. Discuss the factors that control the secretion of gastric juice and gastric emptying.
13. Describe the anatomy and histology of the pancreas and its role in digestion.
14. Explain how pancreatic secretion is regulated.
15. Describe the anatomy and histology of the liver and its role in digestion.
16. Explain how bile secretion is regulated.
17. Discuss the role of the gallbladder in digestion.
18. Discuss the structural features of the small intestine that adapt it for digestion and absorption.
19. Explain how small intestinal secretions are controlled.
20. Define absorption and explain how the end products of digestion are absorbed.
21. Describe the anatomy and histology of the large intestine.
22. Define the processes involved in the formation of feces and defecation.
23. Describe the effects of aging on the digestive system.
24. Describe the developmental anatomy of the digestive system.
25. Describe the clinical symptoms of the following disorders: dental caries, periodontal disease, peritonitis, peptic ulcers, appendicitis, gastrointestinal (GI) tumors, diverticulitis, cirrhosis, hepatitis, gallstones, anorexia nervosa, and bulimia.
26. Define medical terminology associated with the digestive system.

Learning Activities

(pages 590–595)   **A. Regulation of food intake, digestive processes, general organization**

1. Explain why food is vital to life. Give three specific examples of uses of foods in the body.

2. The part of the brain that regulates food intake is the _____

   _____. A low blood glucose level stimulates the *(feeding? satiety?)* center.

1 3. List the five basic activities of the digestive system.

4. Contrast *mechanical digestion* and *chemical digestion*.

5. Identify the organs of the digestive system in Figure LG 24-1. Visualize the locations of these organs on yourself. Relate these to the nine abdominal regions (Figure LG 1-1, page LG 9).

6. List the organs of digestion that are:
   a. Part of the gastrointestinal (GI) tract

   b. Accessory organs

537

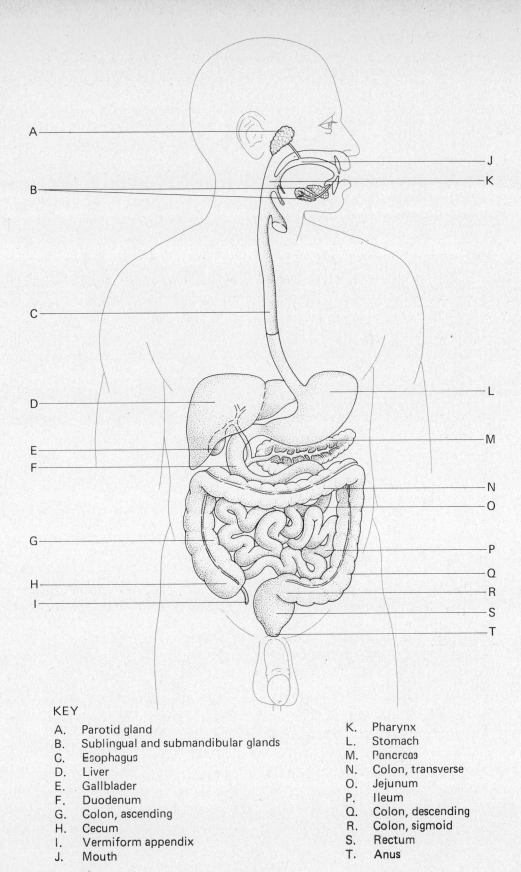

KEY

A. Parotid gland
B. Sublingual and submandibular glands
C. Esophagus
D. Liver
E. Gallblader
F. Duodenum
G. Colon, ascending
H. Cecum
I. Vermiform appendix
J. Mouth

K. Pharynx
L. Stomach
M. Pancreas
N. Colon, transverse
O. Jejunum
P. Ileum
Q. Colon, descending
R. Colon, sigmoid
S. Rectum
T. Anus

Figure LG 24-1 Organs of the digestive system.

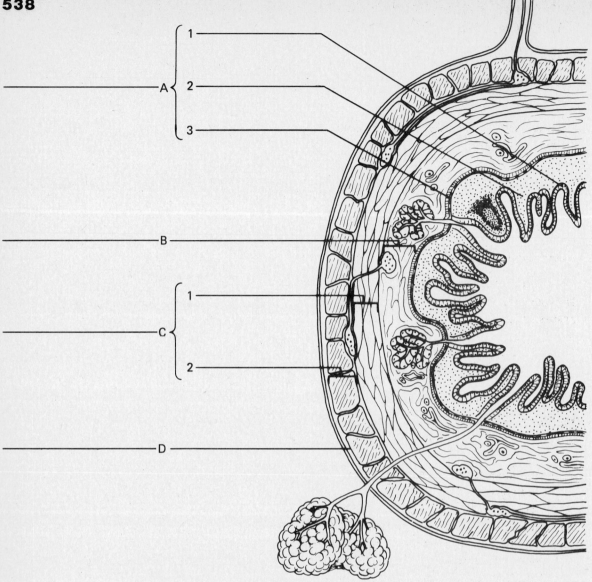

Figure LG 24-2 Gastrointestinal tract seen in cross section. Label as directed in Learning Activity ⬚2⬚ .

⬚2⬚

7. Refer to Figure LG 24-2, which shows a section of the wall of the GI tract. Label the layers on the figure. Then in a few words describe the structure and function of each layer. Use the space below each label line.

8. Contrast the epithelium of the mucosa in the mouth and esophagus with that in stomach and intestine.

9. Is the mucosa of the GI tract keratinized?_____State one advantage of this fact.

10. Contrast *visceral peritoneum* and *parietal peritoneum*. Name the space between these two layers.

Accumulation of fluid in the peritoneal cavity is known as_____

_____.

3

11. In what two ways does the peritoneum differ from the other serous membranes of the body, the pericardium and pleura?

4

12. Match the names of these peritoneal extensions with the correct descriptions.

5

_____ a. Attaches liver to anterior abdominal wall

_____ b. Binds intestines to posterior abdominal wall; provides route for blood and lymph vessels and nerves to reach small intestine

_____ c. Binds part of large intestine to posterior abdominal wall

_____ d. "Fatty apron"; covers and helps prevent infection in snall intestine

_____ e. Suspends stomach and duodenum from liver

F. Falciform ligament
G. Greater omentum
L. Lesser omentum
M. Mesentery
Meso. Mesocolon

13. Note how extensive the peritoneal membrane is. Define *peritonitis* and discuss its clinical significance.

B. Mouth, pharynx, esophagus

(pages 595–603)

1. Describe the following parts of the oral cavity.
   a. Vestibule and oral cavity proper

b. Palates, arches, and fauces

2. Describe these parts of the tongue.
   a. Extrinsic muscles. *For extra review,* refer to Chapter 11, Activity 12 ,
      page LG 205.

   b. Intrinsic muscles

   c. Lingual frenulum (Explain problems which occur if it is too short.)

   d. Papillae

e. Taste zones. *For extra review,* see Chapter 17, Activities 3 and 4, pages LG 338–339.

3. Identify the three salivary glands on Figure LG 24-1 and visualize their locations on yourself.
4. Complete this exercise about salivary glands.

   6

   a. Which glands are largest? *(Parotid? Sublingual? Submandibular?).*
   b. Which secrete the thickest secretion due to presence of much mucus?

   _____

   c. About 1 to 1½ *(tablespoons? cups? liters?)* of saliva is secreted daily.
   d. State three functions of saliva.

   e. The pH of the mouth is appropriate for action of salivary amylase. This is about pH *(2? 6.5? 9?).*
5. Include roles of autonomic nervous system as you briefly describe several conditions in which salivary glands are:
   a. Inhibited

   b. Stimulated

6. Complete the diagram of a typical molar tooth in Figure LG 24-3. On the left side of the diagram, label *crown, cervix (neck),* and *root* regions. Then add to the diagram and label on the right side the following parts: *enamel, cementum, dentin, pulp, cavity, root canal,* and *apical foramen.* Finally, add blood vessels and nerves, and label the *periodontal ligament,* and surrounding bone. *For extra review.* Write a brief description of each structure next to its label.

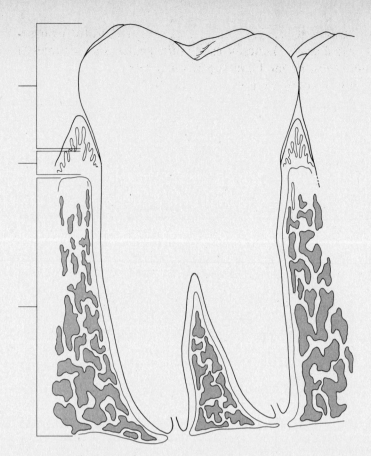

Figure LG 24-3 Outline of a typical molar tooth. Complete the diagram and label as directed.

7. Look at your own teeth in a mirror. Identify the different types of teeth and consider how the structural design of each relates to its function. Then answer these questions. (You may find it helpful to refer to Figure 24-6, page 599, in the text.)

   ☐ 7

   a. How many teeth are in a complete permanent dentition? ____ How many teeth do you have? ____ How many would be in a complete child's dentition, as in a 4-year-old? ____

   b. The four centrally located teeth are named _____. Lateral to these are _____. Posterior to the cuspid teeth are_____and finally _____.

   c. How many cuspids are in an adult set? ____ Premolars? ____ Molars?____

   d. The first permanent tooth to erupt is the _____; the last is the_____.

8. Salivary amylase is an enzyme that digests_____. Most of the starch ingested *(is? is not?)* broken down by the time food leaves the mouth. What inactivates amylase in the stomach?

   ☐ 8

9. Another term for swallowing is _____. The three phases of deglutition are listed below. Describe each of the phases. Include roles of *tongue, soft palate, epiglottis,* and *esophageal sphincters.*

   a. Voluntary (mouth to oropharynx)

   b. Pharyngeal

   c. Esophagus

10. Define peristalsis and explain its role in digestion.

11. Failure of the lower esophageal to close results in the sensation of _____. Consequently, the esophageal lining may be irritated by *(acidic? basic?)* contents of the stomach that enter the esophagus. Failure of this sphincter to relax is a condition known as _____. Resulting distention of the esophagus causes pain that may be confused with _____ pain.

9

12. Summarize digestion in the mouth, pharynx, and esophagus by completing parts a and b of Table LG 24-1.

Table LG 24-1
Summary of Design

Digestive Organs	Carbohydrate	Protein
a. Mouth, salivary glands	Salivary amylase: digests starch to maltose	
b. Pharynx, esophagus		
c. Stomach		
d. Pancreas		
e. Intestinal juices		
f. Liver	No enzymes for digestion of carbohydrates	
g. Large intestine	No enzymes for digestion of carbohydrates	No enzymes for digestion of proteins

Lipid	Mechanical	Other Functions
	Deglutition, peristalsis	
		1. Secretes intrinsic factor 2. Produces hormone stomach gastrin
Pancreatic lipase: digests about 80% of fats		
No enzymes for digestion of lipids		

C. Stomach

1. Describe the position of your stomach relative to these organs: diaphragm, spleen, liver, pancreas.

2. On Figure LG 24-1, identify these regions of the stomach: *cardia, fundus, body,* and *pylorus.* Which is more lateral and inferior in location? *(Greater curvature? Lesser curvature?)* To which curvature is the greater omentum attached? _____ The lesser omentum? _____

| 10 |

_____

3. Describe the pyloric valve according to location and function.

Contrast *pylorospasm* and *pyloric stenosis.*

4. Define the following terms related to the stomach:
a. Rugae

b. Gastroscopy

c. Chyme

5. Complete this table about gastric secretions.

Name of Cell	Type of Secretion	Function of Secretion
a. Zygomatic (chief)		
b. Mucous		
c.	HCl and intrinsic factor	
d.	Gastrin	

6. How does the muscularis of the stomach differ from this layer in the walls [11] of other digestive organs?

7. Explain how food is mixed in the stomach.

8. Answer these questions about chemical digestion in the stomach. [12]

   a. The most important enzyme released by the stomach is _____ _____ . Gastric cells produce the enzyme in the in-active state, called _____ , which is activated by _____ .

b. Pepsin is most active at very *(acid? alkaline?)* pH.

c. State two factors that enable the stomach to digest protein without digesting its own cells (which are composed largely of protein).

d. If mucus fails to protect the gastric lining, the condition known as _____ may result.

e. Another enzyme produced by the stomach is _____ which digests _____. In adults it is quite *(effective? ineffective?)*. Why?

[13] 9. Answer these questions about control of gastric secretion.

a. Name of three phases of gastric secretion: _____, _____, and _____. Which of these causes gastric secretion to begin when you smell or taste food?

_____ .

b. Gastric glands are stimulated mainly by the _____ nerves. Their fibers are *(sympathetic? parasympathetic?)*. Two stimuli which cause vagal impulses to stimulate gastric activity are _____ _____ and _____

_____ .

c. The hormone _____ is released when foods, especially *(carbohydrate? protein? lipid?)* and alcohol, reach the *(fundic? pyloric?)* region of the stomach. This hormone travels through the blood to all parts of the body; its target areas are gastric glands which are *(stimulated? inhibited?)* and the pyloric sphincter which is *(contracted? relaxed?)*.

d. When food (chyme) reaches the intestine, nerves initiate the _____ reflex which *(stimulates? inhibits?)* further gastric secretion. Three hormones released by the intestine also inhibit gastric secretion as well as gastric motility. Name these three hormones:

_____, _____, and

_____ .

10. Refer to Table LG 24-2. Fill in parts related to regulation of gastric function.

Table LG 24-2
Hormones Regulating Digestion

Hormone	Where formed	Stimulated by	Functions
a. Stomach gastrin		Protein and alcohol entering stomach	
b. Enteric gastrin			Stimulates stomach to secrete small amounts of gastric juice
c. GIP	Intestinal wall		
d. Secretin			↓ gastric secretion ↓ GI motility ↑ pancreatic NaHCO₃ ↑ bile production by liver ↑ intestinal juice
e. CCK			

11. Food stays in the stomach for about _____ hours. Which food type leaves the stomach most quickly? _____ Which type stays in the stomach longest? _____

$\boxed{14}$

12. The stomach is responsible for *(much? little?)* absorption of foods. What types of substances are absorbed by the stomach?

$\boxed{15}$ 13. Answer these questions about vomiting (emesis).
   a. Identify the two strongest stimuli for vomiting.

   b. Such stimuli are transmitted to the_____which is the site of the vomiting center. Nerve impulses then convey instructions to *(contract? relax?)* abdominal muscles and *(contract? relax?)* esophageal sphincters.
   c. Explain how prolonged vomiting can lead to serious disturbances in homeostasis.

14. Complete part c of Table LG 24-1, describing the role of the stomach in digestion.

(pages 609–614) **D. Accessory organs: pancreas, liver, gallbladder**

$\boxed{16}$ 1. Study Figures LG 24-1 and LG 24-4. Then complete these statements about the pancreas.

   a. The pancreas lies posterior to the_____.
   b. The pancreas is shaped roughly like a fish, with its head in the curve of the _____ and its tail nudging up next to the

   _____.

   c. The pancreas contains two kinds of glands. Ninety-nine percent of its cells produce *(endocrine? exocrine?)* secretion. One type of these secretions is *(acid? alkaline?)* fluid to neutralize the chyme entering from the stomach.
   d. One enzyme in the pancreatic secretions is trypsin; it digests *(fats? carbohydrates? proteins?)*. Trypsin is formed initially in the inactive form (trypsinogen) and is activated by *(HCl? NaHCO₃? enterokinase?)*.

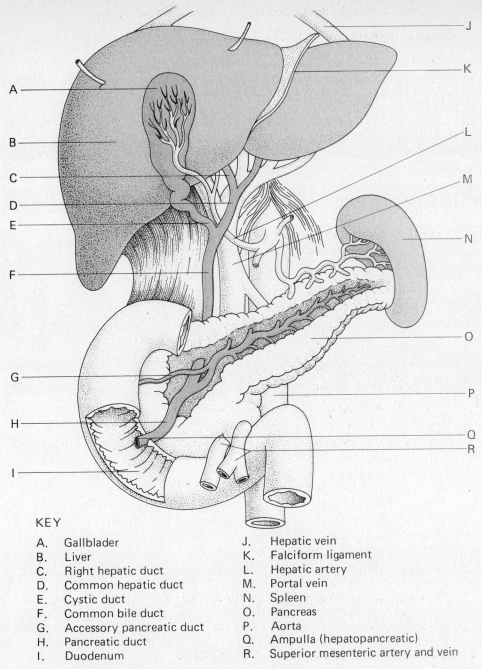

KEY

A. Gallblader
B. Liver
C. Right hepatic duct
D. Common hepatic duct
E. Cystic duct
F. Common bile duct
G. Accessory pancreatic duct
H. Pancreatic duct
I. Duodenum

J. Hepatic vein
K. Falciform ligament
L. Hepatic artery
M. Portal vein
N. Spleen
O. Pancreas
P. Aorta
Q. Ampulla (hepatopancreatic)
R. Superior mesenteric artery and vein

Figure LG 24-4 Liver, gallbladder, pancreas, and duodenum, with associated blood vessels and ducts. (Stomach has been removed.) Complete as directed.

e. Most of the amylase and lipase produced in the body is secreted by the pancreas. Describe the functions of these two enzymes.

f. All exocrine secretions of the pancreas empty into two ducts (_____ and _____ on Figure Lg 24-4). These empty into the_____ _____.

g. *A clinical challenge.* In most persons the pancreatic duct also receives bile flowing through the_____*(F on the figure).* State one possible complication that may occur if a gallstone blocks the pancreatic (as at point *H* on the figure).

h. The endocrine portions of the pancreas are known as the _____ _____. These cells secrete two hormones, _____ and _____. Typical of all hormones, these pass into *(ducts? blood vessels?),* specifically into vessels which empty into the_____vein.

2. Fill in part d of Table LG 24-1, describing the role of the pancreas in digestion.

|17|

3. Describe regulation of the pancreas in this exercise.

a. The pancreas is stimulated by *(sympathetic? parasympathetic?)* nerves, specifically the _____ nerves, and also by hormones produced by cells located in the wall of the_____.

b. The hormone cholecystokinin (CCK) activates the pancreas to secrete fluid rich in *(HCO$_3$$^-$? digestive enzymes such as trypsin, amylase, and lipase?*). What is the nature of the pancreatic fluid produced under the influence of secretin? (Be sure to write this information on Table LG 24-2.)

|18|

4. Answer these questions about the liver.

a. This organ weighs about _____ kg (_____ lb). It lies in the_____ _____ quadrant of the abdomen. Of its _____ lobes, the _____ is the largest.

b. The_____ligament separates the right and left lobes. In the edge of this ligament is the ligamentum teres (round ligament of the liver), which is the obliterated_____vein.

c. Blood enters the liver via vessels named _____

and _____. In liver lobules, blood mixes in channels

called _____ before leaving the liver via vessels

named_____.

5. Complete Figure LG 24-4 as directed.

a. Identify the pathway of bile from liver to intestine by coloring green the following structures in Figure LG 24-4: *C D F H*. The general direction of bile is from *(superior to inferior? inferior to superior?)*.

b. Now draw arrows to show direction of blood through the liver. In order to reach the inferior vena cava, blood must flow *(superiorly? inferiorly?)* through the liver. Color arteries red and veins blue in this figure.

6. Complete this exercise about bile.

a. Two functions of bile are_____of fats and

_____ of fats (and fat-soluble vitamins).

b. The principal pigment in bile is _____ which is

formed largely as a breakdown product of_____cells.

Excessive amounts of bilirubin give skin a yellow color, a condition

known as_____.

7. While blood is in sinusoids of the liver, hepatic cells have ample opportunity to act on this blood, to modify it, to add new substances to it. List here six important functions of the liver. Use these key words:

a. Bile

b. Plasma proteins (Name three that bile makes)

c. Phagocytosis (List three types of cells that may undergo liver phagocytosis)

d. Detoxification (Name two types of chemicals rendered harmless by liver enzymes)

e. Metabolism

f. Vitamins (list two minerals and five vitamins stored in the liver; name one vitamin activated by the liver)

21 8. Complete part f of Table LG 24-1. Note that the liver *(does? does not?)* contribute digestive enzymes.

9. *ACTive learning.* Role play regulation of release of bile. First identify locations of duodenum and gallbladder, bile ducts, and portal vein. Players are needed for all chemicals or structures in CAPITAL letters.

22 a. CHYME strolls in from the stomach (which may be the hallway outside the classroom), entering the duodenum (part of the classroom). Note that

the doorway must therefore be the_____sphincter.

CHYME today is particularly rich in food types such as_____

_____.

b. The presence of these food types serve as a stimulus to the walls of the

_____ to secrete the hormone CHOLECYSTOKI-NIN (CCK). (CCK enters.)

c. CCK then exits from the intestinal area by way of *(common bile duct? portal vein?).* CCK *(goes directly to the gallbladder without passing any other organ? meanders in the bloodstream through all parts of the body.)* (CCK should now take a scenic trip through the blood vessels of arms, legs, ears, heart, as desired.)

d. At last CCK arrives at the GALLBLADDER, "knowing" to attach there since GALLBLADDER has specific receptors for CCK. CCK directs GALLBLADDER to *(contract? relax?).*

e. GALLBLADDER harbors many molecules of BILE SALTS. Contraction of GALLBLADDER expels BILE SALTS. These pass into *(portal vein? cystic duct?)* and via the common bile duct to the duodenum.

f. *For extra review.* A GALLSTONE (biliary calculus) is lodged in the bile pathway, such as in the common bile duct. (Refer to page 610 in your text.) If scenes *a–e* of this role play are now repeated, how will scene *f* differ?

A cholecystectomy may need to be performed. If so, which player must terminate his/her role in your cast?

g. In addition to the bile-related activities already performed by one of our stars, CCK, what other functions might CCK also portray in upcoming productions?

E. Small intestine

(pages 614–621)

1. Describe these aspects of the small intestine.

   a. Its average diameter is_____cm (_____inch); its average length is about_____m (_____ft).

   b. It is divided into three main segments:_____,
   _____, and _____.

2. The small intestine is designed for effective secretion and absorption by modifications of its wall. Describe these by completing the following table.

Structure	Part(s) of Wall Involved	Function(s)
a. Intestinal glands (Crypts of Lieberkuhn)		
b.	Submucosa	Secrete alkaline mucus which neutralizes gastric acid
c. Goblet cells		
d.	Fingerlike projections of plasma membrane of mucosal epithelial cells	
e. Villi		
f.	Deep folds in mucosa and submucosa	

3. Digestion in the small intestine occurs with the aid of enzymes and other secretions from three sources. Name them. (*Hint*: Refer to Table LG 24-1, d–f, pages LG 544–545.)

4. Contrast *segmentation* and *peristalsis,* the two types of movement of the small intestine.

5. Write the main steps in the digestion of each of the three major food types. 23
   a. Carbohydrates

      Polysaccharides → _____ →

      _____ → monosaccharides

   b. Proteins
      Proteins → _____ → _____

      → _____

   c. Lipids

      Neutral fats → emulsified fats → _____

      _____

6. Complete the description of functions of intestinal juices by filling in part
   e of Table LG 24-1.
7. Refer to Table LG 24-1 and review digestion of each of the three major food
   groups by reading your columns vertically.
8. To review roles of GI organs in digestion, identify which chemicals in the
   list at right are made by each of the following organs. 24

   a. Stomach   A. Amylase
                                       B. Bile
                                       CCK. Cholecystokinin
                                       En. Enterokinase
   b. Pancreas   EG. Enteric gastrin
                                       HCl. HCl
                                       L. Lipase
                                       MLS. Maltase, lactase, suc-
   c. Small intestine   rase
                                       P. Pepsin
                                       Pd. Peptidases
                                       S. Secretin
   d. Liver   SG. Stomach gastrin
                                       TCC. Trypsin, chymotryp-
                                            sin, carboxy-peptidase

9. *For extra review* of secretions involved with digestion, match names of
   secretions listed for Activity 24 with descriptions below. Blank lines follow some descriptions. On these indicate whether the secretion is classified
   as an enzyme *(E)*, hormone *(H)*, or neither of these *(N)*. The first one is done
   for you. 25

   _____ a. Causes contraction of gallbladder and relaxation of the sphincter
              of the hepatopancreatic ampulla so that bile enters duodenum
              __H__

   _____ b. Stimulates bile production in the liver ____

   _____ c. Stimulates pancreas to produce secretions rich in enzymes such
              as amylase and lipase ____

_____ d. Stimulates pancreas to secrete alkaline fluid to neutralize stomach acid _____

_____ e. Stimulates intestinal secretions and inhibits gastric secretions and motility (2 answers)

_____ f. Stimulate gastric secretions and motility (2 answers) _____

_____

_____ g. Active form of a proteolytic enzyme formed in the stomach

_____ h. Activates pepsinogen _____

_____ i. Starts protein digestion in the GI tract

_____ j. Protein-digesting enzymes produced by the pancreas

_____ k. Activates trypsinogen to trypsin _____

_____ l. Intestinal enzymes that complete the breakdown of protein to amino acids

_____ m. Most effective of this type of enzyme produced by the pancreas; digests fats

_____ n. Emulsifies fats before they can be digested effectively _____

_____ o. Starch-digesting enzymes secreted by salivary glands and pancreas

_____ p. Intestinal enzymes that complete carbohydrate breakdown, digesting disaccharides to simple sugars

26 10. Describe the absorption of end products of digestion in this exercise.

a. Almost all absorption takes place in the _(large? small?)_ intestine. Glucose and amino acids move into the epithelial cells lining the intestine by _(active transport? diffusion?)._

b. Simple sugars, amino acids, and short-chain fatty acids are absorbed into

(blood? lymph?) capillaries which lead to the_____

vein. These substances can then be stored or metabolized in the

_____.

c. Long-chain fatty acids and monoglycerides first combine with

_____salts to form _(micelles? chylomicrons?)._ This enables fatty acids and monoglycerides to enter epithelial cells in the intestinal lining and soon enter lacteals leading to the _(portal vein? thoracic duct?)._ Most bile salts are ultimately _(eliminated in feces? recycled to the liver?)._

d. _A clinical challenge._ Which vitamins are absorbed with the help of bile? _(Water-soluble? Fat-soluble?)_ Circle the fat-soluble vitamins: A B_{12} C D E K. Obstruction of bile pathways may lead to symptoms related to vitamin deficiency. Name two.

e. About____liters (or quarts) of fluids are ingested or secreted into the GI tract each day. Of this, all but about 0.5 to 1.0 liter is reabsorbed into blood capillaries in the walls of the *(small? large?)* intestine. Several hundred milliliters of fluid are also reabsorbed each day into the *(small? large?)* intestine.

f. When inadequate water reabsorption occurs, as in *(constipation? diar-rhea?)*, then_____such as Na$^+$ and Cl$^-$ are also lost.

F. Large intestine

(pages 622–625)

1. Identify the regions of the large intestine in Figure LG 24-1. Draw arrows to indicate direction of movement of intestinal contents.

2. The total length of the large intestine is about ____ m (____ ft). More than 90 percent of its length consists of the part known as the

_____.

⟨27⟩

3. Match each structure in the large intestine with the related description.

⟨28⟩

_____ a. Valve between small and large in-testine

_____ b. Blind-ended tube attached to cecum

_____ c. Portion of colon located between ascending and transverse colons

_____ d. Pouches that give large intestine its puckered appearance

_____ e. Terminal 3 cm (1 inch) of rectum

AC. Anal columns
H. Haustra
HF. Hepatic flexure
IV. Ileocecal valve
VA. Vermiform appendix

4. Inflammation and enlargement of the rectal veins is a condition called

_____. These may occur as a result of repeated episodes of constipation and are related to *(high? low?)*-fiber diet. Contrast first- and third-degree hemorrhoids.

5. Taeniae coli are parts of the *(submucosa? muscularis?)* wall of the large intestine. These fibers run *(circularly? longitudinally?)*; label them on Figure LG 14-1.

6. Explain why movements of the colon begin immediately following a meal. Include roles of *gastrin* and the *ileocecal valve*.

29 What is the name of the reflex you just described?

7. Define *mass peristalsis.*

8. Describe the roles of bacteria residing in the large intestine. Include these terms in your description: *flatus, vitamins.*

30 9. *A clinical challenge.* Certain groups of individuals lack the normal flora of bacteria in the intestine. Name two such groups.

Name one sign or symptom resulting from inadequate bacterial synthesis of vitamin K.

10. List the chemical components of feces.

11. Describe the process of defecation. Include these terms: *rectal receptors, rectal muscles, sphincters, diaphragm,* and *abdominal muscles.*

12. Complete part g of Table LG 24-1, describing the role of the large intestine in digestion.

G. Aging and developmental anatomy of the digestive system (pages 625–627)

1. *A clinical challenge.* List changes with aging that may lead to a decreased desire to eat among the elderly population.
   a. Related to the upper GI tract (to stomach) | 31 |

   b. Related to the lower GI tract (stomach and beyond)

2. Indicate which germ layer, endoderm (E) or mesoderm (M) gives rise to each of these structures. | 32 |

   _____ a.   Epithelial lining and digestive glands of the GI tract

   _____ b.   Liver, gallbladder, and pancreas

   _____ c.   Muscularis layer and connective tissue of submucosa
3. List GI structures derived from each portion of the primitive gut.
   a. Foregut

562   b. Midgut

c. Hindgut

(pages 627-629) **H.** Disorders, medical terminology

1. Explain how tooth decay occurs. Include the roles of bacteria, plaque, and acid. List the most effective known measures for preventing dental caries.

2. Briefly describe these disorders, stating possible causes of each.
a. Periodontal disease

b. Appendicitis

c. Cirrhosis

3. Explain why a patient may be given large doses of antibiotics prior to abdominal surgery.

563

4. Which of the following is an inflammation? *(Diverticulosis? Diverticulitis?)*

33

5. Describe how these two techniques can be used to help diagnose cancer of the GI tract.
   a. Colonoscopy

   b. Fecal occult blood testing

6. Match the terms with the descriptions.

34

_____ a. Incision of the colon, creating artificial anus		A. Anorexia nervosa
_____ b. Inflammation of the liver		B. Bulimia
_____ c. Inflammation of the colon		Cho. Cholecystitis
_____ d. Burning sensation in region of esophagus and stomach; probably due to gastric contents in lower esophagus		Colit. Colitis
		Colos. Colostomy
		Con. Constipation
		D. Diarrhea
_____ e. Frequent defecation of liquid feces		F. Flatus
_____ f. Inflammation of the gallbladder		Htb. Heartburn
_____ g. Infrequent or difficult defecation		Hem. Hemorrhoids
_____ h. Craterlike lesion in the GI tract due to acidic gastric juices		Hep. Hepatitis
_____ i. Excess air (gas) in stomach or intestine, usually expelled through anus		P. Peptic ulcer
_____ j. Binge-purge syndrome		
_____ k. Loss of appetite and self-imposed starvation		

1   Ingestion, digestion, movement of food, absorption, and defecation.

2   *A,* mucosa: 1, lining epithelium; 2, lamina propria; 3, muscularis mucosae. *B,* submucosa. *C,* Muscularis: 1, circular muscle; 2, longitudinal muscle. *D,* serosa.

3   Ascites.

4   Peritoneum lines the abdomen and covers abdominal viscera; also it forms folds, such as mesentery and omenta.

5   (a) F. (b) M. (c) Meso. (d) G. (e) L.

6   (a) Parotid. (b) Sublingual. (c) Liters. (d) Dissolving medium for foods, lubrication, source of lysozyme and salivary amylase. (e) 6.5.

7   (a) 32, check your own teeth, 20. (b) Incisors, cuspids (canines), premolars (bicuspids), molars. (c) 4, 8, 12. (d) First molar and/or central incisor, third molar. If you had difficulty answering these questions, draw and label sets of teeth on separate paper. Then compare to Figure 24-7, page 600, in the text.)

8   Starch, is not, acidic pH of stomach.

9   Heartburn, acidic, achalasia, heart.

10   Greater curvature, greater curvature, lesser curvature.

11   It has three, rather than two, layers of smooth muscle; the extra one is an oblique layer located deep to the circular layer.

12   (a) Pepsin, pepsinogen, HCl. (b) Acid. (c) Pepsin is released in the inactive state (pepsinogen); mucus protects the stomach lining from pepsin. (d) Ulcer. (e) Gastric lipase; emulsified fats, as in butter; ineffective; its optimum pH is 5 or 6 and most fats have yet to be emulsified (by bile from liver).

13   (a) Cephalic, gastric, intestinal, cephalic. (b) Vagus; parasympathetic; sight, smell, or thought of food; presence of food in the stomach. (c) Stomach gastrin, protein, pyloric, stimulated, relaxed. (d) Enterogastric, inhibits, CCK, GIP, secretin.

14   2-6, carbohydrate, fat.

15   (a) Irritation and distension of the stomach. (b) Medulla, contract, relax. (c) Loss of electrolytes (such as HCl) as well as fluids can lead to fluid/electrolyte imbalances.

16   (a) Stomach. (b) Duodenum, spleen. (c) Exocrine, alkaline. (d) Proteins, enterokinase. (e) Amylase digests carbohydrates, including starch; lipase digests lipids. (f) *G, H,* duodenum. (g) Common bile duct; pancreatic proteases such as trypsin, chymotrypsin and carboxypeptidase may digest tissue proteins of the pancreas itself. (h) Pancreatic islets (of Langerhans), insulin, glucagon, blood vessels, portal.

17   (a) Parasympathetic, vagus small intestine. (b) Digestive enzymes such as trypsin, amylase and lipase; alkaline ($HCO_3{}^-$) fluid.

18   (a) 1.4, 4.0, upper right, 4, right. (b) Falciform, umbilical. (c) Hepatic artery, portal vein, sinusoids, hepatic veins.

19 (a) Superior to inferior. (b) Superiorly.

20 (a) Emulsification, absorption. (b) Bilirubin, red blood, jaundice.

21 Does not.

22 (a) Pyloric, fats and partially digested proteins. (b) Duodenum. (c) Portal vein, meanders in the bloodstream through all parts of the body. (d) Contract. (e) Cystic duct. (f) Obstruction by the stone backs up bile, leading to spasms of smooth muscle (biliary colic) in bile pathways and entrance of new player: INTENSE PAIN; GALLBLADDER. (g) Stimulation of pancreatic and intestinal secretions, inhibition of gastric secretion and motility, and opening of hepatopancreatic sphincter. (Note this on Table LG 24-2.)

23 (a) Shorter chain polysaccharides, disaccharides. (b) Shorter chain polypeptides, dipeptides, amino acids. (c) Fatty acids and glycerol.

24 (a) EG, HCl, P, SG. (b) A, L, TCC. (c) CCK, En, MLS, Pd, S. (d) B.

25 (a) CCK. (b) S, H. (c) CCK, H. (d) S, H. (e) CCK, S. (f) EG, SG, both H. (g) P. (h) HCl, N. (i) P. (j) TCC. (k) En, E. (l) Pd. (m) L. (n) B, N. (o) A. (p) MLS.

26 (a) Small, active transport. (b) Blood, portal, liver. (c) Bile, micelles, thoracic duct, recycled to the liver. (d) Fat-soluble, A D E K; examples, A: night blindness; D: rickets or osteomalacia due to decreased calcium absorption; K: excessive bleeding. See Exhibit 25-4 (pages 652-653) in the text. (e) 9, small, large. (f) Diarrhea, electrolytes.

27 1.5, 5, colon.

28 (a) IV. (b) VA. (c) HF. (d) H. (e) AC.

29 Gastroileal.

30 Newborns, persons who have been on long-term antibiotic therapy; excessive bleeding.

31 Decrease taste sensations, gum inflammation (pyorrhea) leading to loss of teeth and loose-fitting dentures, and difficulty swallowing (dysphagia).

32 (a) E. (b) E. (c) M.

33 Diverticulitis.

34 (a) Colos. (b) Hep. (c) Colit. (d) Htb. (e) D. (f) Cho. (g) Con. (h) P. (i) F. (j) B. (k) A.

MASTERY TEST: Chapter 24

Questions 1–10: Choose the one best answer to each question.

_____ 1. Which of these organs is not part of the GI tract, but is an accessory organ?

A. Mouth   B. Pancreas   C. Stomach
D. Small intestine   E. Esophagus

_____ 2. The main function of salivary and pancreatic amylase is to:

    A. Lubricate foods
    B. Help absorb fats
    C. Digest polysaccharides to smaller carbohydrates
    D. Digest disaccharides to monosaccharides
    E. Digest polypeptides to amino acids

_____ 3. Which enzyme is most effective at pH 1 or 2?

    A. Gastric lipase   B. Maltase   C. Pepsin
    D. Salivary amylase   E. Pancreatic amylase

_____ 4. Choose the TRUE statement about fats.

    A. They are digested mostly in the stomach.
    B. They are the type of food which stays in the stomach the shortest length of time.
    C. They stimulate release of gastrin.
    D. They are emulsified and absorbed with the help of bile.

_____ 5. Most digestion and absorption occurs in the:

    A. Stomach   B. Small intestine   C. Liver
    D. Large intestine   E. Pancreas

_____ 6. Which of the following is under only nervous (not hormonal) control?

    A. Salivation   B. Gastric secretion
    C. Intestinal secretion   D. Pancreatic secretion

_____ 7. All of the following are enzymes involved in protein digestion EXCEPT:

    A. Amylase   B. Trypsin   C. Carboxypeptidase
    D. Pepsin   E. Chymotrypsin

_____ 8. All of the following chemicals are produced by the walls of the small intestine EXCEPT:

    A. Lactase   B. Secretin   C. CCK   D. Trypsin
    E. Peptidases

_____ 9. Which type of movement is used primarily to propel chyme through the intestinal tract, rather than to mix chyme with enzymes.

    A. Peristalsis   B. Rhythmic segmentation
    C. Haustral churning

_____ 10. Choose the FALSE statement about layers of the wall of the GI tract.

    A. Most large blood and lymph vessels are located in the submucosa.
    B. The myenteric plexus is part of the muscularis layer.
    C. Most glandular tissue is located in the layer known as the mucosa.
    D. The mucosa layer forms the peritoneum.

Questions 11–15: Arrange the answers in correct sequence.

_____ _____ _____ 11. From anterior to posterior:

    A. Glossopalatine arch
    B. Pharyngopalatine arch
    C. Palatine tonsils

____ ____ ____ ____ 12. GI tract wall, from deepest to most superficial:

    A. Mucosa
    B. Muscularis
    C. Serosa
    D. Submucosa

____ ____ ____ ____ ____ 13. Pathway of chyme:

    A. Ileum
    B. Jejunum
    C. Cecum
    D. Duodenum
    E. Pylorus

____ ____ ____ ____ ____ 14. Pathway of bile:

    A. Bile canaliculi
    B. Common bile duct
    C. Common hepatic duct
    D. Right and left hepatic ducts
    E. Hepatopancreatic ampulla and duodenum

____ ____ ____ ____ ____ 15. Pathway of wastes:

    A. Ascending colon
    B. Transverse colon
    C. Sigmoid colon
    D. Descending colon
    E. Rectum

Questions 16–20: Circle T (true) or F (false). If the statement is false, change the underlined word or phrase so that the statement is correct.

T F 16. Teeth are composed mostly of a <u>bonelike substance called dentin.</u>

T F 17. The principal chemical activity of the stomach is to begin digestion of <u>protein.</u>

T F 18. The esophagus produces <u>no digestive enzymes or mucus.</u>

T F 19. <u>Villi, microvilli, goblet cells, and rugae</u> are all located in the walls of the small intestine.

T F 20. Cirrhosis and hepatitis are diseases of the <u>liver.</u>

Questions 21–25: fill-ins. Complete each sentence with the word or phrase that best fits.

_____ 21. Stomach motility is ___ creased by the enterogastric reflex, CCK and GIP, and ___-creased by stomach gastrin and parasympathetic nerves.

_____ 22. Mumps involves inflammation of the ___ salivary glands.

_____ 23. Most absorption takes place in the ____, although some substances, such as ____, are absorbed in the stomach.

_____ 24. Mastication is a term which means ____.

_____ 25. Epithelial lining of the GI tract, as well as the
liver and pancreas, are derived from ____-derm.

25

Metabolism

In Chapter 24 you studied how nutrients are ingested, digested, and absorbed into the bloodstream. In this chapter you will see how nutrients are utilized within cells throughout the body. First, you will contrast functions of each type of nutrient (Objective 1). You will define metabolism and compare its two principal aspects: catabolism and anabolism (2–4). Next, you will consider metabolism of the major nutrients: carbohydrates (5–7), fats (8–12), proteins (13–15), minerals and vitamins (18–21). You will learn how metabolism is regulated (16–17). You will also study heat production and heat loss related to metabolic activities (22–26). Finally, you will consider metabolic disorders (27–28).

Topics Summary

A. Metabolism
B. Carbohydrate metabolism
C. Lipid metabolism
D. Protein metabolism
E. Absorptive and postabsorptive states, control of metabolism
F. Minerals, vitamins
G. Metabolism and heat
H. Disorders

Objectives

1. Define a nutrient and list the functions of the six principal classes of nutrients.
2. Define metabolism and contrast the roles of anabolism and catabolism.
3. Define oxidation and reduction and their importance in metabolism.
4. Describe the characteristics and importance of enzymes in metabolism.
5. Explain the fate of absorbed carbohydrates.
6. Describe the complete oxidation of glucose via glycolysis, the Krebs cycle, and the electron transport chain.
7. Define glycogenesis, glycogenolysis, and gluconeogenesis.
8. Explain the fate of absorbed lipids.
9. Discuss fat storage in adipose tissue.
10. Explain how glycerol may be converted to glucose.
11. Explain the catabolism of fatty acids via beta oxidation.
12. Define ketosis and list its effects on the body.
13. Explain the fate of absorbed proteins.
14. Discuss the catabolism of amino acids.
15. Explain the mechanism involved in protein synthesis.
16. Distinguish between the absorptive and postabsorptive states.
17. Explain how various hormones regulate metabolism.
18. Compare the sources, functions, and importance of minerals in metabolism.
19. Define a vitamin and differentiate between fat-soluble and water-soluble vitamins.
20. Compare the sources, functions, and deficiency symptoms and disorders of the principal vitamins.
21. Describe abnormalities associated with megadoses of several minerals and vitamins.
22. Explain how the caloric value of foods is determined.
23. Describe the various mechanisms that produce body heat.
24. Define basal metabolic rate (BMR) and explain several factors that affect it.
25. Describe the various ways that body heat is lost.
26. Explain how normal body temperature is maintained.
27. Describe fever, heat cramp, heatstroke, and heat exhaustion as abnormalities of temperature regulation.
28. Define obesity, phenylketonuria, cystic fibrosis, celiac disease, and kwashiorkor.

(pages 634–635)
A. **Metabolism**

1. List the six classes of nutrients on lines below. Then write one or more function of each.

1

a. _____

b. _____

c. _____

d. _____

e. _____

f. _____

2. Explain why metabolism might be thought of as an "energy-balancing act."

3. Complete this table comparing catabolism with anabolism.

Process	Definition	Releases or Uses Energy	Examples
a. Catabolism			
b. Anabolism			

4. In the final breakdown of glucose (and other products of digestion), hydrogens are *(added? removed?)* and combined with oxygen. This process is known as *(oxidation? reduction?)* of the nutrients. Energy released during these catabolic processes may be stored in the compound _____

_____ or released as heat.

For extra review of ATP, refer to Chapter 2, Activity `26`, page LG 35.

5. In order for molecules to react with one another, they must have a certain amount of energy, known as _____ energy. Answer the following questions about this energy.

    a. One way to attain such energy is to heat the reacting molecules. Why would this be a problem in living systems?

    b. How can enzymes increase chemical reaction rates without requiring an increase in body temperature?

6. Do this exercise about characteristics of enzymes.

    a. Chemically, all enzymes consist (either entirely or mostly) of _____. The molecular weight of most enzymes is *(100–2,000? 10,000–millions?).*

    b. Since enzymes speed chemical reaction rates, they are known as _____. They react with specific molecules called _____.

    c. The rate of production of enzymes depends on the cells' needs at a given moment and *(is? is not?)* controlled by genes.

    d. Names of most enzymes end in the letters _____. Enzyme names also give clues to their functions. For example, enzymes that remove hydrogen are called _____, while those such as amylase or protease that split apart substrates by adding water are known as _____.

    e. Whole enzymes are sometimes known as *(apo? holo?)*-enzymes. These consist of a protein portion called the *(apo? holo?)*-enzyme, as well as a cofactor.

    f. Cofactors are of two types. Some are metal ions, such as _____. So, for example, a protein (apoenzyme) plus calcium ion may serve as a holoenzyme.

g. The other type of cofactor is known as a _____-enzyme. Most coenzymes are derived from *(minerals? vitamins?)*. List three examples of coenzymes: _____. So, for example, NAD may function as a coenzyme by picking up _____ atoms from the substrate.

h. The substrate interacts with a specific region on the enzyme called the _____. An intermediate known as an _____ _____ complex is formed. Very quickly the *(enzyme? substrate?)* is transformed in some way (such as broken down or transferred to another substrate), while the *(enzyme? substrate?)* is recycled for further catalytic action.

(pages 636–643)   **B. Carbohydrate metabolism**

5   1. Answer these questions about carbohydrate metabolism.

a. The story of carbohydrate metabolism is really the story of _____ metabolism, since this is the most common carbohydrate (and in fact the most common energy source) in the human diet.

b. What other carbohydrates besides glucose are ingested? How are these converted to glucose?

c. Just after a meal, the level of glucose in the blood *(increases? decreases?)*.

Cells use some of this glucose; by _____ glucose, they release energy.

d. List three mechanisms by which excess glucose is used or discarded.

e. Increased glucose level of blood is known as _____. This can occur after a meal containing concentrated carbohydrate or in the absence of the hormone _____, since this hormone *(facilitates? inhibits?)* entrance of glucose into cells. Lack of insulin occurs in the condition known as _____.

f. As soon as glucose enters cells, glucose combines with _____

_____ to form glucose-6-phosphate. This process is

known as _____.

2. Describe the process of glycolysis in this exercise.   6

a. Glucose is a _____-carbon molecule (which also has
   hydrogens and oxygens on it). During glycolysis each glucose molecule

   is converted to two _____-carbon molecules named

   _____.

b. *(A lot? A little?)* energy is released from glucose during glycolysis. This
   process requires *(one? many?)* step(s) and *(does? does not?)* require oxygen.

c. The fate of pyruvic acid depends on whether sufficient _____

   _____ is available. If it is, pyruvic acid undergoes
   chemical change in phases 2 and 3 of glucose catabolism (see below).
   These processes are *(aerobic? anaerobic?),* that is, they require oxygen.

d. If the respiratory rate cannot keep pace with glycolysis, insufficient oxygen is available to break down pyruvic acid. In the absence of sufficient

   oxygen, pyruvic acid is temporarily converted to _____

   _____. This is likely to occur during active

   _____.

e. Several mechanisms prevent the accumulation of an amount of lactic acid

   which might be harmful. The liver changes some back to _____

   _____ and _____. Excess PCO_2
   caused by exercise *(stimulates? inhibits?)* respiratory rate, so more oxygen
   is available for breakdown of the pyruvic acid. Lactic acid contributes to

   muscle _____ which will also cause the person to
   want to rest. When exercise has slowed or stopped, sufficient oxygen is

   inspired so that the oxygen _____ is paid.

3. Review glucose catabolism up to this point and summarize the remaining
   phases by doing this exercise. Refer to Figure LG 25-1.   7

A six-carbon compound named (a) _____ is converted

to two molecules of (b) _____ via a number of steps,

together called (c) _____, occurring in the (d)

_____ of the cell.

   In order for pyruvic acid to be further broken down, it must undergo a

(e) _____ step. This involves removal of the one-carbon

molecule, (f) _____. The remaining two-carbon (g)

_____ group is attached to a carrier called (h)

_____, forming (i) _____.

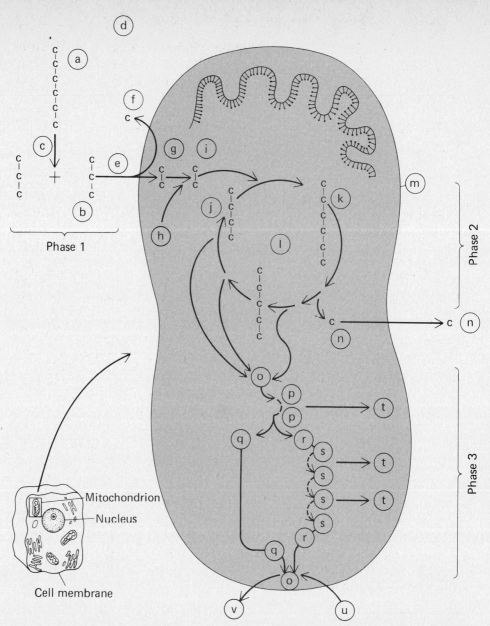

Figure LG 25-1 Summary of glucose catabolism. Letters refer to Learning Activity 7 .

This compound hooks on to a four-carbon compound called (j) _____. The result is a six-carbon molecule of (k) _____. The name of this compound is given to the cycle of reactions called the (1) _____ cycle, occurring in the (m) _____ of the cell.

Two main types of reactions occur in the Krebs cycle. One is the decarboxylation reaction, in which (n) _____ molecules are removed. (Notice these locations on Figure 25-5, page 640, of your text.) (What happens to the CO_2 molecules that are removed?) As a result the six-carbon citric acid is eventually shortened to regenerate oxaloacetic, giving the cyclic nature to this process.

The other major type of reaction involves removal of (o) _____

_____ atoms during oxidation of compounds such as isocitric and α-ketoglutaric acids. (See Figure 25-5, page 640, in your text.) These hydrogen atoms are carried off by two coenzymes named (p)

_____ and "trapped" for use in phase 3.

In phase 3, hydrogen atoms on NAD and FAD (and later coenzyme

Q) are ionized to (q) _____ and (r) _____

_____ . Electrons are shuttled along a chain of (s)

_____ . During electron transport energy in these electrons (derived from hydrogen atoms in ingested foods) is "tapped"

and stored in (t) _____ (Figure 25-6, page 641, of your text).

Finally, the electrons, depleted of some of their energy, are reunited

with hydrogen ions and with (u) _____ to form (v)

_____ . Notice why your body requires oxygen—to keep drawing hydrogen atoms off of nutrients so that energy from them can be stored in ATP for use in cellular activities.

4. Write a chemical equation showing the total catabolism of glucose. `8`

5. Most of the 38 ATPs generated from the total oxidation of glucose derive from *(glycolysis? Krebs cycle?)*. Of those generated by the Krebs cycle, most are formed from the oxidation of hydrogens carried by *(NAD? FAD?)*. `9`

6. About what percent of the energy originally in glucose is stored in ATP after glucose is completely catabolized? *(10%? 43%? 60%? 99%?)* `10`

7. Complete the exercise about the equation shown below. `11`

$$\text{Glucose} \; \underset{(b)}{\overset{(a)}{\rightleftharpoons}} \; \text{Glycogen}$$

a. You learned earlier that excess glucose may be stored as glycogen. In other words, glycogen consists of large branching chains of

_____ . Name the process of glycogen formation. Label (a).

b. Between meals, when glucose is needed, glycogen can be broken down again to release glucose. Label (b) with the name of this process.

c. Which of these two reactions is anabolic? _____

d. Where is most (80 percent) of glycogen in the body stored?

576

e. Identify a hormone that stimulates reaction (a). Write its name on line (a). As more glucose is stored in the form of glycogen, the blood level of glucose ____creases. Now write below line (b) two hormones that stimulate glycogenolysis.

f. In order for glycogenolysis to occur, a _____ group must be added to glucose as it breaks away from glycogen. The enzyme catalyzing this reaction is known as _____. Does the reverse reaction (shown in b) require phosphorylation also? *(Yes? No?)*

8. Refer to Figure 25-8, page 643, in your text. Define *gluconeogenesis* and briefly discuss how it is related to other metabolic reactions.

9. Circle the processes at left and the hormones at right that lead to increased blood glucose level.

| 12 |

Glycogenesis   Insulin
Glycogenolysis   Glucagon
Glycolysis   Epinephrine
Gluconeogenesis   Cortisol
     Thyroxine
     Growth hormone

10. The two molecules that appear to be most involved in interconversions among amino acids, fatty acids, and carbohydrates are _____

| 13 |

_____ and _____.

(pages 644–646) **C. Lipid metabolism**

1. Name four examples of structural or functional roles of fats in the body.

2. List areas where fat is stored.

Is the fat stored in your subcutaneous tissue "the same fat" that was in that location two years ago? Explain.

3. Complete the exercise about fat catabolism. $\boxed{14}$

   a. Fats must first be broken into _____
   and _____. Glycerol is converted into
   _____ and enters glycolysis pathways.

   b. Recall that fatty acids are long chains of carbons, with attached hydrogens and a few oxygens. Two-carbon pieces are "snipped off" of fatty acids by a process called _____, occurring in the
   _____.

   c. Some of these two-carbon pieces, called _____, can enter Krebs cycle reactions; in this way fats can release energy just as carbohydrates do.

   d. When fat catabolism is excessive, large numbers of acetyl CoA's form and tend to pair chemically:

   Acetic acid + acetic acid → _____

   (2C)   (2C)   (4C)

   The presence of these acids in blood *(raises? lowers?)* blood pH. Since these acids are called "keto" acids, this condition is known as _____
   _____.

   e. A slight alteration of acetoacetic acid (removal of CH_3) forms acetone. Collectively, acetoacetic acid and acetone are known as _____
   _____ bodies. Formation of ketone bodies occurs in the liver; the process is known as _____.

   f. Ketogenesis occurs when cells are forced to turn to fat catabolism. State two reasons why cells might carry out excessive fat catabolism leading to ketogenesis and ketosis.

   g. Explain why a diabetic might have sweet breath. Tell how this relates to fat metabolism.

4. Briefly describe how a diet that is excessive in carbohydrates or proteins can lead to formation of fat deposits in the body.

(page 646) **D. Protein metabolism**

15   1. Which of the three major nutrients (carbohydrates, lipids, and proteins) fulfills each function?

    a. Most direct source of energy; stored in body least: _____

    _____

    b. Second as source of energy; used more for body-building: _____

    _____

    c. Used least for energy; used most for body-building: _____

    _____

16   2. Contrast the caloric values of the three major food types.

    a. Carbohydrates and proteins each produce _____ Cal/g or (since a pound contains 454 g) _____ Cal/lb.

    b. Fats produce _____ Cal/g or _____ Cal/lb.

3. Throughout your study of systems of the body so far, you have learned about a variety of roles of proteins. In one or two words each, list at least six functions of proteins in the body. Include some structural and some regulatory roles.

4. Name a hormone that enhances transport of amino acids into cells. (For help, see Figure LG 18–1, page LG 368.)

17   5. Refer to Figure LG 25–2 and describe the uses of protein.

    Two sources of proteins are shown: (a) _____

    _____ and (b)

    _____. Amino acids from these sources may undergo a

    process known as (c) _____

    _____ to form new body proteins.

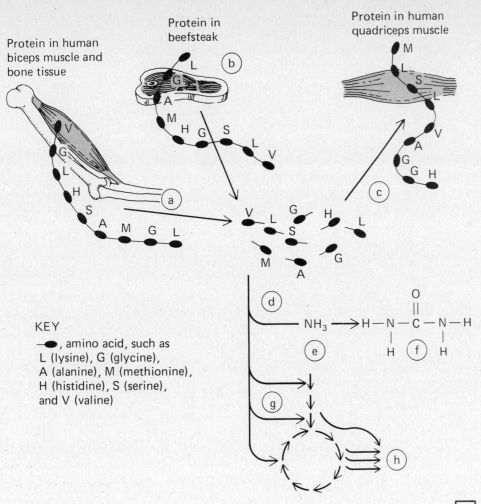

KEY

━●, amino acid, such as
L (lysine), G (glycine),
A (alanine), M (methionine),
H (histidine), S (serine),
and V (valine)

Figure LG 25-2 Metabolism of protein. Lowercase letters refer to Learning Activity ⌐17⌐.

If other energy sources are used up, amino acids may undergo catabolism. The first step is (d) _____, in which an amino group (NH₂) is removed and converted (in the liver) to (e) _____ _____, a component of (f) _____ _____ which exists in urine. The remaining portion of the amino acid may enter (g) _____ _____ pathways at a number of points (Figure 25-11, page 647, in your text). In this way, proteins can lead to formation of (h) _____.

6. Contrast *essential amino acids* and *nonessential amino acids.*

580 (pages 647–649)

E. Absorptive and postabsorptive states, control of metabolism

1. Contrast these two states by writing A if the description refers to the absorptive state and P if it refers to the postabsorptive state.

[18]

_____ a. Period when the body is fasting
_____ b. Time when the body is absorbing nutrients from the GI tract
_____ c. State during which the principal concern is formation of stores of glucose (as glycogen) and fat
_____ d. State during which glucose (stored in glycogen in liver and muscle) is released
_____ e. State during which most systems (excluding nervous system) switch over to fat as energy source

2. Now discuss utilization of amino acids in the absorptive state.

[19]

3. Do this exercise about maintenance of glucose level between meals.
   a. The normal blood glucose level in the postabsorptive state is about *(30–50? 80–100? 150–170? 600–700?)* mg glucose/100 ml blood. Adequate blood glucose is especially important to the brain and nerves. Explain why.

   b. Now list four sources of glucose that may be called upon during the postabsorptive state.

   c. Can fatty acids serve as a source of glucose? *(Yes? No?)* Explain.

   Fats can be used to generate ATP by being broken down into _____ which can then enter the Krebs cycle. Therefore fatty acids are said to provide a glucose *(sparing? utilizing?)* mechanism between meals.

4. Complete the table describing how each hormone assist in regulation of metabolism. (Refer to Figures LG 18-1 and LG 18-5, pages LG 368 and 380, and Exhibit 25–2, page 650 of your text.)

Hormone	Source	Action
a.		Stimulates glycogenesis in liver; stimulates glucose catabolism in other cells
b.	Alpha cells of islets of Langerhans in pancreas	

Hormone	Source	Action
c. Growth hormone (GH)		
d.		Stimulates adrenal cortex to produce glucocorticoids
e. Glucocorticoids		
f.	Thyroid gland	
g.	Adrenal medulla	
h. Testosterone		

5. Circle the hormones in the activity above that are hyperglycemic, that is, tend to *increase* blood glucose? $\boxed{20}$

F. Minerals, vitamins

(pages 649–655)

1. Define *minerals.*

Minerals make up about _____ percent of body weight and are concentrated $\boxed{21}$

in the _____.

2. Study Exhibit 25-3, page 651 in your text. Then check your understanding of minerals by doing this matching exercise. $\boxed{22}$

_____ a. Main anion in extracellular fluid, part of HCl in stomach; component of table salt

_____ b. Involved in generation of nerve impulse, helps to regulate osmosis, acts in buffer systems

_____ c. Most abundant cation in the body, found mostly in bones and teeth; necessary for normal muscle contraction and for blood clotting

_____ d. Important component of hemoglobin and cytochromes

_____ e. Main cation inside of cells; used in nerve transmission

_____ f. Essential component of thyroxin

_____ g. Constituent of vitamin B_{12}, so necessary for red blood cell formation

_____ h. Important component of amino acids, vitamins, and hormones

_____ i. Improves tooth structure

Ca. Calcium
Cl. Chlorine
Co. Cobalt
F. Fluorine
Fe. Iron
I. Iodine
K. Potassium
Mg. Magnesium
Na. Sodium
P. Phosphorus
S. Sulfur

_____ j. Found mostly in bones and teeth; important in buffer system and in ATP processes; component of DNA and RNA.

3. Vitamins are *(organic? inorganic?)*. Most vitamins *(can? cannot?)* be synthesized in the body. In general, what are the functions of vitamins?

|23|

4. Name three sources of vitamins.

|24|

5. Contrast the two principal groups of vitamins, and list the main vitamins in each group.
   a. Fat-soluble

   b. Water-soluble

|25|

6. Match the vitamins with the descriptions.

_____ a. This serves as a coenzyme that is essential for blood clotting, so it is called the antihemorrhagic vitamin; synthesized by intestinal bacteria. _____		A
		B_1
		B_2
		B_{12}
_____ b. Its formation depends upon sunlight on skin and also on kidney and liver activation; necessary for		C
		D
		E
		K

calcium absorption. _____

_____ c. Riboflavin is another name for it; a component of FAD; necessary for normal integrity of skin, mucosa and eye. _____

_____ d. This vitamin acts as an important coenzyme in carbohydrate metabolism; deficiency leads to beriberi.

_____

_____ e. Formed from carotene, it is necessary for normal bones and teeth; it prevents night blindness. _____

_____ f. This substance is also called ascorbic acid; deficiency causes anemia, poor wound healing, and scurvy.

_____

_____ g. This coenzyme, the only B vitamin not found in vegetables, is necessary for normal erythropoiesis; absorption from GI tract depends on intrinsic factor. _____

_____ h. Also known as tocopherol, it is necessary for normal red blood cell membranes; deficiency is associated with sterility in some animals. _____

G. Metabolism and heat

(pages 655–658)

1. Define *heat*.

2. Contrast *calorie* and *kilocalorie*.

26

If you ingest, digest, absorb, and metabolize a slice of bread, the energy released from the bread equals about 80–100 *(Cal? kCal?)*.

3. Explain how a calorimeter determines calorie values of foods.

584

4. Answer these questions about catabolism.

   a. Catabolism of foods *(uses? releases?)* energy. Most of the heat produced by the body comes from catabolism by *(oxidation? reduction?)* of nutrients.

   [27]

   b. Discuss factors that affect how rapidly your body catabolizes, that is, your metabolic rate. Include: *exercise, nervous system, hormones,* and *body temperature, specific dynamic action (SDA), age,* and *pregnancy.*

5. Define *basal metabolic rate (BMR).*

   List four conditions that are necessary in order for the body to be in a basal state.

[28] 6. Calculate a typical BMR in this exercise.

   a. BMR may be determined by measuring the amount of _____ _____ consumed within a period of time, since oxygen is necessary for the metabolism of foods. Normally for each liter of oxygen consumed, the body metabolizes enough food to release about _____ kilocalories.

   b. Suppose you consume 15 liters of oxygen in an hour. Your body would release _____ kCal of heat in that hour.

c. BMR also factors in differences in body size. A typical body surface (as if you could lay carpet over all of your skin is about *(0.5? 2? 5? 10?)* square meters. If you used 72.4 kCal in an hour, and your body surface is 2 m², your BMR is ____ kCal/m²/hr.

d. If a normal value for BMR for a person your age is 40.2 kCal/m²/hr, and your BMR is 36.2 kCal/m²/hr, then your BMR is about 10% *(above? below?)* normal. This would be expressed as a BMR of *(+10? −10?)*.

e. A BMR of −20 is more likely to be due to *(hyper? hypo)*-thyroidism.

7. Your body temperature usually *(does? does not?)* rise much as a result of catabolism. Why? `29`

8. Write the percent of heat loss by each of the following routes (at room temperature). Then write an example of each of these types of heat loss. One is done for you.
   a. Radiation ____

   b. Conduction ____

   c. Convection ____ cooling by draft while taking a shower

   d. Evaporation ____

9. During an hour of active exercise Bill produces 1 liter of sweat. The amount of heat loss (cooling) that accompanies this much evaporation is ____ Cal. On a humid day, Bill would sweat *(more? less?),* so Bill would be cooled *(more? less?)* on such a day. `30`

10. Explain how hypothalamus serves as the body's thermostat.

11. Lil is outside on a snowy day without a warm coat. Her skin is pale and chilled and she begins to shiver. Explain how these responses are attempts of her body to maintain homeostasis of temperature. What other responses might raise her body temperature? Include these words in your answers: *sympathetic, metabolism, adrenal medulla, thyroid,* and *coat.*

12. Define *pyrogens*.

31 13. Arrange in order the events believed to occur in the production of fever of 39.4°C (103°F) and recovery from this state. ___ ___ ___ ___ ___

A. 39.4°C (103°F) temperature is reached and maintained.
B. Pyrogens cause the hypothalamic "thermostat" to be reset from normal 37.5°C (98.6°F) to a higher temperature such as 39.4°C (103°F).
C. Vasoconstriction and shivering occur as heat is conserved in order to raise body temperature to 39.4°C (103°F) level directed by the hypothalamus.
D. Source of pyrogens is removed (for example, bacteria are killed by antibiotics), lowering hypothalamic "thermostat" to 37.5°C (98.6°F).
E. "Crisis" occurs. Obeying hypothalamic orders, the body shifts to heat-loss mechanisms such as sweating and vasodilation, returning body temperature to normal.

14. In what ways is a fever beneficial?

15. Describe each of these temperature abnormalities.
    a. Heat cramp

    b. Heatstroke

    c. Heat exhaustion

(pages 658–660) **H. Disorders**

1. Define *obesity*.

Contrast *regulatory obesity* and *metabolic obesity*.

2. Contrast advantages and disadvantages of three surgical methods for weight reduction.

3. Individuals with PKU are unable to convert _____ to _____. What are results of toxic levels of phenylalanine? [32]

4. Cystic fibrosis *(is? is not?)* an inherited disorder. It affects *(endo? exo?)*-crine glands. List three regions of the body in which glands are affected. [33]

5. *A clinical challenge.* Explain why the child with cystic fibrosis may require a low fat diet and is at risk for developing tetany. [34]

6. Persons with celiac disease require a diet that is lacking in most *(meats? vegetables? grains?)*. The problem stems from a water-insoluble protein named _____ which causes destruction of the intestinal lining. Name two grains that are acceptable in the diet of a celiac disease patient. [35]

36

7. *A clinical challenge.* Explain why an individual with kwashiorkor may have a large abdomen even though the person's diet is inadequate.

Answers to Numbered Questions in Learning Activities

1. Carbohydrates, proteins, lipids, minerals, vitamins, and water.

2. Removed, oxidation, ATP.

3. Activation. (a) Temperature increases could kill living cells. (b) Increase frequency of collisions, orient molecules so that they are more likely to react, lower activation energy.

4. (a) Protein, 10,000–millions. (b) Catalysts, substrates. (c) Is. (d) -ase, dehydrogenases, hydrolases. (e) Holo, apo. (f) Magnesium, zinc, calcium. (g) Co; vitamins; NAD, NADP and FAD; hydrogen. (h) Active site, enzyme-substrate, substrate, enzyme.

5. (a) Glucose. (b) Starch, sucrose, lactose; by enzymes in the liver. (c) Increases, oxidizing. (d) Stored as glycogen, converted to fat and stored, excreted in urine. (e) Hyperglycemia, insulin, facilitates, diabetes mellitus. (f) Phosphate, phosphorylation.

6. (a) Six, three, pyruvic acid. (b) A little, many, does not. (c) Oxygen, aerobic. (d) Lactic acid, exercise. (e) Pyruvic acid, glucose, stimulates, fatigue, debt.

7. (a) Glucose. (b) Pyruvic acid. (c) Glycolysis. (d) Cytoplasm. (e) Transition. (f) CO_2. (g) Acetyl. (h) Coenzyme A. (i) Acetyl coenzyme A. (j) Oxaloacetic acid. (k) Citric acid. (l) Citric acid (or Krebs). (m) Mitochondria. (n) CO_2. (o) Hydrogen. (p) NAD and FAD. (q) H^+. (r) Electrons (e^-). (s) Cytochromes or electron transfer system. (t) ATP. (u) Oxygen. (v) Water.

8. $C_6H_{12}O_6 + 6 O_2 \rightarrow 38 \text{ ATP} + 6 CO_2 + 6 H_2O$

9. Krebs, NAD.

10. 43.

11. (a) Glucose, glycogenesis. (b) Glycogenolysis. (c) Glycogenesis. (d) Muscles. (e) Insulin, de, glucagon and epinephrine. (f) Phosphate, phosphorylase, yes.

12. Glycogenolysis and gluconeogenesis; all of the hormones listed EXCEPT insulin.

13. Pyruvic acid and acetyl coenzyme A (see Figure 25-9, page 644 of the text).

14   (a) Fatty acids, glycerol, glyceraldehyde-3-phosphate. (b) Beta oxidation, liver. (c) Acetyl coenzyme A. (d) Acetoacetic acid, lowers, ketosis. (e) Ketone, ketogenesis. (f) Examples of reasons: starvation, fasting diet, lack of glucose in cells due to lack of insulin (diabetes mellitus), excess growth hormone (GH) which stimulates fat catabolism. (g) As acetone (formed in ketogenesis) passes through blood in pulmonary vessels, some is exhaled and detected by its sweet aroma.

15   (a) Carbohydrates. (b) Lipids. (c) Proteins.

16   (a) 4, about 1,800. (b) 9, about 4,000.

17   (a) Worn-out cells as in bone or muscle. (b) Ingested foods. (c) Anabolism. (d) Deamination. (e) Ammonia. (f) Urea. (g) Glycolytic or Krebs cycle. (h) $CO_2 + H_2O + ATP$.

18   (a) P. (b) A. (c) A. (d) P. (e) P.

19   (a) 80–100, the nervous system cannot use any carbohydrate other than glucose. (b) Liver glycogen (4-hour supply), glycerol from fats, muscle glycogen and lactic acid during vigorous exercise, amino acids from tissue proteins. (c) No, since acetyl CoA cannot be readily converted into pyruvic acid and then to glucose; acetyl CoA, sparing.

20   (b) Glucagon, (c) GH, (d) ACTH, (e) Glucocorticoids, (f) Thyroxine, (g) Epinephrine.

21   4, skeleton.

22   (a) Cl. (b) Na. (c) Ca. (d) Fe. (e) K. (f) I. (g) Co. (h) S. (i) F. (j) P.

23   Organic; cannot; most serve as coenzymes, maintaining growth and metabolism.

24   (a) A D E K. (b) B complex and C.

25   (a) K. (b) D. (c) B_2. (d) B_1. (e) A. (f) C. (g) B_{12}. (h) E.

26   A kilocalorie is the amount of heat required to raise 1,000 ml (1 liter) of water 1°C; a kilocalorie (kCal) = 1,000 calories (Cal); kCal.

27   Releases, oxidation.

28   (a) Oxygen, 4.825. (b) 72.4 (= 4.825 kCal/l O_2 × 15 l O_2/hr). (c) 2. 36.2 (= 72.4/2). (d) Below, −10. (e) Hypo.

29   Does not. About half of the energy released by catabolism is stored in ATP and used for body activities. The remaining energy is released as heat and regulated by heat loss mechanisms.

30   580 (= 0.58 Cal/ml water × 1000 ml/liter), less, less.

31   B C A D E.

32   Phenylalanine, tyrosine; toxicity to the brain with possible mental retardation.

33   Is: exo; respiratory, pancreas, salivary, and sweat glands.

590

34 Pancreatic ducts are blocked, so lipases do not reach the intestine to digest fats. So fats and fat-soluble vitamins (including D) are not absorbed. Vitamin D is necessary for calcium absorption; hypocalcemia can cause overstimulation of nerves to tetany muscles.

35 Grains, gluten, rice, and corn.

36 Protein deficiency (due to lack of essential amino acids) decreases plasma proteins. Blood has less osmotic pressure, so fluids exit from blood. Movement of these into the abdomen (ascites) increases the size of the abdomen. Fatty infiltration of the liver also adds to the abdominal girth.

MASTERY TEST: Chapter 25

Questions 1–13. Choose the one best answer to each question.

_____ 1. Which of these processes is anabolic?

    A. Pyruvic acid $\rightarrow CO_2 + H_2O + ATP$
    B. Glucose $\rightarrow$ pyruvic acid $+$ ATP
    C. Protein synthesis
    D. Digestion of starch to maltose
    E. Glycogenolysis

_____ 2. All of these statements are true of coenzyme A EXCEPT:

    A. It acts as a carrier.
    B. It is nonprotein.
    C. It functions in the transition step between glycolysis and the Krebs cycle.
    D. It carries a two-carbon acetyl group.
    E. It is used in glycolysis.

_____ 3. Which hormone is said to be hypoglycemic since it tends to lower blood sugar?

    A. Glucagon   B. Glucocorticoids   C. Growth hormone
    D. Insulin   E. Epinephrine

_____ 4. The complete oxidation of glucose yields all of these products EXCEPT:

    A. ATP   B. Oxygen   C. Carbon dioxide
    D. Water

_____ 5. All of these processes occur exclusively or primarily in liver cells except one, which occurs in virtually all body cells. This one is:

    A. Gluconeogenesis
    B. Beta oxidation of fats
    C. Ketogenesis
    D. Deamination   E. Krebs cycle

_____ 6. The process of forming glucose from fats or proteins is called:

    A. Gluconeogenesis   B. Glycogenesis   C. Ketogenesis
    D. Deamination   E. Glycogenolysis

_____ 7. All of the following are parts of fat absorption and fat metabolism EXCEPT:

     A. Lipogenesis   B. Beta oxidation
     C. Chylomicron formation   D. Ketogenesis
     E. Glycogenesis

_____ 8. Which of the following vitamins is water-soluble: A, C, D, E, K?

_____ 9. At room temperature most body heat is lost by:

     A. Evaporation   B. Convection
     C. Conduction   D. Radiation

_____ 10. Choose the FALSE statement about vitamins.

     A. They are organic compounds.
     B. They regulate physiological processes.
     C. Most are synthesized by the body.
     D. Many act as parts of enzymes or coenzymes.

_____ 11. Choose the FALSE statement about temperature regulation.

     A. Some aspects of fever are beneficial.
     B. Fever is believed to be due to a "resetting of the body's thermostat."
     C. Vasoconstriction of blood vessels in skin will tend to conserve heat.
     D. Heat-producing mechanisms which occur when you are in a cold environment are primarily parasympathetic.

_____ 12. Which of the following processes involves a cytochrome chain?

     A. Glycolysis
     B. Transition step between glycolysis and Krebs cycle
     C. Krebs cycle
     D. Electron transport system

_____ 13. Which of the following can be represented by the equation: Glucose → pyruvic acids + small amount ATP?

     A. Glycolysis
     B. Transition step between glycolysis and Krebs cycle
     C. Krebs cycle
     D. Electron transport system

Question 14: Arrange the answers in correct sequence.

_____ _____ _____ _____ _____ _____ 14. Steps in complete oxidation of glucose:

     A. Glycolysis takes place.
     B. Pyruvic acid is converted to acetyl coenzyme A.
     C. Hydrogens are picked up by NAD and FAD; hydrogens ionize.
     D. Krebs cycle releases CO_2 and hydrogens.

592

E. Oxygen combines with hydrogen ions and electrons to form water.

F. Electrons are transported along cytochromes and energy from electrons is stored in ATP.

Questions 15–20: Circle T (true) or F (false). If the statement is false, change the underlined word or phrase so that the statement is correct.

T F 15. Catabolism of carbohydrates involves <u>oxidation</u> which is a process of <u>addition of hydrogens.</u>

T F 16. The metabolic rate of a typical child is about <u>half</u> that of a typical elderly person.

T F 17. Nonessential amino acids are those that <u>are not used in the synthesis of human protein.</u>

T F 18. The complete oxidation of glucose to CO_2 and H_2O yields <u>4</u> ATPs.

T F 19. Anabolic reactions are <u>synthetic reactions that release energy.</u>

T F 20. Regulatory obesity seems to be far <u>more</u> common than metabolic obesity.

Questions 21–25: fill-ins. Complete each sentence with the word or phrase that best fits.

_____ 21. Catabolism of each gram of carbohydrate or protein results in release of about ____ kCal, while each gram of fat leads to about ____ kCal.

_____ 22. The body's "favorite" source of energy (since it is most easily catabolized) is ____, while the least favorite is ____.

_____ 23. ____ is the mineral that is most common in extracellular fluid (ECF); it is also important in osmosis, buffer systems, and in nerve impulse conduction.

_____ 24. Aspirin and acetaminophen (Tylenol) reduce fever by inhibiting synthesis of ____ so that the body's "thermostat" located in the ____ is reset to a lower temperature.

_____ 25. In uncontrolled diabetes mellitus, the person may go into a state of acidosis due to the production of ____ acids resulting from excessive breakdown of ____.

The Urinary System

In the last chapter you saw that metabolism of nutrients results in production of some wastes, such as carbon dioxide, water, and urea. In this chapter you will examine the urinary system and its mechanisms for eliminating wastes and controlling composition and volume of body fluids. You will study the urine-producing organs, the kidneys: their structure (Objectives 1–3) and their functions (4–11). You will also look at excretory functions of organs of other systems. You will identify the components of urine and factors affecting its composition (12–15). You will trace the pathway of urine through other urinary organs leading to the exterior of the body (16–20). You will study developmental anatomy and effects of aging upon the urinary system (21–22). Finally, you will consider some disorders and medical terminology associated with the urinary system (23–24).

Topics Summary

A. Kidneys: anatomy
B. Kidneys: physiology
C. Homeostasis: other excretory organs, urine
D. Ureters, urinary bladder, urethra
E. Aging and developmental of the urinary system
F. Disorders, medical terminology

Objectives

1. Identify the external and internal gross anatomical features of the kidneys.
2. Define the structural adaptations of a nephron for urine formation.
3. Describe the blood and nerve supply to the kidneys.
4. Describe the structure and function of the juxtaglomerular apparatus.
5. Discuss the process of urine formation through glomerular filtration, tubular reabsorption, and tubular secretion.
6. Define the forces that support and oppose the filtration of blood in the kidneys.
7. Explain the mechanism and importance of tubular reabsorption.
8. Compare the obligatory and facultative reabsorption of water.
9. Explain the mechanism and importance of tubular secretion.
10. Explain how kidneys produce dilute and concentrated urine.
11. Discuss the operational principle of hemodialysis.
12. Compare the effects of blood pressure, blood concentration, temperature, diuretics, and emotions on urine volume.
13. List and describe the physical characteristics of urine.
14. List the normal chemical constituents of urine.
15. Define albuminuria, glycosuria, hematuria, pyuria, ketosis, bilirubinuria, urobilinogenuria, casts, and renal calculi.
16. Discuss the structure and physiology of the ureters.
17. Describe the structure and physiology of the urinary bladder.
18. Explain the physiology of the micturition reflex.
19. Compare the causes of incontinence and retention.
20. Explain the structure and physiology of the urethra.
21. Describe the effects of aging on the urinary system.
22. Describe the development of the urinary system.
23. Discuss the causes of gout, glomerulonephritis, pyelitis, pyelonephritis, cystitis, nephrosis, polycystic disease, renal failure, and urinary tract infections (UTIs).
24. Define medical terminology associated with the urinary system.

Learning Activities

(pages 664–670)   **A. Kidneys: anatomy**

[1] 1. Describe functions of the urinary system in this exercise.
        a. Name waste products eliminated through urine.

        b. Kidneys produce _____ which regulates blood pressure, and _____ which is vital to normal red blood cell formation. Kidneys also activate vitamin ____.

     2. Identify the organs that make up the urinary system on Figure LG 26-1 and
[2]   answer the following questions about them.

        a. The kidneys are located at about *(waist? hip?)* level, between _____ and _____ vertebrae.

        b. The kidneys are in an extreme *(anterior? posterior?)* position in the abdomen. They are described as _____ since they are posterior to the peritoneum.

     3. What are the dimensions of the kidneys?

     4. Identify the parts of the internal structure of the kidney on Figure LG 26-1A–G. Then check your understanding of these structures by coloring the parts of the *cortex* red, *medulla* green, and *pelvis* yellow.

     5. The functional unit of the kidney is called a _____. Part of a nephron is shown in Figure LG 26-2. Label: glomerular (Bowman's) capsule (visceral and parietal layers), and the five parts of the renal tubule (proximal convoluted tubule, descending and ascending limbs of loop of the nephron, distal convoluted tubule, and collecting duct.

[3] 6. Name the structures which form a renal corpuscle.

     7. Describe the endothelial-capsular membrane.
        a. Arrange in order the layers that fluid or solutes pass through as they move
[4]   from blood into the forming urine.
           A. Filtration slits between podocytes of visceral layer of glomerular capsule
           B. Basement membrane of the glomerulus
           C. Endothelial pores of capillary membrane

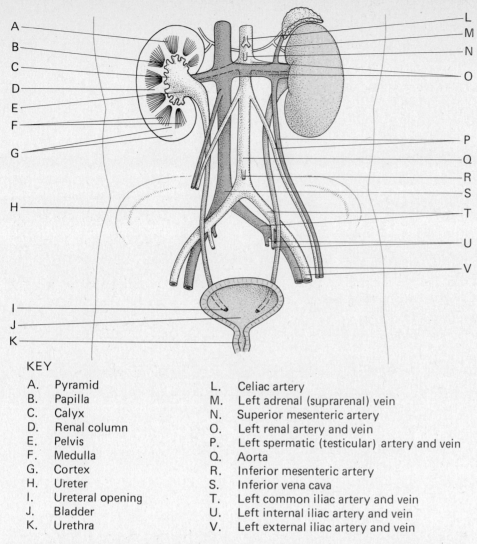

A. Pyramid
B. Papilla
C. Calyx
D. Renal column
E. Pelvis
F. Medulla
G. Cortex
H. Ureter
I. Ureteral opening
J. Bladder
K. Urethra

L. Celiac artery
M. Left adrenal (suprarenal) vein
N. Superior mesenteric artery
O. Left renal artery and vein
P. Left spermatic (testicular) artery and vein
Q. Aorta
R. Inferior mesenteric artery
S. Inferior vena cava
T. Left common iliac artery and vein
U. Left internal iliac artery and vein
V. Left external iliac artery and vein

KEY

Figure LG 26-1 *(Left side of diagram)* Organs of the urinary system. Structures *A* to *G* are parts of the kidney. *(Right side of diagram)* Blood vessels of the abdomen. Color as directed.

b. List the functional advantages of these structures: *endothelial pores* and *filtration slits.*

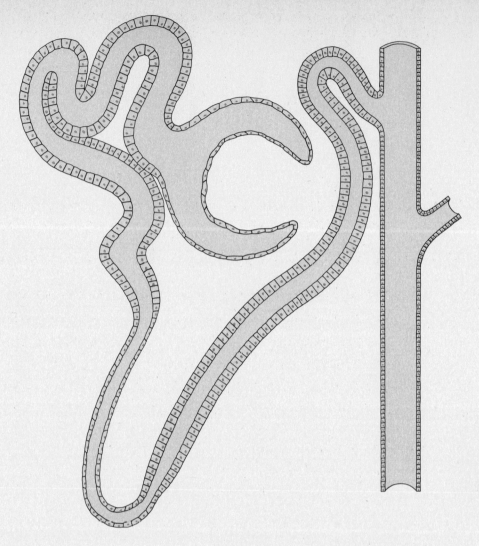

Figure LG 26-2 Diagram of a nephron (partial) to be completed as directed.

8. Contrast the locations of the *cortical nephron* and the *juxtaglomerular nephron*. Note which parts of the nephrons lie in the cortex and which parts are in the medulla.

5   9. The walls of the proximal and distal convoluted tubules consist of

_____ epithelium. Cells of the *(proximal? distal?)* tubules contain more microvilli. This factor ____-creases absorptive ability of proximal tubule cells.

6   10. About what percent of cardiac output passes through the kidneys each minute?_____

11. Describe the pathway blood takes as it courses through the kidneys.
    a. Name the vessels in order. (Consult Figure 26-6, page 670, in the text.)

    b. On Figure LG 26-2, draw and label these vessels: *afferent arteriole, glomerular capillaries, efferent arteriole, peritubular capillaries, vasa recta,* and *veins.* (See Figure 26-3, page 667, in your text.)
12. Locate the juxtaglomerular apparatus on Figure 26-7, (page 671) in the text.

    □ 7

    a. Describe its two components. First, smooth muscle cells of the *(afferent? efferent?)* arteriole; these are called the _____ cells. And also adjacent epithelial cells of the _____ convoluted tubule. These cells are known as the _____.

    (More about function of this apparatus in Activity □19.)
    b. Circle and label these structures on Figure LG 26-2.

B. Kidneys: physiology

(pages 670–680)

1. In what main ways is blood modified as it passes through the kidneys?   □ 8

2. Describe the first step of urine production by completing this exercise.   □ 9

    a. The three major steps in urine production are _____ _____ , _____ , and _____ _____ .

    b. Glomerular filtration is a process of *(pushing? pulling?)* fluids and solutes out of _____ and into the fluid known as _____ _____ .

    c. *(All types? All small substances? Only selected small substances?)* are forced out of blood. Refer to Figure LG 26-3, area A (plasma in glomeruli) and area B (filtrate just beyond the capsule). Compare composition of these two fluids (described in the first two columns of Exhibit 26-1 in your textbook). Note that during filtration all solutes are freely filtered from blood except _____. Why is so little protein filtered?

d. Blood pressure (or hydrostatic pressure) in glomerular capillaries is about

_____ mm Hg. This value is *(higher? lower?)* than that in other capillaries of the body. This extra pressure is accounted for by the fact that the diameter of the efferent arteriole is *(larger? smaller?)* than that of the afferent arteriole. Picture three garden hoses connected to each other. The third one is extremely narrow; it creates such resistance that pressure builds up in the first two. Fluids will be forced out of pores in the middle (glomerular) hose.

3. Blood hydrostatic pressure is not the only force determining the amount of filtration occurring in glomeruli.

[10]

a. Name two other forces (Figure 26-8, page 672, in your textbook).

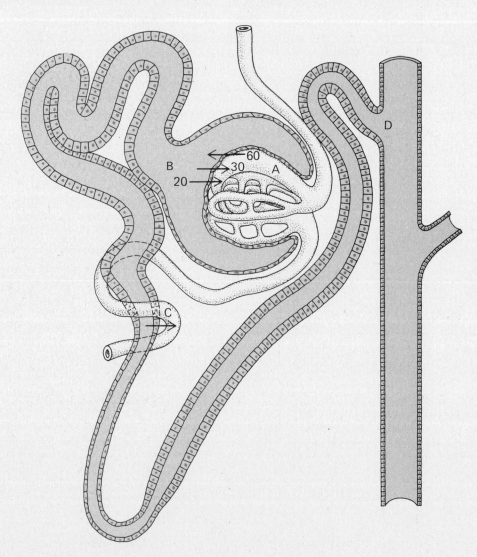

Figure LG 26-3 Part of a nephron. Renal tubule is abbreviated. Numbers refer to Activity [10] ; letters refer to Learning Activities [9] , [13] , and [14] .

b. Normal values for these three forces are given on Figure LG 26-3. Write the formula for calculating effective filtration pressure (P_{eff}); draw an arrow beneath each term to show the direction of the force. Then calculate a normal P_{eff} using the values given.

$$P_{eff} = \left(\qquad\right) - \left(\qquad + \qquad\right)$$

$= (\underline{\qquad}$ mm Hg$) - (\underline{\qquad}$ mm Hg $+ \underline{\qquad}$ mm Hg$)$
$= \underline{\qquad}$ mm Hg

Notice that blood (hydrostatic) pressure pushes *(out of? into?)* glomerular blood and is largely counteracted by the other two forces.

c. *For extra review* of these types of forces, refer to Figure LG 21-2, page LG 454.

4. *A clinical challenge.* Show the effects of alterations of these pressures in pathological situations. Determine P_{eff} of the following patients. Note which values are abnormal and suggest causes. ⏹11

	Patient A	Patient B
Glomerular blood Hydrostatic pressure	50	73
Osmotic pressure	30	25
Capsular filtrate Hydrostatic pressure	20	20

a. Patient A P_{eff} =

b. Patient B P_{eff} =

5. Explain why a person under severe stress might go into a state of oliguria ⏹12 or anuria. Under stress the *(sympathetic? parasympathetic?)* system dominates. It particularly affects the _____ arteriole, vaso-constricting it. This *(increases? decreases?)* glomerular blood pressure and *(increases? decreases?)* urine output.

[13] 6. Do this exercise.

a. Glomerular filtration is a(n) *(passive, nonselective? active, selective?)* process. If this were the only step in urine formation, all of the substances in the filtrate would leave the body in urine. Note from Exhibit 26-1 (page 672) that the body would produce _____ liters of urine each day! And it would contain many valuable substances. Obviously some of these "good" substances must be drawn back into blood and saved.

b. Recall that blood in most capillaries flows into vessels named

_____, but blood in glomerular capillaries flows into

_____ and _____ (Figure 26-3, page 667, in your text and Figure LG 26-2). This unique arrangement permits blood to recapture some of the substances indiscriminately pushed out during filtration. This occurs during the second step of urine

formation, called _____. This process moves substances in the *(same? opposite?)* direction as glomerular filtration as shown by the direction of the arrow at area C of Figure LG 26-3.

7. Define *tubular maximum (Tm)*.

8. Refer to areas *C* and *D* of Figure LG 26-3 and to the third and fourth columns of Exhibit 26-1 (page 672) in your textbook. Discuss the effects of
[14] tubular reabsorption in this learning activity.

a. Which solutes are 100 percent reabsorbed into area *C,* so that virtually none remains in urine (area *D*)?

b. Which substances are mostly, but not entirely, reabsorbed?

About what percentage of filtered water is reabsorbed?

c. Which solute is approximately 50 percent resorbed into blood?

d. Identify the solute which is filtered from blood, but none of which is reabsorbed. _____ State the clinical significance of this fact.

e. You can conclude that tubular reabsorption is a *(nonselective? discriminating?)* process which enables the body to save valuable nutrients, ions, and water.

9. *A clinical challenge.* List possible causes of glycosuria in the following two situations. Plasma concentration of glucose is shown for each person. `15`
    a. Toni (240 mg/100 ml)

    b. Marlene (80 mg/100 ml)

10. Do this exercise about reabsorption of electrolytes. `16`
    a. Na^+ is *(actively? passively?)* reabsorbed mostly from the *(proximal? distal?)* convoluted tubule. Its reabsorption in the distal tubule and collecting duct is enhanced by *(high? low?)* levels of aldosterone.
    b. Reabsorption of Na^+ from the proximal tubule is accompanied by *(active? passive?)* transport of Cl^- back into blood, thus achieving electrochemical balance.
    c. In the ascending limb of the loop of the nephron, *(Na^+? Cl^-?)* is actively reabsorbed, while ____ follows passively.
    d. The net effect of the processes described in (b) and (c) is *(the same? different?).*

11. Low Na^+ concentration in ECF is one factor that stimulates release of `17` _____ from juxtaglomerular cells of the nephrons. Briefly summarize the *renin-angiotensin pathway. (For extra review,* see Chapter 18, Activity `19d-e`, pages LG 376–378.)

18

12. Identify statements that describe facultative water reabsorption (F) or obligatory water reabsorption (O).

_____ a. Water follows sodium according to osmotic pressure factors.

_____ b. The process is influenced by presence of hormone ADH.

_____ c. About 80 percent of water reabsorption is of this type.

_____ d. Reabsorption occurs specifically across distal and collecting tubules.

19

13. Explain how the juxtaglomerular apparatus regulates glomerular filtration rate (GFR).

a. Recall that approximately _____ ml of blood (or about 650 ml plasma) enter kidneys via renal arteries each minute. Of this, about 125 ml/min of fluids and solutes are removed from blood in the glomeruli and enter the renal tubule. This value (125 ml/min) is a typical _____

_____ rate (GFR).

b. GFR is maintained in two ways. Both are based on the fact that when GFR is low, *(more? less?)* filtrate is passing through renal tubules. Consequently, filtrate flows more *(slowly? rapidly?)*, permitting *(more? less?)* reabsorption of electrolytes such as Cl^- back into blood.

c. Less Cl^- is then present in tubule cells. Distal tubule cells (called the macula _____) respond to the low Cl^- levels by causing dilation of *(afferent? efferent?)* arterioles. This ____-creases GFR.

d. Juxtaglomerular cells (which you may recall are smooth muscle cells of _____) also respond to the lowered Cl^- level by secreting _____. Renin is a vaso-*(dilator? constrictor?)*. It especially constricts the ____-ferent arteriole, which increases glomerular blood pressure, again ____-creasing GFR.

e. A term closely related to GFR is filtration _____ which means the percentage of plasma entering glomeruli that does form glomerular filtrate. Calculate a typical filtration fraction using values in (a) *(1–3%? 15–20%? 50–60%?)*. If the endothelial-capsular membrane is especially porous, as in some kidney conditions, filtration fraction would ____-crease.

14. The third step in urine production is _____. It involves movement of substances from *(blood to urine? urine to blood?)*. In other words, tubular secretion is movement of substances in *(the same? opposite?)* direction as movement occurring in filtration.

20

15. List four substances that are secreted by the process of tubular secretion.

16. Explain how tubular secretion helps to control pH. Refer to Figure 26-11 (page 676) in the text and do this exercise. 21

   a. Deborah's respiratory rate is slow. Her blood therefore contains high levels of _____. It is also somewhat *(acidic? alkaline?)*. Explain why.

   b. Her kidneys can assist in restoring normal pH (Figure 26-11a). The presence of acid (H^+) stimulates the kidneys to eliminate H^+. In order for this to happen, another cation present in urine must exchange places with H^+. What ion enters kidney tubule cells? _____

      It joins HCO_3^- to form _____.

   c. Figure 26-11b shows a second possible mechanism for eliminating acid.

      H^+ combines with ammonia (_____) to form

      _____ (_____) ion. This cation

      displaces _____ from NaCl in urine. In this way

      _____ is conserved and the H^+ is lost as part of the

      salt _____.

   d. In both of these cases the body eliminated acid (H^+) by exchanging it for another cation, _____. Consequently, blood pH is raised. Since the body is constantly forming acids during metabolism, this process occurs continually, and so urine is slightly *(acidic? alkaline?)*.

17. State an example of a situation in which your body needs to produce dilute urine in order to maintain homeostasis.

   Now describe how your body can do this by doing this learning activity. 22

   a. To form dilute urine, kidney tubules must reabsorb *(more? fewer?)* solutes than water. Two factors facilitate this. One is permeability of the ascending limb and part of the distal tubule to Na^+ and Cl^- *(and also to? but not to?)* water. Thus solutes enter interstitium, but water stays in urine.

   b. The second requirement for dilute urine is a *(high? low?)* level of ADH.

      The function of ADH is to ____-crease permeability of distal and collecting tubules to water. Less ADH forces more water to stay in urine, leading to a *(hyper? hypo?)*-osmotic urine.

604

18. Describe the role of the vasa recta in the concentration of urine.

23| 19. Summarize factors that result in concentrated urine in this exercise. The countercurrent mechanism provides a means for producing *(hypotonic? hypertonic?)* urine during times when the body is _____

_____. This occurs because the *(ascending? descending?)* limb of the loop of Henle is completely impermeable to H_2O, and also related to vasa recta permeability. As a result, the interstitial fluid of the renal _____ is extremely hypertonic, that is, has a *(high? low?)* osmolality, so pulls H_2O from urine in the _____

_____ tubules. These tubules are made more permeable at this time under the influence of the hormone _____

_____. As a result, a *(large? small?)* volume of concentrated urine is produced.

20. Explain how *hemodialysis* can help a person with impaired kidney function to maintain homeostasis.

21. Describe *peritoneal dialysis* as an alternative method of cleansing blood.

(pages 680–684) **C. Homeostasis: other excretory organs, urine**

1. Besides the urinary organs, what other body structures perform excretory functions? List three and name the substances they eliminate.

2. As an example of how the activities of the kidneys and other excretory organs are coordinated, describe the adjustments made by the kidneys when the skin increases its output of water (as by sweating on a July day).

3. Summarize factors that influence urine volume. $\boxed{24}$

   a. *Blood pressure.* When blood pressure drops, _____ cells of the kidney release _____ which leads to formation of _____. This substance *(increases? decreases?)* blood pressure in two ways: directly, since it serves as a *(vasoconstrictor? vasodilator?)* and, indirectly, since it stimulates release of _____ which causes retention of both _____ and _____. Increase in fluid volume will *(increase? decrease?)* blood pressure.

   b. *Blood concentration.* The hypothalamic hormone _____ _____ helps to *(conserve? eliminate?)* fluid. When body fluids are hypertonic, ADH is released, causing *(increased? decreased?)* tubular reabsorption of water back into blood.

   c. *Medications.* A chemical that mimics secretion of ADH is called a(n) _____. One that inhibits (or is antagonistic to) ADH is called a(n) _____. Diuretics cause diuresis which means a *(large? small?)* volume of urine. Removal of extra ECF in this manner is likely to *(increase? decrease?)* blood pressure.

   d. *Temperature.* On hot days skin gives up *(large? small?)* volumes of fluid via sweat glands. As blood becomes more concentrated, the posterior pituitary responds by releasing *(more? less?)* ADH, so urine volume *(increases? decreases?)*.

4. Describe the following characteristics of urine. Discuss variations from the normal in each case.
   a. Volume

   b. Color

c. Turbidity

d. Odor

e. pH

f. Specific gravity

25 5. Match the names of tests for renal function with descriptions below.

_____ a. Assesses glomerular filtration rate: low value indicates low GFR. Example is creatinine clearance test.   BUN. Blood urea nitrogen
RC. Renal clearance
SC. Serum creatinine

_____ b. Measures levels of urea in blood; level increases in kidney failure.

_____ c. Measures level of a by-product of muscle metabolism that remains in blood; value increases in kidney failure.

6. The following substances are not normally found in urine. Explain what the presence of each might indicate.
a. Albumin

b. Red blood cells

c. Ketone bodies

d. Bilirubin

7. Contrast *casts* and *calculi*. Explain what their presence in urine indicates.

8. Discuss two nonsurgical methods used for elimination of renal calculi (kidney stones).

D. Ureters, urinary bladder, urethra (pages 684–686)

1. Refer to Figure LG 26-1 as you complete the following exercise. 26

a. Ureters connect _____ to _____

_____ . Ureters are about ____ cm (_____ inches) long.

b. These tubes enter the urinary bladder at two of the three corners of the

_____ . The third corner marks the opening into the

_____ .

c. The urinary bladder is located in the *(abdomen? pelvis?)*. Two sphincters lie just inferior to it. The *(internal? external?)* sphincter is under voluntary control.

d. Urine leaves the bladder through the _____. In females the length of this tube is about _____ cm; in males it is about _____ cm.

2. Ureters, urinary bladder, and urethra are lined with _____

_____ membrane. What is the clinical significance of that fact?

3. Define *micturition reflex.*

In this reflex *(sympathetic? parasympathetic?)* nerves stimulate the _____ muscle of the urinary bladder and cause relaxa-

tion of the internal sphincter.

4. What is the function of a cystoscope?

5. Contrast *incontinence* and *retention.*

E. Aging and development of the urinary system.

1. GFR ____-creases with age. Two other common problems associated with |29| aging are inability to control voiding (_____) and increased frequency of _____. Explain why UTIs may occur more often in elderly males.

Nocturia may also occur more among the aged population. Define *nocturia.*

2. Do this exercise about urinary system development. |30|
   a. Kidneys form from _____-derm, beginning at about the _____ week of gestation.
   b. Which develops first? *(Pro? Meso? Meta?)*-nephros. Which extends most superiorly in location? _____-nephros. Which one ultimately forms the kidney? _____
   c. The urinary bladder develops from the original _____ _____ .

F. Disorders, medical terminology

(pages 688–689)

1. Match each of the terms with the correct description. |31|

_____ a. Painful urination
_____ b. Inflammation of the urinary bladder
_____ c. Inflammation of the pelvis and calyces of kidney
_____ d. Urea in blood
_____ e. Excessive urine
_____ f. Inflammation of the kidney involving glomeruli; may follow strep infection
_____ g. Floating kidney; slipping of the kidney from its normal position
_____ h. High uric acid level of blood; painful inflammation of joints caused by deposits of the crystallized acid

C. Cystitis
D. Dysuria
Gl. Glomerulonephritis
Go. Gout
Po. Polyuria
Pt. Ptosis
Py. Pyelitis
U. Uremia

2. Discuss and contrast two types of renal failure in this learning activity.

    a. *(Acute? Chronic?)* kidney failure is an abrupt cessation (or almost) of kidney function. Write one example of each type of cause of acute renal failure (ARF): prerenal, renal, postrenal.

    b. *(Acute? Chronic?)* renal failure is progressive and irreversible. Describe changes that occur during the three stages of CRF.

3. Urinary tract infections (UTIs) are most often caused by Gram-*(positive? negative?)* bacteria named _____. List three signs or symptoms of UTIs.

1   (a) Nitrogen-containing products of protein catabolism, such as ammonia and urea; also certain ions and excessive water. (b) Renin, renal erythropoietic factor, D.

2   (a) Waist, T12, L3. (b) Posterior, retroperitoneal.

3   Glomerular (Bowman's) capsule with enclosed glomerular capillaries.

4   C B A.

5   Simple cuboidal, proximal, in.

6   20 to 25 (1,200 ml/min).

7   Afferent, juxtaglomerular, distal, macula densa.

8   Its volume, electrolyte content, and pH are adjusted; toxic wastes are removed.

9   (a) Glomerular filtration, tubular reabsorption, tubular secretion. (b) Pushing, blood, filtrate. (c) All small substances, protein molecules, too large to pass through filtration slits of healthy endothelial-capsular membranes. (d) 70, higher, smaller.

10   (a) Glomerular blood osmotic pressure and capsular filtrate hydrostatic pressure. (b) (Glomerular blood hydrostatic pressure) $\leftarrow$ (capsular hydrostatic pressure $\rightarrow$ + blood osmotic $\rightarrow$ pressure); (60) $\leftarrow$ − (20 $\rightarrow$ + 30) $\rightarrow$; 10 $\leftarrow$.

11   (a) P_{eff} = 0 mm Hg, (50) − (20 + 30). Anuria due to low blood pressure could be related to hemorrhage or stress. (b) P_{eff} = 28 mm Hg $\leftarrow$, (73) − (20 + 25). Possibly related to kidney disease in which protein is allowed to pass out of blood into filtrate, thus lowering blood osmotic pressure.

12   Sympathetic, afferent, decreases, decreases (such as patient A above).

13   (a) Passive, nonselective; 180. (b) Venules, peritubular capillaries and vasa recta, tubular reabsorption, opposite.

14   (a) Glucose and proteins. (b) Cl^-, Na^+, HCO_3^-, K^+, uric acid; about 99 percent of water. (c) Urea. (d) Creatinine; creatinine clearance is often used as a measure of glomerular filtration rate (GFR). 100 percent of the creatinine (a breakdown product of muscle protein) filtered or "cleared" out of blood will show up in urine since 0 percent is reabsorbed. (e) Discriminating.

15   (a) Ingestion of excessive amounts of carbohydrates or hormone imbalance, such as insulin deficiency (diabetes mellitus) or excess growth hormone (GH) may lead to Toni's hyperglycemia. Such high blood glucose levels exceed capacity of glucose reabsorption mechanisms even in the normal kidney. (b) Although Marlene's blood glucose is in the normal range, kidney malfunction may prevent adequate reabsorption of glucose.

16   (a) Actively, proximal, high. (b) Passive. (c) Cl^-, Na^+. (d) The same.

612

17 Renin.

18 (a) O. (b) F. (c) O. (d) F.

19 (a) 1,200, glomerular filtration. (b) Less, slowly, more. (c) Densa, afferent, in. (d) Afferent arteriole, renin, constrictor, ef. in. (e) Fraction, 15–20% (650/125), in.

20 Tubular secretion, blood to urine, the same.

21 (a) CO_2, acidic, increasing levels of CO_2 tend to cause increase of H^+: $CO_2 + H_2O \rightarrow H_2CO_3 \rightarrow H^+ + HCO_3^-$. (Review content on transport of respiratory gas in Chapter 23.) (b) Na^+, $NaHCO_3^-$ (sodium bicarbonate). (c) NH_3, ammonium (NH_4^+), Na^+, $NaHCO_3$, NH_4Cl. (d) Na^+, acidic.

22 (a) More, but not to. (b) Low, in, hypo.

23 Hypertonic, dehydrated, ascending, medulla, high, collecting, ADH, small.

24 (a) Juxtaglomerular, renin, angiotensin II, increases, vasoconstrictor, aldosterone, Na^+ and H_2O, increase. (b) ADH, conserve, increased. (c) Antidiuretic, diuretic, large decrease. (d) Large, more, decreases.

25 (a) RC. (b) BUN. (c) SC.

26 (a) Kidneys to urinary bladder, 25–30 (10-12). (b) Trigone, urethra. (c) Pelvis, external. (d) Urethra, 3.8, 20.

27 Mucous, microbes can spread infection from exterior of body (at urethral orifice) along mucosa to kidneys.

28 Parasympathetic, detruser.

29 De, incontinence, urinary tract infections (UTIs); increased frequency of prostatic hypertrophy which leads to urinary retention; excessive urination at night.

30 (a) Meso, third. (b) Pro, pro, meta. (c) Cloaca (urogenital sinus portion).

31 (a) D. (b) C. (c) Py. (d) U. (e) Po. (f) Gl. (g) Pt. (h) Go.

MASTERY TEST: Chapter 26

Questions 1–6: Arrange the answers in correct sequence.

_____ _____ _____ 1. From superior to inferior:
A. Ureter
B. Bladder
C. Urethra

_____ _____ _____ _____ 2. From most superficial to deepest:
A. Renal capsule
B. Renal medulla
C. Renal cortex
D. Renal pelvis

_____ _____ _____ _____ _____ 3. Pathway of glomerular filtrate:

     A. Ascending limb of loop
     B. Descending limb of loop
     C. Collecting tubule
     D. Distal convoluted tubule
     E. Proximal convoluted tubule

_____ _____ _____ _____ _____ 4. Pathway of blood:

     A. Arcuate arteries
     B. Interlobular arteries
     C. Renal arteries
     D. Afferent arteriole
     E. Interlobar arteries

_____ _____ _____ _____ _____ 5. Pathway of blood:

     A. Afferent arteriole
     B. Peritubular capillaries and vasa recta
     C. Glomerular capillaries
     D. Venules and veins
     E. Efferent arteriole

_____ _____ _____ _____ _____ _____ 6. Order of events to restore low blood pressure to normal:

     A. This substance acts as a vasoconstrictor and stimulates aldosterone.
     B. Renin is released and converts angiotensinogen to angiotensin I.
     C. Angiotensin I is converted into angiotensin II
     D. Decrease in blood pressure stimulates juxtaglomerular cells.
     E. Under influence of this hormone Na^+ and H_2O reabsorption occurs.
     F. Increased blood volume raises blood pressure.

_____ 7. Which of these is a normal constituent of urine?

    A. Albumin   B. Urea   C. Glucose   D. Casts
    E. Acetone

_____ 8. Which is a normal function of the bladder?

    A. Oliguria   B. Nephrosis   C. Calculi
    D. Micturition

_____ 9. Which parts of the nephron are composed of simple squamous epithelium?

    A. Ascending and descending limbs of loop of the nephron
    B. Glomerular capsule (parietal layer) and descending limb of loop of the nephron
    C. Proximal and distal convoluted tubules
    D. Distal convoluted and collecting tubules

_____ 10. Choose the one FALSE statement.

    A. The efferent arteriole normally has a larger diameter than the afferent arteriole.
    B. Glomerular capillaries have higher blood pressure than other capillaries of the body.
    C. Blood in glomerular capillaries flows into arterioles, not into venules.
    D. Vasa recta pass blood from peritubular capillaries toward veins.

T   F   11. Antidiuretic hormone (ADH) <u>decreases</u> permeability of distal and collecting tubules to water.

T   F   12. A ureter is a tube which carries urine from <u>the bladder to the outside of the body.</u>

T   F   13. As you run on a hot summer day, you are likely to <u>increase</u> your ADH secretion and therefore to <u>decrease</u> urine production.

T   F   14. The juxtaglomerular apparatus consists of cells of the <u>afferent arteriole and distal convoluted tubule.</u>

T   F   15. The renin-angiotensin mechanism tends to <u>increase</u> blood pressure.

T   F   16. A person taking diuretics is likely to urinate <u>less</u> than a person taking antidiuretics.

T   F   17. Blood osmotic pressure is a force which tends to <u>push substances from blood into filtrate.</u>

T   F   18. The ascending limb of the loop of the nephron is permeable to <u>Na^+ and Cl^-, but not to water.</u>

T F 19. The countercurrent multiplier mechanism permits production of <u>small volumes of concentrated urine.</u>

T F 20. In infants 2 years old and under, <u>retention</u> is normal.

Questions 21–25: fill-ins. Complete each sentence with the word or phrase that best fits.

_____ 21. The glomerular capsule and its enclosed glomerular capillaries constitute a ___.

_____ 22. A ___ consists of a renal corpuscle and a renal tubule.

_____ 23. Write a value for specific gravity of dilute urine. ___

_____ 24. Sympathetic impulses cause greater constriction of ___ arterioles with resultant ___-crease in GFR and ___-crease in urinary output.

_____ 25. A person with chronic renal failure is likely to be in acidosis since kidneys fail to excrete ___, may be anemic since kidneys do not produce ___, and may have symptoms of hypocalcemia since kidneys do not activate ___.

Fluid, Electrolyte, and Acid–Base Dynamics

27

In this chapter you will correlate much of the information discussed so far in this unit as you consider interrelationships of systems working together to maintain homeostasis of fluids, electrolytes, and pH. First, you will review the fluid compartments of the body and look at avenues and regulation of fluid intake and output (Objectives 1–4). You will examine the role of electrolytes in fluids (5–8). You will consider factors involved in the movement of body fluids (9–10). Lastly, you will study acid-base dynamics and consider the profound effects of imbalances such as acidosis and alkalosis (11–13).

Topics Summary

A. Fluid compartments, water
B. Electrolytes
C. Movements of body fluids
D. Acid-base balance and imbalance

Objectives

1. Define a body fluid.
2. Distinguish between intracellular fluid (ICF) and extracellular fluid (ECF).
3. Define the processes available for fluid intake and fluid output.
4. Compare the mechanisms regulating fluid intake and fluid output.
5. Compare the osmotic effects of nonelectrolytes and electrolytes on body fluids.
6. Calculate the concentration of ions in a body fluid.
7. Contrast the electrolytic concentration of the three major fluid compartments.
8. Explain the functions and regulation of sodium, chloride, potassium, calcium, phosphate, and magnesium.
9. Define the factors involved in the movement of fluid between plasma and interstitial fluid and between interstitial fluid and intracellular fluid.
10. Define the relationship between electrolyte imbalance and fluid imbalance.
11. Compare the role of buffers, respirations, and kidney excretion in maintaining body pH.
12. Define acid-base imbalances and their effects on the body.
13. Explain appropriate treatments for acidosis and alkalosis.

Learning Activities

(page 693) **A.** **Fluid compartments, water**

1. The body contains two great fluid compartments. Describe them, naming the components of each.
   a. ICF

   b. ECF

   What approximate proportion of body fluid is located in each? ICF: _____
   ECF: _____

2. What does it mean to say that the body is in *fluid balance?* Explain why *water balance* implies *electrolyte balance.*

3. About what percent of body weight consists of water? _____ Circle the person in each pair who is more likely to have a higher proportion of water content.
   a. 37-year-old woman/67-year-old woman
   b. Female/male
   c. Lean person/fat person

4. Daily intake and output of fluid usually both equal _____ ml (_____ liters).

5. List the three main sources of fluids and the average daily amounts of each.

6. Now list four systems of the body that eliminate water. Indicate the amounts ⬛3 **619**
lost by each system on an average day. One is done for you.
a. Integument (skin): 500 ml/day

b. _____ : _____ ml/day

c. _____ : _____ ml/day

d. _____ : _____ ml/day
On a day when you exercise vigorously, how would you expect these values ⬛4
to change?

7. Briefly describe the mechanisms regulating:
a. Fluid intake (Which part of the brain regulates thirst?)

b. Fluid output (Which two hormones regulate urine production?)

B. Electrolytes (pages 694–697)

1. Contrast *electrolytes* with *nonelectrolytes.*

5
2. Circle all of the answers which are classified as electrolytes:
   A. Proteins in solutions   B. Anions
   C. Glucose   D. K^+ and Na^+
   E. Compounds containing at least one ionic bond
3. List three general functions of electrolytes.

(Electrolytes? Nonelectrolytes?) exert a greater osmotic effect. In one or two sentences, explain why.

4. The three main compartments of the body are shown in Figure LG 1-2 (page LG 11). Concentrations of ions in these compartments are graphed in Figure 27-2 (page 695) in your textbook. After you have studied this graph, complete the table below. List the electrolytes in greatest concentration in each
6
   area.

	Major Cations	Major Anions
a. Intracellular fluid	(two) K^+, Mg^{2+}	(three)
b. Interstitial fluid	(one)	(two) Cl^-, HCO_3^-
c. Plasma	(one)	(three)

7
5. Analyze the information in the above table by answering these questions.
   a. Which two compartments are most similar in electrolyte concentration?

   b. What is the one major difference in electrolyte content of these two similar compartments?

c. Most body protein is located in which compartment?

d. Protein anions are more likely to be bound to the cation *(Na$^+$? K$^+$?)* in intracellular fluid.

6. Refer again to Figure 27-2 (page 695) in your text. Notice that concentra- $\boxed{8}$ tions of electrolytes are expressed in units called _____

_____, abbreviated _____. Write a formula which can be used to determine meq/liter.

meq/liter = (_____ × _____)/
_____

7. Calculate meq/liter of each of the plasma electrolytes listed in the table below. Use information in that table and in Figure 27-2 (page 695) in the text for help. $\boxed{9}$

Electrolyte	mg/liter	Number of Charges	Atomic Weight	meq/liter
a. Sodium (Na$^+$)	3,300	1	23	
b. Chloride (Cl$^-$)	3,670		35	
c. Magnesium (Mg^{2+})	30	2	24	

Notice how your answers correlate with values for plasma shown in the graph (Figure 27-2, page 695 in your text); much less Mg^{2+} is present than either Na$^+$ or Cl$^-$.

8. Complete the following table describing important electrolytes.

Electrolyte	Principal Functions	Disorders and Some Symptoms
a. Sodium (Na$^+$)		
b.		Hypochloremia; muscle spasms, alkalosis, depressed respiration, coma

Electrolyte	Principal Functions	Disorders and Some Symptoms
c.	Most abundant cation in ICF; helps maintain fluid volume of cells; functions in nerve transmission	
d. Calcium (Ca^{2+})		
e.	Helps form bones, teeth; important component of DNA, RNA, ATP, and buffer system	
f. Magnesium (Mg^{2+})		

9. *A clinical challenge.* Norine is having her serum electrolytes analyzed. Complete this exercise about her results.

   10

   a. Her Na^+ level is 128. Refer back to Learning Activity **9** and identify a normal value for serum Na^+. Norine's value indicates that she is more likely to be in a state of *(hyper? hypo?)*-natremia. Write two symptoms of this electrolyte imbalance.

   b. A diagnosis of hypochloremia would indicate that Norine's blood level of

      _____ ion is lower than normal. One cause of this is excessive

      _____.

   c. A normal range for K^+ is 3.5–5.0 meq/liter. This range is considerably *(higher? lower?)* than that for Na^+. This is reasonable since K^+ is the main cation in *(ECF? ICF?)*, yet lab analysis of serum electrolytes is examining *(ECF? ICF)* K^+.

d. Norine's potassium (K^+) level is 5.6 meq/liter, indicating that she is in

a state of hyper-_____. Hyperkalemia and hypona-
tremia (as described in *a*) may be caused by *(high? low?)* levels of aldosterone. Explain why.

e. A normal range for Mg^{2+} level of blood is 1.5–2.5 meq/liter. Norine's
electrolyte report indicates a value of 1.3 meq/liter for Mg^{2+}. This elec-

trolyte imbalance is known as _____. Hypomagnese-
mia, like hyponatremia, may be caused by *(high? low?)* aldosterone levels.

f. Norine's Ca^{2+} level is 7.5 meq/liter. A normal range is 9.0–10.6 meq/

liter. This electrolyte imbalance, known as _____, is

most closely linked to ___-creased levels of parathyroid hormone (PTH)

and ___-creased secretion of the thyroid hormone, _____

_____. Low blood levels of both Ca^{2+} and Mg^{2+}
cause overstimulation of the central nervous system and muscles. List two
symptoms of these electrolyte imbalances.

C. Movement of body fluids

(page 697)

1. Blood constantly moves through your vessels. The water, oxygen, and nutrients it carries must pass out of blood to interstitial fluid so that they can be
delivered to cells. Wastes must move out of cells (via interstitial fluid) to
plasma so that they can reach excretory organs. Four forces are involved in
the movement of water between plasma and interstitial compartments and
between interstitial and intracellular compartments. (They include, incidentally, the same types of forces involved in glomerular filtration.) Movement
across capillary membranes due to these four forces is according to

_____ law of capillaries.

2. Define *effective filtration pressure* (P_{eff}).

In Chapter 21 (Activity 14, page LG 454) you calculated P_{eff} across the
arterial end of a capillary. Now refer to the table below and calculate P_{eff}
in each situation (a-c). In order to visualize the forces, write each one (with
an arrow to indicate the direction of the force) on parts a-c of Figure LG
27-1. Write the P_{eff} in each case on spaces provided on the figure. (*Hint:*
It may help to show the comparative strength of the forces by the relative
length of the arrows you draw.) 11

Situation	Fluid	Hydrostatic Pressure (HP) in mm Hg	Osmotic Pressure (OP) in mm Hg
a. At arterial end of capillary (normal)	Blood plasma Interstitial fluid	30 0	25 5
b. After a hemorrhage (average pressure in capillary)	Blood plasma Interstitial fluid	18 0	22 5
c. In edema (average pressure in capillary)	Blood plasma Interstitial fluid	44 0	25 6

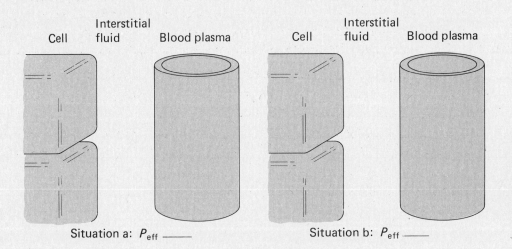

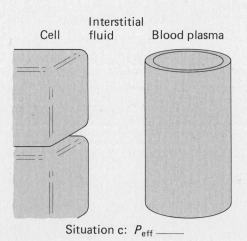

Figure LG 27-1 Practice diagrams for determining P_{eff} under four different circumstances. See Learning Activity 11 .

3. *A clinical challenge.* Now discuss situations b and c of the above activity by answering these questions. `12`
   a. After a hemorrhage (situation b) which value is appreciably lower than

   normal? _____ What is the effect of this?
   b. In a condition of edema (situation c), which value is much higher than

   normal? _____ What factors besides hypertension might be causes of edema? Name and describe reasons for three factors.

4. Now consider movement of fluid between interstitial fluid and cells. Such movement is controlled by the same pairs of forces: osmotic pressure and hydrostatic pressure. Examine the following situation which demonstrates these forces. Laurie runs 10 kilometers on a hot, humid day; she loses much fluid through perspiration. She stops to sip some water during her run, and also drinks water after finishing the run. Answer these questions. `13`
   a. The sweat Laurie lost contained not only water, but also _____

   _____ . As a result, plasma level of sodium will drop,

   a condition known as _____ . This will lead to decreased sodium in interstitial fluid also.
   b. Since extracellular fluids will be low in the solute sodium, they will be *(hypertonic? hypotonic?)* to fluid in cells (ICF). As a result, water will

   move *(into? out of?)* cells by the process of _____ .
   c. Describe results of this movement. (See Figure 27-4, page 698, in your textbook.)

   d. Note that loss of solute from blood—whether salt or protein—tends to

   lead to low blood pressure and _____ .
   e. List symptoms of water intoxication that may occur if a runner like Laurie loses excessive amounts of sweat.

   f. How can hyponatremia and water intoxication be avoided?

D. Acid–base balance and imbalance

14 1. The pH of blood and other extracellular fluids should be maintained between
    _____ and _____.

15 2. Name the three major mechanisms that work together to maintain acid-base
       balance.

3. Define *strong acid*. Explain how strong acids affect pH.

16 4. Name the four principal buffer systems of the body.

17 5. Consider the components of the carbonic acid-bicarbonate system by doing
       this exercise.
    a. This buffer system is extremely important in the body since so much of
       each chemical forms. Carbonic acid forms readily from two products of

       metabolism: _____ and _____.

       Write the reaction using chemical formulas:

       _____ + _____ → _____
                                  (Carbonic acid)
    b. Show the ionization of carbonic acid and the two products formed. In-
       clude names and chemical formulas.

       _____ → _____ + _____

       (Carbonic acid)   (   )   (Hydrogen ion)
    c. When a cation such as Na^+ or K^+ replaces the ionized H^+, the com-
       pound sodium bicarbonate ($NaHCO_3$) or potassium bicarbonate
       ($KHCO_3$) will be formed. Which of these is more likely to form in extra-

       cellular fluid (ECF), such as plasma? _____ Explain.
       (*Hint:* Refer to Figure 27-2, page 695.)

d. When a person actively exercises, *(much? little?)* CO_2 forms leading to H_2CO_3 production. Consequently, *(much? little?)* H^+ will be added to body tissues. In order to buffer this excess H^+ (to avoid drop in pH and damage to tissues), the *(H_2CO_3? $NaHCO_3$?)* component of the buffer system will be called upon. The excess H^+ will then replace the _____ ion of $NaHCO_3$, in a sense "tying up" the potentially harmful H^+.

e. *For extra review* of the mechanism of this buffer system, look again at Chapter 2, Learning Activity $\boxed{17}$, using Figure LG 2-2, page LG 30.

6. Phosphates are more concentrated in *(ECF? ICF?)*. (For help, see Figure 27-2, page 695, in the text.) Name two types of cells in which the phosphate buffer system is most important? $\boxed{18}$

Write the reaction showing how phosphate buffers strong acid.

7. Explain how the hemoglobin buffer system buffers carbonic acid, so that an acid even weaker than carbonic acid (that is, hemoglobin), is formed.

For extra review. See Chapter 23, Learning Activity $\boxed{24}$ (page LG 525) and Figure 27-5 (page 699) in the text.

8. Which is the most abundant buffer system in the body? _____

_____ Draw the structural formula of an amino acid. $\boxed{19}$

20

What portion of the amino acid buffers acids? *(—COOH?—NH₂?)* Which portion buffers base? *(—COOH?—NH₂?)*

9. Explain how respirations can help to regulate pH.

Hyperventilation will tend to *(raise? lower?)* blood pH, since as the person exhales CO_2, less CO_2 is available for formation of _____

21

_____ acid and free hydrogen ion.

22

10. A slight decrease in pH will tend to *(stimulate? inhibit?)* the respiratory center and so will *(increase? decrease?)* respirations.

23

11. The kidneys regulate acid-base by increasing or decreasing their tubular secretion of _____ or _____ ions.

24

12. Complete this exercise about acid-base imbalances.

a. An arterial blood pH of 7.15 indicates _____-osis. In such a case, the bicarbonate–carbonic acid buffer system would buffer the excess H^+, and an increase in *(NaHCO₃? HHCO₃ = H₂CO₃?)* would result.

b. Normal blood pH is maintained by balancing components of this buffer system such that the *(NaHCO₃? H₂CO₃?)* member of the system is in much greater abundance. In fact, the normal ratio of bicarbonate (or sodium bicarbonate) to carbonic acid is ____ : ____. Write that ratio here:

$$\frac{NaHCO_3}{H_2CO_3} = \rule{3cm}{0.4pt}$$

c. Any condition that increases the ratio (so that this buffer system consists of more bicarbonate or less carbonic acid) results in _____-osis. A condition that decreases the ratio, for example, by raising *(bicarbonate? carbonic acid?)* in blood, will lead to acidosis.

d. Tell whether each of these ratios indicates alkalosis or acidosis.

A. 30:1 _____

B. 20:0.5 _____

C. 12:1 _____

25

13. An acid-base imbalance caused by abnormal alteration of the respiratory system is classified as *(metabolic? respiratory?)* acidosis or alkalosis. Any other cause of acid-base imbalance, such as a urinary or digestive tract disorder, is identified as *(metabolic? respiratory?)* acidosis or alkalosis.

14. A *clinical challenge.* Indicate which of the four categories of acid–base imbalances at right is most likely to occur as a result of each condition described below.

26

_____ a. Decreased blood level of CO_2 as a result of hyperventilation

_____ b. Decreased respiratory rate in a patient taking overdose of morphine

_____ c. Excessive intake of antacids

_____ d. Prolonged vomiting of stomach contents

_____ e. Ketosis in uncontrolled diabetes mellitus

_____ f. Decreased respiratory minute volume in a patient with emphysema or fractured rib

_____ g. Excessive loss of bicarbonate from the body, as in prolonged diarrhea or renal dysfunction

MAcid. Metabolic acidosis
MAlk. Metabolic alkalosis
RAcid. Respiratory acidosis
RAlk. Respiratory alkalosis

15. Briefly describe treatments that may be used for:
   a. Respiratory acidosis

   b. Metabolic acidosis

   c. Respiratory alkalosis

   d. Metabolic alkalosis

Answers to Numbered Questions in Learning Activities

1   (a) 37-year-old woman. (b) Male. (c) Lean person.

2   2,500 (2.5).

3 (Any order) (b) Kidney, 1,500. (c) Lungs, 300. (d) GI, 200

4 Output of fluid increases greatly via sweat glands and slightly via respirations. To compensate, urine output decreases and intake of fluids increases.

5 A, B, D, E (all except glucose).

6 (a) Anions: HPO_4^{2-}, protein anions, SO_4^{2-}. (b) Cations: Na^+. (c) Cations: Na^+; anions: Cl^-, HCO_3^-, protein anions.

7 (a) Interstitial fluid and plasma (the two extracellular fluids). (b) Plasma contains more protein. (c) Intracellular. (d) K^+ (since it is more concentrated inside of cells).

8 Milliequivalents/liter, meq/liter,

$$\frac{\text{mg of ion/liter solution} \times \text{number of valence charges}}{\text{atomic weight}}$$

9 (a) $143.0 = (3,300 \times 1)/23$. (b) $105 = (3,670 \times 1)/35$. (c) $2.5 = (30 \times 2)/24$.

10 (a) Normal serum Na^+ range is 137–150 meq/liter, hypo; examples: headache, confusion, lethargy, eventually coma. (b) Cl^-, vomiting of gastric contents (HCl). (c) Lower, ICF, ECF. (d) Kalemia, low, aldosterone promotes reabsorption of Na^+ and H_2O and also Mg^{2+} into blood in kidneys, and secretion of K^+ into urine. (e) Hypomagnesemia, low. (f) Hypocalcemia, de, in, calcitonin; tremor, tetany and possible convulsions.

11 (a) 10. (b) 1. (c) 25

12 (a) BHP (due to loss of blood, as in hypovolemic shock); kidney shutdown (no filtration occurs). (b) BHP; inflammation, burns, and nephrosis (increase in glomerular permeability) all lead to loss of blood protein and decreased blood osmotic (pulling) pressure.

13 (a) Electrolytes, hyponatremia. (b) Hypotonic, into, osmosis. (c) Cells become overhydrated; blood volume decreases, causing low blood pressure and shock. (d) Circulatory shock. (e) See Norine's signs and symptoms of hyponatremia in Activity **10a**. Others include abdominal cramps and diarrhea which may also signal the runner of the onset of hyponatremia. (f) Replace electrolytes with salt tablets or oral solutions (such as Gatorade by Stokely Van Camp or Sportade Punch Mix by Becton, Dickinson) or stop running.

14 7.35–7.45.

15 Buffers, respirations, and kidney excretion.

16 Carbonic acid–bicarbonate, phosphate, hemoglobin–oxyhemoglobin, protein.

17 (a) Carbon dioxide, water; $CO_2 + H_2O \rightarrow H_2CO_3$. (b) $H_2CO_3 \rightarrow HCO_3^-$ (bicarbonate) $+ H^+$. (c) $NaHCO_3$; Na^+ is the major cation in ECF. (d) Much, much, $NaHCO_3$, Na^+.

18 ICF; inside red blood cells and in cells of kidney.

19 Protein.

|20| NH$_2$ buffers acids; —COOH buffers base.

|21| Raise, carbonic.

|22| Stimulate, increase.

|23| H$^+$, HCO$_3$$^-$.

|24| (a) Acid, HHCO$_3$ = H$_2$CO$_3$. (b) NaHCO$_3$, 20:1, NaHCO$_3$/H$_2$CO$_3$ = 20/1. (c) Alkal, carbonic acid. (d) A, alkalosis; B, alkalosis; C, acidosis.

|25| Respiratory, metabolic.

|26| (a) RAlk. (b) RAcid. (c) MAlk. (d) MAlk. (e) MAcid. (f) RAcid. (g) MAcid.

MASTERY TEST: Chapter 27

Questions 1–16: Circle T (true) or F (false). If the statement is false, change the underlined word or phrase so that the statement is correct.

T F 1. Hyperventilation will tend to raise pH.

T F 2. Carbonic acid is a weaker acid than reduced hemoglobin (H·Hb).

T F 3. Under normal circumstances fluid intake each day is greater than fluid output.

T F 4. During edema there is increased movement of fluid out of plasma and into interstitial fluid.

T F 5. Parathyroid hormone causes a high blood level of calcium.

T F 6. When aldosterone is in high concentrations, sodium is conserved (in blood) and potassium is excreted (in urine).

T F 7. In general, electrolytes cause greater osmotic effects than non-electrolytes.

T F 8. The body of an infant contains a higher percentage of water than the body of an adult.

T F 9. Starch, glucose, HCO$_3$$^-$, Na$^+$, and K$^+$ are all electrolytes.

T F 10. The cation in highest concentration inside of cells is Na$^+$.

T F 11. Protein is the most abundant buffer system in the body.

T F 12. Sodium plays a much greater role than magnesium in osmotic balance of ECF since more meq's of sodium are in ECF.

T F 13. The carbonic acid portion of the bicarbonate-carbonic acid system buffers base.

T F 14. A bicarbonate-carbonic acid ratio of 10:1 indicates acidosis.

T F 15. The pH of arterial blood should be <u>higher</u> than that of venous blood and should be about pH <u>7.45.</u>

T F 16. Hydrostatic pressure of blood and hydrostatic pressure of interstitial fluid tend to <u>move substances in the same direction: out of blood and into interstitial fluid.</u>

Questions 17–20: Choose the one best answer to each question.

_____ 17. On an average day the greatest volume of fluid output is via:

   A. Feces   B. Sweat   C. Lungs   D. Urine

_____ 18. Which term refers to a lower than normal blood level of sodium?

   A. Hyperkalemia   B. Hypokalemia
   C. Hypernatremia   D. Hyponatremia
   E. Hypercalcemia

_____ 19. Which two electrolytes are in greatest concentration in ICF?

   A. Na^+ and Cl^- B. Na^+ and K^+ C. K^+ and Cl^-
   D. K^+ and HPO_4^{2-} E. Na^+ and HPO_4^{2-}
   F. Ca^{2+} and Mg^{2+}

_____ 20. All of the following factors will tend to increase fluid output EXCEPT:

   A. Fever
   B. Vomiting and diarrhea
   C. Hyperventilation
   D. Increased glomerular filtration rate
   E. Decreased blood pressure

Questions 21–25: fill-ins. Complete each sentence with the word or phrase that best fits.

_____ 21. Interstitial fluid, blood plasma and lymph are all ____-cellular fluids.

_____ 22. About ____ percent of body fluids form part of ____-cellular fluid, and ____ percent are in ECF.

_____ 23. ____ are chemicals that have at least one ionic bond and dissociate into positive and negative ions.

_____ 24. To compensate for acidosis, kidneys will increase secretion of ____, and in alkalosis kidneys will eliminate more ____.

_____ 25. Normally blood plasma contains about ____ meq/liter of Na and ____ meq/liter of K.

UNIT V
Continuity

The Reproductive Systems

In all of the previous chapters you have studied homeostatic mechanisms that promote the health of the individual. In Unit V you will consider how the human organism is adapted for continuity of the species. In this chapter you will define reproduction (Objective 1) and you will examine the special processes of division which result in gametes (3, 12). You will study reproductive organs in the male (2, 4–10) and in the female (11, 13–14, 17–19). You will discuss hormonally regulated cycles in the female (15–16). You will consider roles of the male and female in sexual intercourse (20) and you will examine birth control methods (21). You will learn about age-related changes and development of the reproductive system (22–23). Finally you will discuss disorders and medical terminology associated with reproduction (24–26).

Topics Summary

A. Male reproductive system
B. Female reproductive system: ovaries, uterine tubes, uterus
C. Female reproductive system: endocrine relations
D. Female reproductive system: vagina, external genitalia, mammary glands
E. Sexual intercourse and birth control
F. Aging and development of the reproductive systems
G. Disorders, medical terminology

Objectives

1. Define reproduction and classify the organs of reproduction by function.
2. Explain the structure, histology, and functions of the testes.
3. Define meiosis and explain the principal events of spermatogenesis.
4. Describe the physiological effects of testosterone and inhibin.
5. Describe the seminiferous tubules, straight tubules, and rete testis as components of the duct system of the testis.
6. Describe the location, structure, histology, and functions of the ductus epididymis, ductus (vas) deferens, and ejaculatory duct.
7. Describe the three anatomical subdivisions of the male urethra.
8. Explain the locations and functions of the seminal vesicles, prostate gland, and bulbourethral (Cowper's) glands, the accessory reproductive glands.
9. Discuss the chemical composition of semen (seminal fluid).
10. Explain the structure and functions of the penis.
11. Describe the location, histology, and functions of the ovaries.
12. Describe the principal events of oogenesis.
13. Explain the location, structure, histology, and functions of the uterine (Fallopian) tubes.
14. Explain the histology and blood supply of the uterus.
15. Compare the principal events of the menstrual and ovarian cycles.
16. Discuss the physiological effects of estrogens, progesterone, and relaxin.
17. Describe the location, structure, and functions of the vagina.
18. Describe the components of the vulva and explain their functions.
19. Explain the structure, development, and histology of the mammary glands.
20. Explain the roles of the male and female in sexual intercourse.
21. Contrast the various kinds of birth control and their effectiveness.
22. Describe the effects of aging on the reproductive systems.
23. Describe the development of the reproductive systems.
24. Explain the symptoms and causes of sexually transmitted diseases (STDs) such as gonorrhea, syphilis, genital herpes, trichomoniasis, and nongonococcal urethritis (NGU).
25. Describe the symptoms and causes of male disor-

ders (prostate dysfunctions, impotence, and infertility) and female disorders [amenorrhea, dysmenorrhea, premenstrual syndrome (PMS), toxic shock syndrome (TSS), ovarian cysts, endometriosis, infertility, breast tumors, cervical cancer, and pelvic inflammatory disease (PID)].

26. Define medical terminology associated with the reproductive systems.

Learning Activities

(pages 706–717)   **A.** **Male reproductive system**

1. Define each of these groups of reproductive organs.
   a. Gonad

   b. Ducts

   c. Accessory glands

   d. Supporting structures

1   As you go through the chapter, note which structures of the male reproductive system are included in each category.

2. Describe the scrotum and testes in this exercise.
   a. Describe the structure and contents of the scrotum.

2   b. The temperature in the scrotum is about 3°C *(higher? lower?)* than the temperature in the abdominal cavity. Of what significance is this?

c. What is the function of cremaster muscles?

d. Where do testes develop in early fetal life?

e. Define *cryptorchidism*.

3. Match the parts of the testes listed at right with descriptions below. | 3 |

_____ a. 200 to 300 of these per testis
_____ b. Tightly coiled tubes (1 to 3 per lobule) composed of cells that develop into sperm
_____ c. Cells located between seminiferous tubules that produce nutrients for sperm and also the hormone inhibin forms blood-testis barrier.
_____ d. Cells located between seminiferous tubules that secrete testosterone

IE. Interstitial endocrinocytes
L. Lobule
SC. Sustantacular cells
ST. Seminiferous tubules

4. Explain how the *blood-testis barrier* helps assure adequate sperm production.

5. Discuss spermatogenesis in this exercise. | 4 |
   a. Spermatozoa are one type of gamete; name the other type: _____. Gametes are *(haploid? diploid?);* human gametes contain ____ chromosomes.

b. Fusion of gametes produces a cell called a _____.
This cell, and all cells of the organism derived from it, contain the *(haploid? diploid?)* number of chromosomes, written as *(n? 2n?)*. These chromosomes, received from both sperm and ovum, are said to exist in

_____ pairs.

c. Gametogenesis assures production of haploid gametes from otherwise

diploid humans. In males, this process is called _____

_____ -genesis. It occurs in the _____

_____ , and requires about 2–3 *(days? weeks? years?)* for full maturation of spermatazoa. As cells proceed through stages of spermatogenesis, they move *(toward? away from?)* the lumen of the tubule.

d. Refer to Figure LG 3-2, page LG 55. The process of spermatogenesis involves several stages. Certain spermatogonia may continue to undergo mitosis throughout a lifetime, assuring a reservoir of sperm-producing cells. Which type of spermatogonia have that capacity?

e. Other spermatogonia, known as type *(A? B?)*, (as at *D* in the figure), undergo meiosis. This process involves *(1? 2?)* division(s). Cells resulting

from the reduction division are called _____ *(F)*,

while spermatids *(G)* are products of the _____ division.

f. Maturation of spermatids *(G)* to spermatozoa *(H)* occurs in the process

known as _____ . Cells in *G* and *H* each contain

____ chromosomes. Once formed, mature spermatazoa *(do? do not?)* divide again.

6. *For extra review* of the process of meiosis, look again at Chapter 3 Learning Activity 20 , pages 54–56.

7. Explain how cytoplasmic bridges between cells formed during spermatogenesis may facilitate survival of Y-bearing sperm, and therefore permit generation of male offspring.

8. Describe each of these aspects of spermatozoa.

a. Number produced daily in a normal male: _____

b. Life expectancy of sperm once inside the female reproductive tract: _____ hours

c. Function of each of these parts of a sperm: *head, acrosome, midpiece, tail.*

9. Match the hormones listed at right with descriptions below.

a. Released by the hypothalamus; stimulates release of two anterior pituitary hormones: _____

b. Stimulates seminiferous tubules to start sperm production; also stimulates sustantacular cells to nourish sperm: _____

c. Stimulates release of testosterone: _____

d. Principal male hormone: _____

e. Inhibits release of GnRF: _____

f. Testicular hormone that inhibits FSH production: _____

FSH
GnRF
Inhibin
LH
Testosterone

10. List five functions of testosterone.

11. Refer to Figure LG 28-1. Structures in the male reproductive system are numbered in order along the pathway that sperm take from their site of origin to the point where they exit from the body. Match the structures in the figure (using numbers 1 to 9) with descriptions below.

_____ a. Ejaculatory duct

_____ b. Epididymis

_____ c. Prostatic urethra

_____ d. Spongy (cavernous) urethra

_____ e. Membranous urethra

_____ f. Ductus (vas) deferens (portion within the abdomen)

_____ g. Ductus (vas) deferens (portion within the scrotum)

_____ h. Urethral orifice

_____ i. Testis

12. Which structures numbered 1 to 9 in Figure LG 28-1 are paired? Circle their numbers: 1 2 3 4 5 6 7 8 9. Which are singular? Write their numbers:

7 _____

8 13. Answer these questions about the ductus (vas) deferens.

  a. The ductus deferens is about _____ cm (____ inches) long.
  b. It is located *(entirely? partially?)* in the abdomen. It enters the abdomen

    via the _____. Protrusion of abdominal contents

    through this weakened area is known as _____.
  c. What structures besides the vas compose the seminal cord?

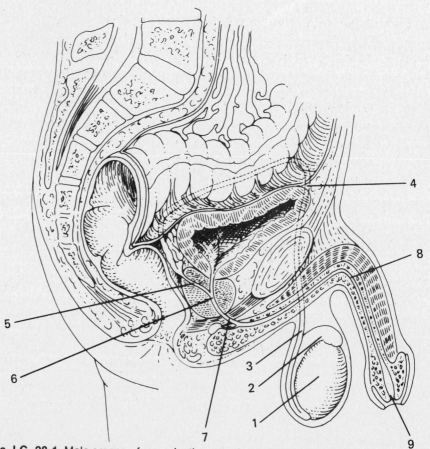

Figure LG 28-1 Male organs of reproduction seen in sagittal section. Structures are numbered in order along the pathway taken by sperm. Refer to Activities 6 and 7 . Label and color as directed.

d. A vasectomy is performed at a point along the section of the vas that is

within the _____. Must the peritoneal cavity be entered during this procedure? *(Yes? No?)* Will the procedure affect testosterone level of the male? *(Yes? No?)* Why?

14. Which portion of the male urethra is longest? *(Prostatic? Membranous? Spongy,* or *cavernous?)* | 9 |

Now that you have studied all structures in the pathway of sperm within the male, color that pathway in red.

15. Label and color the three accessory glands on Figure LG 28-1: seminal vesicles, blue; prostate, purple; and bulbourethral, green. Identify other pelvic organs in the figure by coloring the bladder yellow and rectum brown.

Now match these structures with their descriptions. | 10 |

_____ a. Paired pouches posterior to the urinary bladder

_____ b. Structures that empty secretions (along with contents of vas deferens) into ejaculatory duct

_____ c. Single doughnut-shaped gland located inferior to the bladder; surrounds and empties secretions into urethra

_____ d. Contributes about 60 percent of seminal fluid

_____ e. Pair of pea-sized glands located in the urogenital diaphragm

B. Bulbourethral glands
P. Prostate gland
S. Seminal vesicles

16. *A clinical challenge.* Explain the clinical significance of the shape and location of the prostate gland. | 11 |

17. Complete the following exercise about semen.

a. The average amount per ejaculation is _____ ml.

b. The average range of number of sperm is _____/ml. When the count falls below _____/ml, the male is likely to be sterile. Only one sperm fertilizes the ovum. Explain why such a high sperm count is required for fertility.

c. The pH of semen is slightly *(acid? alkaline?)*. State the advantage of this fact.

d. What is the function of *seminalplasmin?*

18. Discuss characteristics of semen studied in a semen analysis.

12 19. Refer to Figure 28-10 (page 716) in the text and do this exercise.

a. The _____, or foreskin, is a covering over the _____ of the penis. Surgical removal of the foreskin is known as _____.

b. The names *corpus* _____ and *corpus* _____ _____ indicate that these bodies of tissue in the penis contain spaces that distend in the presence of excess _____ _____. As a result, blood is prohibited from leaving the penis, since *(arteries? veins?)* are compressed. This temporary state is known as *(erect? flaccid?)* state.

c. The urethra passes through the corpus _____. The urethra functions in the transport of _____ and _____. During ejaculation, what prevents sperm from entering the urinary bladder and urine from entering the urethra?

20. Label these parts of the penis on Figure LG 28-1: *prepuce, glans, corpus spongiosum, corpora cavernosa, urethra, bulb of penis.* Color the corpus spongiosum (including glans) pink and the corpora cavernosa orange.

B. Female reproductive system: ovaries, uterine tubes, uterus

(pages 717–722)

1. Refer to Figure LG 28-2a. Structures are numbered in order along the pathway taken by ova from the site of formation toward the point at which an unfertilized ovum would exit from the body. Match those five structures (using numbers 1 to 5) with their descriptions.

   13

   _____ a. Uterus (body)

   _____ b. Uterus (cervix)

   _____ c. Ovary

   _____ d. Uterine (fallopian) tube

   _____ e. Vagina

2. Indicate which of those five structures in Figure LG 28-2 are paired. Circle the numbers: 1 2 3 4 5. Which structures are singular? Write their numbers:

   _____

   14

3. Describe the following aspects of the ovaries.
   a. Shape and size

   b. Location

   c. Ligaments. Label on Figure LG 28-2b: *suspensory ligaments, ovarian ligament, mesovarian.*

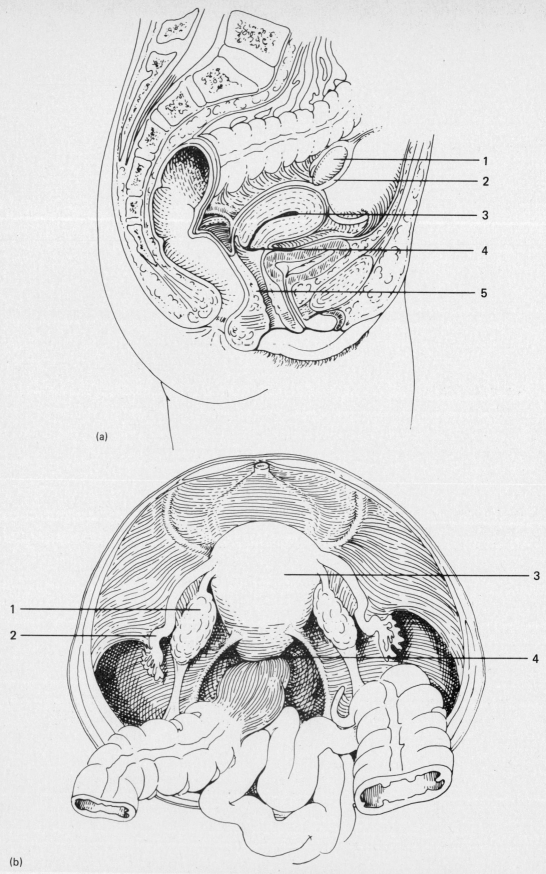

(a)

(b)

Figure LG 28-2 Female organs of reproduction. Structures are numbered in order along the pathway taken by ova. Refer to Learning Activities 13 and 14. Label and color as directed. (a) Sagittal section. (b) Viewed from above.

4. Define these parts of the ova.
   a. Germinal epithelium

   b. Stroma

   c. Vesicular ovarian (Graafian) follicle

   d. Corpus luteum

5. Answer these questions about aspects of oogenesis which differ from spermatogenesis.
   a. When do oogonia terminate mitosis so that no new potential female gametes can be generated?

   b. What are polar bodies?

   c. In which stage of meiosis is the developing ovum at the time that the cell is discharged from the ovary? 15

6. *For extra review,* contrast mitosis with meiosis (as in Figure LG 3-2, page 55) by completing this table.

Process	Identical or Different Resulting Cells	Number of Chromosomes in Resulting Cells	Number of Cells Resulting from One Cell
a. Mitosis			
b. Meiosis in spermatogenesis			
c. Meiosis in oogenesis			

7. Complete this exercise about the uterine tubes.

a. Uterine tubes are also known as _____ tubes. They are about _____ cm (_____ inches) long.

b. The funnel-shaped open end of each tube is known as the _____. Its fingerlike projections, called _____, are close to (but not in direct contact with) the _____. Label the *infundibulum* and *fimbriae* on Figure LG 28-2b.

c. What structural features of the tube enhance passage of ova into and through the tubes?

d. List two functions of uterine tubes.

Draw arrows in Figure LG 28-2b showing direction of movement of ova in these tubes from ovaries toward the uterus.

e. Define *ectopic pregnancy* and list three possible causes of this condition.

8. On Figure LG 28-2a label these three parts of the uterus: *fundus, body,* and *cervix.*

9. Answer these questions about uterine position. $\boxed{18}$
   a. The organ that lies anterior and inferior to the uterus is the

   _____. The _____ lies posterior to it.

   Color the pelvic organs as follows: bladder, yellow; uterus, red; rectum, brown.
   b. The rectouterine pouch lies *(anterior? posterior?)* to the uterus, while the

   _____-uterine pouch lies between the urinary bladder and the uterus.
   c. The fundus of the uterus is normally tipped *(anteriorly? posteriorly?)*. If it is malpositioned posteriorly, it would be *(anteflexed? retroflexed?)*.
   d. The structures that hold the uterus in position are referred to as

   _____. The _____ ligaments

   attach the uterus to the sacrum. The _____ ligaments pass through the inguinal canal and anchor into external genitalia (labia

   majora). The _____ ligaments are broad, thin sheets of peritoneum extending laterally from the uterus. Label these ligaments on Figure LG 28-2.
   e. The most important ligaments in prevention of drooping (prolapse) of the

   uterus are the _____ ligaments which are attached to the base of the uterus. (*Note:* These are not visible on Figure LG 28-2.)

10. *A clinical challenge.* Name the procedures described. $\boxed{19}$
    a. Dilation of the cervix of the uterus with scraping of the uterine lining:

    _____.

    b. Relatively painless procedure in which a small number of cells are removed from the cervical area and examined microscopically for possible

    changes indicating malignancy: _____.

    c. Surgical removal of the uterus: _____.

11. Refer to Figure LG 28-2a and do this exercise about layers of the wall of $\boxed{20}$ the uterus.
    a. Most of the uterus consists of _____-metrium. This layer is *(smooth muscle? epithelium?)*. It should appear red in the figure (as directed above).
    b. Now color the peritoneal covering over the uterus green. This is a *(mucous? serous?)* membrane.
    c. The innermost layer of the uterus is the _____-metrium. Which portion of it is shed during menstruation? Stratum *(basalis? functionalis?)*. Which arteries supply the stratum functionalis? *(Spiral? Straight?)*

C. Female reproductive system: endocrine relations

1. Contrast *menstrual cycle* and *ovarian cycle*.

2. Indicate where each of the following hormones is produced. Write H for hypothalamus, AP for anterior pituitary, and O for ovary.

21

_____ a. GnRF

_____ b. FSH

_____ c. LH

_____ d. Estrogens

_____ e. Progesterone

Now write one or more principal function next to each hormone. (Try to list five functions of estrogen.)

3. After carefully studying Figure 28-18 (page 724) and the related description in your text, summarize the events of menstrual and ovarian cycles in this exercise.

   a. The average duration of the menstrual cycle is _____ days. The

   first phase is the _____ phase, usually about

   _____ days in length. What uterine changes take place during

   this time?

   b. Following the menstrual phase is the _____ phase.

   During both of these phases, _____ begin to develop in the ovary. At birth of a female infant, about *(20? 400? 5,000? 200,000?)*

   primary follicles are present in her ovaries. Each month about _____ of

   these follicles become secondary follicles, while only _____ usually matures each month.

   c. Follicle development occurs under the influence of the hypothalamic

   hormone _____, which regulates the anterior

   pituitary hormone _____.

   d. What are the two main functions of follicles?

   e. How does estrogen affect the endometrium?

   For this reason, the preovulatory phase is also known as the

   _____ phase. Since follicles reach their peak during this

   phase, it is also called the _____ phase.

   f. Toward the end of this phase estrogen level is very high. What effect does a high level of a target hormone exert? (See Figure LG 18-2, page LG 370.)

   g. The principle of this negative feedback effect is utilized in oral contraceptives. By starting to take "pills" (consisting of target hormones estrogen and progesterone) early in the cycle (day 5), and continuing until day 25, a woman will maintain a very low level of the tropic hormone

   _____. Without FSH, follicles and ova will not

   develop, and so the woman will not _____.

   h. The second tropic hormone released during the cycle is _____

   _____. LH level surges just *(before? after?)* ovulation. LH stimulates release of the ovum from the follicle. This event is

   known as _____.

i.   Following ovulation is the _____ phase. It lasts until day _____ of a 28-day cycle. Under the influence of tropic hormone _____, follicle cells are changed into the corpus _____. These cells secrete two hormones: _____ and _____. Both prepare the endometrium for _____. What preparatory changes occur?

j.   What effect do rising levels of target hormones estrogen and progesterone have upon the tropic hormone LH?

k.   One function of LH is to form and maintain the corpus luteum. So as LH decreases, the corpus luteum _____, forming a _____ on the ovary which is known as the _____.

l.   The corpus luteum had been secreting _____ and _____. With the demise of the corpus luteum, the levels of these hormones rapidly *(increase? decrease?)*. Since these hormones were maintaining the endometrium, endometrial tissue now deteriorates and will be shed during the next _____ phase.

m.   If fertilization should occur, estrogens and progesterone are needed to maintain the endometrial lining. The _____ around the developing embryo secretes a hormone named _____. It functions much like LH in that it maintains the corpus _____, even though LH has been inhibited (step j above). The corpus luteum continues to secrete estrogen and progesterone for several months until the placenta itself can secrete sufficient amounts. Incidentally, HCG is present in the urine of pregnant women only, and so is routinely used to detect _____.

4. To check your understanding of the events of these cycles, list letters of the major events in chronological order beginning on day 1 of the cycle. To do this exercise it may help to list on a separate piece of paper days 1 to 28 and write each event next to the approximate day on which it occurs. |23|

____ ____ ____ ____ ____ ____ ____ ____ ____ ____ ____

A. Ovulation occurs.
B. Estrogen and progesterone levels drop.
C. FSH begins to stimulate follicles.
D. LH begins to be secreted.
E. High levels of estrogens and progesterone inhibit LH.
F. Estrogens are secreted for the first time during the cycle.
G. Corpus luteum dies and becomes corpus albicans.
H. Endometrium begins to deteriorate.
I. Follicle is converted into corpus luteum.
J. Rising level of estrogen inhibits FSH secretion.
K. Corpus luteum secretes estrogens and progesterone.

5. Contrast *menarche, menopause,* and *climacteric.*

6. *ACTive learning.* Role play the menstrual cycle. Lab groups prepare such "Reproductions" and present them in lab or lecture periods. Students assume the following roles and wear or carry signs to make clear their identities:
   a. Hormones (GnRF, FSH, LH, estrogens, progesterone) which emerge from the correct endocrine gland. Hormones are on stage and active only during the appropriate phases or scenes.
   b. Organs such as hypothalamus and anterior pituitary; a uterus (with arms extended as uterine tubes and fingers as fimbriae, possibly wearing layers of red clothing which can be shed during menstrual phase); and ovaries (crouched below extended uterine tubes, possibly holding balloons or balls or small persons to represent ova released).
   c. Spermatozoa may be represented by a number of students. The group can determine whether one sperm will fertilize the ovum, in which case roles of other hormones (like HCG, HCS, and relaxin) and a placenta may be added.
   Special effects may include taped segments of music with appropriate lyrics and a creative selection of attire and props.

D. Female reproductive system: vagina, external genitalia, mammary glands (pages 726–730)

1. Refer to Figure LG 28–2a. In the normal position, at what angle does the |24| uterus join the vagina? _____
2. Describe these vaginal structures.
   a. Fornix

652   b. Rugae

c. Hymen

25   3. The pII of vaginal mucosa is *(acid? alkaline?)*. What is the clinical significance of this fact?

4. On Figure LG 28-3 label the following parts of the vulva: *mons pubis, labia majora, labia minora, clitoris,* and *vaginal orifice.*

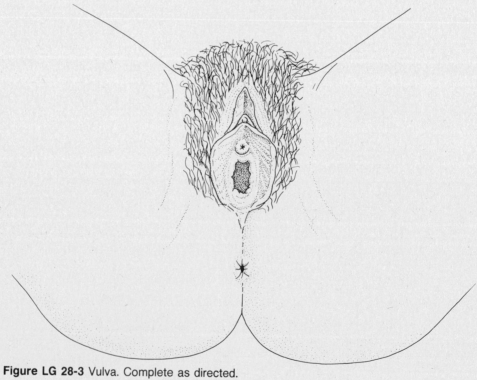

Figure LG 28-3 Vulva. Complete as directed.

5. Identify male homologues for each of the following. $\boxed{26}$ **653**

    a. Labia majora: _____

    b. Clitoris: _____

    c. Lesser vestibular (Skene's) glands: _____

    d. Greater vestibular (Bartholin's) glands: _____

6. Name the structures which form the four corners of the *perineum* in both sexes. (Refer to Figure 28-22 (page 728) in the text.) $\boxed{27}$

    Draw the borders of this diamond-shaped area on Figure LG 28-3. Then draw a line between the ischial tuberosities. The two triangles formed are

    called the _____ and the _____.

    All of the female genitalia are located in the _____ triangle.

7. Locate the clinical perineum in your diagram. Draw a red line where the episiotomy incision would be performed.

8. Each mammary gland consists of about 15–20 _____. Each lobe contains several lobules. These contain clusters of milk-secreting

    cells called _____. How does the size of a breast correlate to the amount of milk produced by it?

9. Number in correct sequence the structures in the pathway of milk from its site of formation to its exit point from the body. $\boxed{28}$

    _____ a. Alveoli

    _____ b. Ampullae

    _____ c. Lactiferous ducts

    _____ d. Mammary ducts

    _____ e. Secondary tubules

10. Which has the larger diameter? *(Areola? Nipple?)* $\boxed{29}$

11. Match the hormones with their roles in breast development and lactation. $\boxed{30}$

    _____ a. Causes duct development in the breast beginning at puberty

    _____ b. Stimulates alveolar (gland) development of the breast beginning during adolescence

    _____ c. Stimulates secretion of milk in alveoli of lactating mother

    _____ d. Stimulates ejection of milk from alveoli to ducts

E. Estrogen
OT. Oxytocin
P. Progesterone
PRL. Prolactin

31   12. Do this exercise about breast cancer detection.
     a. Identify the most promising method for increasing the survival rate for breast cancer.

At what point in the monthly cycle should self-breast examination be performed?

     b. _____ is a screening technique that shows small calcium deposits by low-dose radiation. Ultrasound procedures utilize *(radiation? sound waves?)* to distinguish malignant tumors from _____ tumors or cysts.

(pages 730–732)   **E. Sexual intercourse and birth control**

1. Define *sexual intercourse* or *coitus.*

32   2. Answer these questions about the process.
     a. *(Sympathetic? Parasympathetic?)* impulses from the *(brain stem? sacral cord?)* cause dilation of arteries of the penis, leading to the *(flaccid? erect?)* state.
     b. *(Sympathetic? Parasympathetic?)* impulses cause sperm to be propelled into the urethra. This process is called *(emission? ejaculation?)*. Expulsion from the urethra to the exterior of the body is called _____.

3. Describe the processes of erection, lubrication, and orgasm in the female.

4. Most of the lubricating fluid that facilitates sexual intercourse is produced by the *(male? female?)*. What structure produces most of this fluid? $\boxed{33}$

5. Match the methods of birth control listed at right with descriptions below. $\boxed{34}$
(Note: extra lines at right of descriptions will be used in Activity 35.)

_____ a. Removal of testes __S__

_____ b. Removal of uterus _____

_____ c. Removal of a portion of the ductus (vas) deferens _____

_____ d. Tying off of the uterine tubes _____

_____ e. A natural method of birth control involving abstinence during the period when ovulation is most likely _____

_____ f. A mechanical method of birth control in which a dome-shaped structure is placed over the cervix _____

_____ g. Small object, an intrauterine device, placed in the uterus by physician _____

_____ h. A rubber sheath over the penis _____

_____ i. Spermicidal foams, jellies, and creams _____

_____ j. "The pill," combination of progesterone and estrogens, causing decrease in FSH and LH so ovulation does not occur _____

Ca. Castration
Ch. Chemical methods
Co. Condom
D. Diaphragm
H. Hysterectomy
IUD. IUD
O. Oral contraceptive
R. Rhythm
T. Tubal ligation
V. Vasectomy

6. Contrast two general categories of birth control methods by indicating which of the descriptions in Learning Activity 34 above refer to sterilization techniques (S) and which refer to contraceptive methods (C). The first one is done for you.

35

7. Describe the laparoscopic technique of sterilization.

8. List several risks of oral contraceptives.

9. What is *gossypol* and how does it work as a contraceptive?

Explain why the development of such a contraceptive is significant.

(pages 732–735) **F. Aging and development of the reproductive systems**

1. *(Many? Relatively few?)* age-related changes involve the reproductive system. Describe these changes in:
a. Females

b. Males

2. Which type of cancer is more common among women over 65 years? *(Cervical? Uterine?)*

3. Name the reproductive structures (if any) that develop from the following embryonic ducts.

36

a. Mesonephric duct in females: _____

b. Mesonephric duct in males: _____

c. Paramesonephric (Müller's) duct in females: _____

d. Paramesonephric (Müller's) duct in males: _____

4. Write *E* (endoderm) or *M* (mesoderm) next to the structures below to indicate their embryonic origins.

37

_____ a. Ovaries and testes

_____ b. Prostate and bulbourethral glands

_____ c. Greater (Bartholin's) and lesser vestibular glands

G Disorders, medical terminology

(pages 735–738)

1. Define *sexually transmitted disease* and *venereal disease*.

2. Describe the main characteristics of these sexually transmitted diseases by completing the table.

Disease	Causative Agent	Main Symptoms
a. Gonorrhea		
b.	*Treponema pallidum*	
c.	Type II herpes simplex virus	

658

Disease	Causative Agent	Main Symptoms
d.	*Trichomonas vaginalis*	
e. NGU		

3. *For extra review.* Describe the three stages of syphilis.

4. Discuss effects of syphilis on children born to mothers with untreated syphilis.

5. Check your understanding of the disorders listed at right by matching them with the correct description.

38

_____ a. Painful menstruation, partly due to contractions of uterine muscle

_____ b. Possible cause of blindness in newborns if bacteria transmitted to eyes during birth

_____ c. Spreading of uterine lining into abdominopelvic cavity via uterine tubes

_____ d. Caused by bacterium Treponema pallidum; may involve many systems in tertiary stage

_____ e. Absence of menstrual periods

_____ f. Benign tumor with firm, rubbery consistency; easily moved within breast

_____ g. Caused by flagellated protozoan

A. Amenorrhea
D. Dysmenorrhea
E. Endometriosis
F. Fibroadenoma
G. Gonorrhea
S. Syphilis
T. Trichomoniasis

6. Explain why prostate enlargement can lead to urinary tract infections (UTIs).

7. Fill in the term *(impotence* or *infertility)* that answers the description. Then 39 list several causes of each condition.

   a. Inability to fertilize an ovum: _____.

   b. Inability to attain and maintain an erection: _____

      _____.

8. List six symptoms of *premenstrual syndrome (PMS).*

9. Name the microorganism associated with *toxic shock syndrome (TSS).*
   _____ Relate this condition to use of tampons.

10. Choose the correct answers about incidence of breast cancer. It occurs most commonly among women:
    a. *(Under 30? Between 40–60?)* years old.
    b. Who *(do? do not?)* have a mother or sister with a history of breast cancer.
    c. Who have given birth to *(0? 1 or more?)* children.

11. Contrast these surgical procedures for the breast cancer patient: *lumpectomy, radical mastectomy.*

12. What is PID?

    It is most commonly caused by the sexually transmitted disease (STD) known as _____.

13. Oophorectomy is removal of a(n) _____, while _____ refers to removal of a uterine (Fallopian) tube.

40

Answers to Numbered Questions in Learning Activities

1   (a) Testes. (b) Seminiferous tubules, epididymis, vas deferens, ejaculatory duct, urethra. (c) Prostate, seminal vesicles, bulbourethral glands. (d) Scrotum, penis.

2   The lower temperature is required for normal sperm production and survival.

3   (a) L. (b) ST. (c) SC. (d) IE.

4   (a) Ova, haploid, 23. (b) Zygote, diploid, 2n, homologous. (c) Spermato, seminiferous tubules of testes, weeks, toward. (d) Pale type A (or, if necessary, dark type A). (e) B, 2, secondary spermatocytes, equatorial. (f) Spermiogenesis, 23, do not.

5   (a) GnRF. (b) FSH. (c) LH. (d) Testosterone. (e) Testosterone. (f) Inhibin.

6   (a) 5. (b) 2. (c) 6. (d) 8. (e) 7. (f) 4. (g) 3. (h) 9. (i) 1.

7   Paired: 1–5; singular: 6–9.

8   (a) 45, 18. (b) Partially, inguinal canal and rings, inguinal hernia. (c) Testicular artery, veins, lymphatics, nerves, and cremaster muscle. (d) Scrotum, no, no; hormones exit through testicular veins which are not cut during vasectomy.

9   Spongy, or cavernous.

10 (a) S. (b) S. (c) P. (d) S. (e) B.

11 Since it surrounds the urethra (like a doughnut), an enlarged prostate can cause painful and difficult urination (dysuria) and bladder infection (cystitis) due to incomplete voiding.

12 (a) Prepuce, glans, circumcision. (b) Cavernosum, spongiosum, blood, veins, erect. (c) Spongiosum, urine, sperm; sphincter action at base of bladder.

13 (a) 3. (b) 4. (c) 1. (d) 2. (e) 5.

14 Paired: 1–2; singular: 3–5.

15 Secondary oocyte (K in Figure LG 3-2, page 55); it undergoes the final (equatorial) division only if fertilization occurs.

16 (a) Identical, 46, 2. (b) Different, 23, 4 spermatozoa. (c) Different, 23, 4 (1 ovum and 3 polar bodies).

17 (a) Fallopian, 10, 4. (b) Infundibulum, fimbriae, ovary. (c) Cilia lining tubes and contractions of smooth muscle in wall of tubes. (d) Passage of ovum to uterus and site of fertilization. (e) Pregnancy following implantation at any location other than in the body of the uterus, for example, tubal pregnancy.

18 (a) Bladder, rectum. (b) Posterior, vesico-. (c) Anteriorly, retroflexed. (d) Ligaments, uterosacral, round, broad. (e) Cardinal.

19 (a) D & C (dilation and curettage). (b) Papanicolaou smear. (c) Hysterectomy.

20 (a) Myo, smooth muscle. (b) Serous. (c) Endo, functionalis, spiral.

21 (a) H. (b) AP. (c) AP. (d) O. (e) O.

22 (a) 28, menstrual, 5; functionalis layer of the endometrium is shed. (b) Preovulatory, follicles, 200,000, 20, 1. (c) GnRF, FSH. (d) Secretion of estrogens and development of the ovum. (e) Cause cells of endometrium to grow (proliferate); proliferative, follicular. (f) Negative feedback effect inhibits GnRF and FSH. (g) FSH, ovulate. (h) LH, before, ovulation. (i) Postovulatory, 28, LH, luteum, progesterone, estrogens, implantation; thickening, increased secretory activity and vascularization, increased amount of tissue fluid. (j) Inhibition of LH. (k) Disintegrates, white scar, corpus albicans. (l) Progesterone, estrogens, decrease, menstrual. (m) Placenta, human chorionic gonadotropin (HCG), luteum, pregnancy.

23 C F J D A I K E G B H.

24 Close to 90°.

25 Acid, decreases risk of infection but may injure sperm.

26 (a) Scrotum. (b) Penis. (c) Prostate. (d) Bulbourethral glands.

27 Symphysis pubis, the two ischial tuberosities, and coccyx.

28 (a) 1. (b) 4. (c) 5. (d) 3. (e) 2.

29 Areola.

30 (a) E. (b) P. (c) PRL. (d) OT.

31 (a) Early detection, as by self-breast examination (SBE), just after the menstrual phase. (b) Mammography, sound waves, benign.

32 (a) Parasympathetic, sacral cord, erect. (b) Sympathetic, emission, ejaculation.

33 Female, vaginal lining.

34 (a) Ca. (b) H. (c) V. (d) T. (e) R. (f) D. (g) IUD. (h) Co. (i) Ch. (j) O.

35 (a–d) All S. (e–j) All C.

36 (a) None. (b) Ducts of testes and epididymides, ejaculatory ducts and seminal vesicles. (c) Uterus and vagina. (d) None.

37 (a) M. (b) E. (c) E.

38 (a) D. (b) G. (c) E. (d) S. (e) A. (f) F. (g) T.

39 (a) Infertility. (b) Impotence.

40 Ovary, salpingectomy.

MASTERY TEST: Chapter 28

Questions 1–10: Arrange the answers in correct sequence.

_____ _____ _____ 1. From most external to deepest:

    A. Endometrium
    B. Serosa
    C. Myometrium

_____ _____ _____ 2. Portions of the urethra, from proximal (closest to bladder) to distal (closest to outside of the body):

    A. Prostatic
    B. Membranous
    C. Spongy

_____ _____ _____ 3. Thickness of the endometrium, from thinnest to thickest:

    A. On day 5 of cycle
    B. On day 13 of cycle
    C. On day 23 of cycle

_____ _____ _____ 4. From anterior to posterior:

    A. Uterus and vagina
    B. Bladder and urethra
    C. Rectum and anus

_____ _____ _____ _____ 5. Order in which hormones begin to increase level, from day 1 of menstrual cycle:

    A. FSH
    B. LH
    C. Progesterone
    D. Estrogen

_____ _____ _____ _____ 6. Order of events in the female monthly cycle, beginning with day 1:

    A. Ovulation
    B. Formation of follicle
    C. Menstruation
    D. Formation of corpus luteum

_____ _____ _____ _____ 7. From anterior to posterior:

    A. Anus
    B. Vaginal orifice
    C. Clitoris
    D. Urethral orifice

_____ _____ _____ _____ _____ 8. Pathway of sperm:

    A. Ejaculatory duct
    B. Testis
    C. Urethra
    D. Ductus (vas) deferens
    E. Epididymis

_____ _____ _____ _____ _____ 9. Pathway of milk in breasts:

    A. Ampullae
    B. Alveoli
    C. Secondary tubules
    D. Mammary ducts
    E. Lactiferous ducts

_____ _____ _____ _____ _____ _____ 10. Pathway of sperm entering female reproductive system:

    A. Uterine cavity
    B. Cervical canal
    C. External os
    D. Internal os
    E. Uterine tube
    F. Vagina

Questions 11–12: Choose the one best answer to each question.

_____ 11. Eighty to 90 percent of seminal fluid (semen) is secreted by the combined secretions of:

    A. Prostate and epididymis
    B. Seminal vesicles and prostate
    C. Seminal vesicles and seminiferous tubules
    D. Seminiferous tubules and epididymis
    E. Bulbourethral glands and prostate

_____ 12. In the normal male, there are two of each of the following structures EXCEPT:

    A. Testis   B. Seminal vesicle   C. Prostate
    D. Ductus deferens   E. Epididymis   F. Ejaculatory duct

Questions 13–15: Choose ALL correct answers to each question.

_____ 13. Which of the following are produced by the testes?

    A. Spermatozoa   B. Testosterone
    C. Inhibin   D. GnRF
    E. FSH   F. LH

_____ 14. Which of the following are produced by the ovaries and then *leave* the ovaries?

    A. Follicle   B. Ovum   C. Corpus luteum
    D. Corpus albicans   E. Estrogen
    F. Progesterone

_____ 15. Which of the following are functions of LH?

    A. Begin the development of the follicle.
    B. Stimulate change of follicle cells into corpus luteum cells.
    C. Stimulate release of ovum (ovulation).
    D. Stimulate corpus luteum cells to secrete estrogens and progesterone.
    E. Stimulate release of GnRF.

Questions 16–20: Circle T (true) or F (false). If the statement is false, change the underlined word or phrase so that the statement is correct.

T   F   16. <u>Spermatogonia and oogonia both</u> continue to divide throughout a person's lifetime to produce new <u>primary spermatocytes or primary oocytes.</u>

T   F   17. The menstrual cycle refers to a series of changes in the <u>uterus,</u> whereas the ovarian cycle is a series of changes in the <u>ovary.</u>

T   F   18. <u>Meiosis is a part</u> of both oogenesis and spermatogenesis.

T   F   19. Each oogonium beginning oogenesis produces <u>four</u> mature ova.

T   F   20. A high level of a target hormone (such as estrogen) will <u>inhibit</u> release of its related releasing factor and target hormone (GnRF and FSH).

Questions 21–25: fill-ins. Complete each sentence with the word or phrase that best fits.

_____ 21. _____ is a term that means undescended testes.

_____ 22. Most oral contraceptives consist of the hormones _____ and _____.

_____ 23. _____ cells in testes phagocytose degenerating spermatogenic cells and secrete androgen-binding proteins that concentrate testosterone in seminiferous tubules.

_____ 24. Menstrual flow occurs as a result of a sudden _____-crease in the hormones _____ and _____.

_____ 25. The three main functions of estrogens are promotion of growth and maintenance of _____, _____ balance, and stimulation of protein _____-bolism.

29

Development and Inheritance

In the last chapter you learned about the systems involved in production of sperm and ova. In this chapter you will learn about the developmental processes which lead to the formation of a new individual. You will consider the union of gametes and the implantation of the fertilized egg (1–2). You will study development through embryonic and fetal periods (3–10). You will discuss labor and birth, adjustments required of the newborn, the process of lactation, and possible effects on development of harmful environmental factors (11–14). Then you will study several aspects of inheritance and medical terminology associated with development and inheritance (15–16).

Topics Summary

A. Development during pregnancy
B. Hormones of pregnancy, gestation, prenatal diagnostic techniques
C. Parturition and labor, adjustments at birth, potential hazards during development, lactation
D. Inheritance and medical terminology

Objectives

1. Explain the activities associated with fertilization, morula formation, blastocyst development, and implantation.
2. Describe how external human fertilization and embryo transfer gamete intrafallopian transfer (GIFT) and transvaginal oocyte retrieval are accomplished.
3. Discuss the formation of the primary germ layers, embryonic membranes, placenta, and umbilical cord as the principal events of the embryonic period.
4. List representative body structures produced by the primary germ layers.
5. Discuss the function of the embryonic membranes.
6. Compare the roles of the placenta and umbilical cord during embryonic and fetal growth.
7. Discuss the principal body changes associated with fetal growth.
8. Compare the sources and functions of the hormones secreted during pregnancy.
9. Describe some of the anatomical and physiological changes associated with gestation.
10. Explain amniocentesis and chorionic villi sampling (CVS) as procedures for diagnosing diseases in the newborn.
11. Explain the events associated with the three stages of labor.
12. Explain the respiratory and cardiovascular adjustments that occur in an infant at birth.
13. Discuss potential hazards to the embryo and fetus associated with chemicals and drugs, irradiation, alcohol, and cigarette smoking.
14. Discuss the physiology and control of lactation.
15. Define inheritance and describe the importance of PKU, sex and color blindness, and hemophilia.
16. Define medical terminology associated with development and inheritance.

666   **Learning Activities**

(pages 743-752)   **A.** Development during pregnancy

1. 1. Answer these questions about the process of fertilization.
   a. How many sperm are ordinarily introduced into the vagina during sexual
      intercourse? _____About how many reach the area
      where the ovum is located?_____
   b. What two mechanisms may be involved in sperm transport within the
      female reproductive organs?

   c. Name the usual site of fertilization.

   d. What reproductive structure secretes hyaluronidasc? What is the function
      of this enzyme?

   e. How many sperm fertilize an ovum? What prevents further sperm from
      entering the ovum?

2. Describe the roles of these structures in fertilization. Tell how many
   chromosomes are in each.
   a. Male pronucleus

   b. Female pronucleus

d. Zygote

3. Contrast types of multiple births by choosing the answer that best fits each description. Not all answers will be used. | 2 |

_____ a. Also known as fraternal twins

_____ b. Produced by three ova fertilized by three different spermatozoa

_____ c. Two infants derived from a single fertilized ovum that divides once

_____ d. Three "identical" infants developed from a single fertilized ovum that splits twice

DTwin. Dizygotic twins
DTrip. Dizygotic triplets
MTwin. Monozygotic twins
MTrip. Monozygotic triplets
TTrip. Trizygotic triplets

4. Summarize the main events during the week following fertilization. Describe each of the following stages. (For help, refer to Figures 29-3 and 29-4, pages 744 and 745 in the text.)
a. Cleavage

b. Blastomere

c. Morula

668   d. Blastocyst

5. Draw a diagram of a blastocyst. Label: *blastocoel, inner cell mass,* and *ectodermal cells (trophectoderm).* Color red the cells which will become the embryo. Color blue the cells which will become part of the placenta. (Refer to Figure 29-4c, page 745, for help.)

3 6. Attachment of the blastocyst to the endometrium is a process known as _____; this occurs about 7–8 days after ovulation, or about day ____ of a 28-day cycle. Note that the developing baby is implanted in the uterus even before mother "misses" the first day of her next menstrual cycle. Cells of the *(inner cell mass? trophectoderm?)* secrete enzymes that enable the blastocyst to burrow into the uterine lining. Following implantation, the blastocyst can absorb nutrients from _____

_____.

4 7. *Hyperemesis gravidarum* is a term meaning _____. List possible causes.

8. Describe the following aspects of external human fertilization.
   a. In what year was the first live birth as a result of this procedure? *(1962?*
5   *1970? 1978? 1986?)*

b. Briefly describe this procedure.

c. State the significance of this procedure.

d. Contrast *external human fertilization* with *embryo transfer.*

9. A human embryo refers to the developing individual during the *(first two? last seven?)* months of development in utero. Define fetus.

6

10. Complete this exercise about the embryonic period. Refer to Figure 29-6 (pages 748–749) in your text.

7

   a. Identify what structures will ultimately form from these two parts of the implanted blastocyst. Trophectoderm: ＿＿＿＿＿＿＿＿＿＿＿ Inner cell mass: ＿＿＿＿＿＿＿＿＿＿＿ Between these two portions of the blastocyst is the ＿＿＿＿＿＿＿＿＿＿＿ cavity.

b. The inner cell mass first forms the embryonic _____ consisting of two layers. The upper cells (closer to the implantation site toward the right on Figure 29-b, page 748 of the text) are called

_____ -derm cells. The lower cells (towards the left in the figure) are called _____ -derm cells. Eventually, a layer of _____ -derm will form between these. Collectively the three layers are called _____ layers.

c. The process of formation of the primary germ layers is known as

_____ and it occurs by the end of the

_____ week following fertilization.

d. _____ -derm cells line the amniotic cavity. The delicate fetal membrane formed from these cells is called the

_____ . It is sometimes called the "_____

_____ " surrounding the fetus and it is broken at birth. (Ectoderm cells also form fetal structures: see Learning Activity

8 .)

e. The lower cells of the embryonic disc, called _____

_____ cells, line the primitive _____

_____ and for a while extend out to form the

_____ sac. This sac (is? is not?) as significant in humans as it is in birds.

f. Some mesoderm cells move around to line the original trophectoderm cells (Figure 29–4, page 745 of the text). Together these cells form the

_____ layer of the placenta. Which membrane is more superficial? (Amnion? Chorion?) The space between these layers is

called the _____ .

11. Each of the primary germ layers is involved with formation of both fetal organs and fetal membranes. Name the major structures derived from each primary germ layer: endoderm, mesoderm, and ectoderm. (Consult Exhibit 29–1, page 747, in your text.)

Endoderm   Mesoderm   Ectoderm

12. *For extra review.* Do this matching exercise on primary germ layer derivatives. 8

_____ a. Epithelial lining of all of digestive, respiratory, and genitourinary tracts except near openings to the exterior of the body

_____ b. Epidermis of skin, epithelial lining of entrances to the body (such as mouth, nose, and anus), hair, nails

_____ c. All of the skeletal system (bone, cartilage, joint cavities)

_____ d. All muscle (skeletal, smooth, and cardiac)

_____ e. Blood and all blood and lymphatic vessels

_____ f. Entire nervous system, including posterior pituitary

_____ g. Thyroid, parathyroid, thymus, and pancreas

Ecto. Ectoderm
Endo. Endoderm
Meso. Mesoderm

13. Write the name of each fetal membrane next to its description. 9
    a. Originally formed from ectoderm, this membrane encloses fluid that acts as a shock absorber for the developing baby: _____

    _____.

    b. Derived from mesoderm and trophectoderm; it becomes the principal part

    of placenta: _____.

    c. Endoderm-lined membrane serving as exclusive nutrient supply for embryos of some species: _____.

    d. A small membrane that forms umbilical blood vessels: _____

    _____.

14. Further describe the placenta in this exercise. 10

    a. The placenta has the shape of a _____ embedded in

    the wall of the uterus. Its fetal portion is the _____

    fetal membrane. Fingerlike projections, known as _____

    _____, grow into the uterine lining. Consequently maternal and fetal blood are *(allowed to mix? brought into close proximity but do not mix?).* Fetal blood in umbilical vessels is bathed by

    maternal blood in _____ spaces.

    b. The maternal aspect of the placenta is called the *(chorion? decidua basalis?).* This is a portion of the _____-metrium of the uterus.

    c. What is the ultimate fate of the placenta?

11   15. *A clinical challenge.* Following a birth, vessels of the umbilical cord are carefully examined. A total of _____ vessels should be found, surrounded by a mucous jelly-like connective tissue. The vessels should include *(1? 2?)* umbilical artery(-ies) and *(1? 2?)* umbilical vein(s). Lack of a vessel may indicate a congenital malformation.

16. List three uses of human placentas following birth.

12   17. Study Exhibit 29-2 (page 752) in your textbook. Write the number of the month when each of the following events occurs.

_____ a. Heart starts to beat.

_____ b. Heartbeat can be detected.

_____ c. Backbone and vertebral canal form.

_____ d. Eyes are almost fully developed; eyelids are still fused.

_____ e. Eyelids separate; eyelashes form.

_____ f. Limb buds develop.

_____ g. Ossification begins.

_____ h. Limb buds become distinct as arms and legs; digits are well formed.

_____ i. Nails develop.

_____ j. Fine (lanugo) hair covers body.

_____ k. Lanugo hair is shed.

_____ l. Fetus is capable of survival.

_____ m. Subcutaneous fat is deposited.

_____ n. Testes descend into scrotum.

(pages 752–756)   **B. Hormones of pregnancy, gestation, prenatal diagnostic techniques.**

13   1. Complete this exercise about the hormones of pregnancy.

a. During pregnancy, the level of progesterone and estrogens must remain *(high? low?)* in order to support the endometrial lining. During the first two or three months or so, these hormones are produced principally by the _____ located in the _____, under the influence of the tropic hormone _____. (It might be helpful to review the section on endocrine relations in Chapter 28.)

b. HCG reaches its peak between the _____ and _____ month. Since it is present in blood, it will be filtered into _____ where it can readily be detected as an indication of pregnancy.

c. The factor pLRF is produced by the *(ovary? hypothalamus? placenta? anterior pituitary?)*. It is similar chemically to *(GnRF? FSH? LH?)*. Its function appears to be to stimulate the placenta to produce

_____.

d. You have learned about two hormones produced by the placenta: pLRF and also the hormone that pRLF evokes, _____. In addition, the chorion produces HCS; these initials stand for _____ chorionic _____. Describe functions of HCS.

e. The placenta also produces progesterone and estrogens from about the _____ month of pregnancy until _____ _____. Therefore it serves as a temporary endocrine organ.

f. _____ is a hormone that relaxes pelvic joints and helps dilate the cervix near the time of birth. This hormone is produced by the _____ and _____.

2. The normal human gestation period is ____ days. This period is defined as the time from *(the beginning of the last menstrual period? conception?)* until birth. |14|

3. Complete this exercise about maternal adaptations during pregnancy. Describe normal changes that occur. |15|

   a. Pulse ____-creases.

   b. Blood volume ____-creases by about ____ percent.

   c. Tidal volume ____-creases.

   d. Gastrointestinal (GI) motility ____-creases, which may cause *(diarrhea? constipation?)*.

   e. Pressure upon the inferior vena cava ____-creases which may lead to

   _____.

4. Describe uses of the following techniques in pregnancy. State one reason why each may be utilized.
   a. Ultrasonography

674

b. Doppler detector

16 5. Complete this exercise about amniocentesis and CVS.

a. Amniocentesis involves withdrawal of _____ fluid, usually at about *(2–4? 8–10? 16–20?)* weeks after conception. Amniotic fluid is constantly recycled (by drinking and excretion) through the fetus; it *(does? does not?)* contain fetal cells and products of fetal metabolism. For what purposes is amniotic fluid collected and examined?

b. CVS (meaning _____) is a procedure most often performed at about *(2–4? 8–10? 16–20?)* weeks. It *(does? does not?)* involve penetration of the uterine cavity. Cells of the _____ layer of the placenta embedded in the uterus are studied. These *(are? are not?)* derived from the same sperm and egg that is forming the fetus. Therefore this procedure, like amniocentesis, can be used to diagnose _____.

(pages 756–760) **C. Parturition and labor, adjustments at birth, potential hazards during development, lactation**

1. Contrast these pairs of terms.
a. Labor/parturition

b. False labor/true labor

2. Do the following hormones stimulate (S) or inhibit (I) uterine contractions during the birth process?

_____ a. Progesterone

_____ b. Estrogens

_____ c. Oxytocin

3. What is the "show" produced at the time of birth?

4. Identify descriptions of the three phases of labor (first, second, or third). |17|

_____ a. Stage of expulsion: from complete cervical dilation through delivery of the baby

_____ b. Time after the delivery of the baby until the placenta ("afterbirth") is expelled; the placental stage

_____ c. Time from onset of labor to complete dilation of the cervix; the stage of dilation

5. Define these terms:
a. Pudendal nerve block

b. Dystocia

6. A premature infant ("preemie") is considered to be one who weighs less than _____ gm (_____ lb) at birth. These infants are at high risk for the RDS (or _____). *For extra review* of RDS, refer to Chapter 23, Activity |29|, page LG 529.

7. What adjustments must the newborn make at birth as it attempts to cope with its new environment? Describe adjustments of the following:
a. Respiratory system (What serves as a stimulus for the respiratory center of the medulla?)

The respiratory rate of a newborn is usually about *(one-third? three times?)* that of an adult.

b. Heart and blood vessels. *(For extra review* go over changes in fetal circulation, Chapter 21, pages LG 467–468.)

c. Blood cell production

8. Explain why perhaps 50 percent of newborns may experience a temporary state of jaundice soon after birth.

9. Complete the exercise about potential hazards to the embryo and fetus. Match the terms at right with the related descriptions.

|18|

_____ a. An agent or influence that causes defects in the developing embryo

_____ b. May cause defects such as small head, facial irregularities, defective heart, retardation

_____ c. Linked to a variety of abnormalities such as lower infant birth weight, higher mortality rate, GI disturbances, and SIDS

C. Cigarette smoking
FAS. Fetal alcohol syndrome
T. Teratogen

|19| 10. Complete this exercise about lactation.

a. The major hormone promoting lactation is _____ which is secreted by the _____. During pregnancy the level of this hormone increases somewhat, but is inhibited by the hormone _____ which is produced during pregnancy while levels of estrogens and progesterone are *(high? low?)*.

b. Following birth and the loss of the placenta, a major source of estrogens and progesterone is gone, so prolactin *(increases? decreases?)* dramatically.

c. The sucking action of the newborn facilitates lactation in two ways. What are they?

d. What hormone stimulates milk letdown? _____

11. What is colostrum?

12. Explain how lactation may inhibit ovulation and therefore decrease likelihood of pregnancy in a breast-feeding mother.

13. List five advantages offered by breast-feeding.

D. Inheritance and medical terminology (pages 760–763)

1. Define these terms.
   a. Inheritance

   b. Genetics

c. Homologous chromosomes

2. Contrast:
   a. Genotype/phenotype

   b. Homozygous/heterozygous

3. Answer these questions about genes controlling PKU. (See Figure 29-16, page 761, in the text.)
   a. The dominant gene for PKU is represented by *(P? p?)*. The dominant gene is for the *(normal? PKU?)* condition.
   b. The letters PP are an example of a genetic makeup or *(genotype? phenotype?)*. A person with such a genotype is said to be *(homozygous dominant? homozygous recessive? heterozygous?)*. The phenotype of that individual would be *(normal? normal, but serve as a "carrier" of the PKU gene? PKU?)*.

   c. The genotype for a heterozygous individual is _____. The phenotype is

   _____.

   d. Use a Punnett square to determine possible genotypes of offspring when both parents are Pp.

4. In most traits the normal gene is *(dominant? recessive)*. Name several exceptions to this rule. (See Exhibit 29-3, page 761, in the text.)

5. Contrast *autosome* and *sex chromosome.*

6. Answer these questions about sex inheritance.
   a. All *(sperm? ova?)* contain the X chromosome. About half of the sperm produced by a male contain the _____ chromosome, and half contain the _____.
   b. Every cell (except those forming ova) in a normal female contains *(XX? XY?)* genes on sex chromosomes. All male cells (except those forming sperm) are said to be *(XX? XY?)*.
   c. Who determines the sex of the child? *(Mother? Father?)* Show this by using a Punnett square. (See Figure 29–17, page 762, in your text.)

7. Complete this exercise about X-linked inheritance.
   a. Sex chromosomes contain other genes besides those determining sex of an individual. Such traits are called _____ traits. Y chromosomes are shorter than X chromosomes and lack some genes. One of these is the gene controlling ability to _____ colors. Thus this ability is controlled entirely by the _____ gene.
   b. Write the genotype for females of each type: color blind, _____; carrier, _____; normal, _____.

c. Now write possible genotypes for males: color blind, _____; normal, _____.

d. Determine the results of a cross between a color-blind male and a normal female.

e. Now determine the results of a cross between a normal male and a carrier female.

f. Color blindness and other X-linked, recessive traits are much more common in (males? females?).

8. Name another X-linked trait. _____

Answers to Numbered Questions in Learning Activities

1   (a) Hundreds of millions, probably several hundred to thousands. (b) Peristaltic contractions and cilia of uterine (Fallopian) tubes. (c) Uterine tube. (d) Acrosomes of spermatozoa, dissolve covering around ovum. (e) One, electrical charges and enzymes of the ovum.

2   (a) DTwin. (b) TTrip. (c) MTwin. (d) MTrip.

681

3 Implantation, 21–23, trophectoderm, mother's uterine blood and glands.

4 Morning sickness in early months of pregnancy.

5 1978.

6 First two; developing individual during the last seven months of human gestation.

7 (a) Trophectoderm: part of chorion of placenta; inner cell mass: embryo and parts of placenta; amniotic. (b) Disc, ecto, endo, meso, primary germ. (c) Gastrulation, third. (d) Ecto, amnion, bag of waters. (e) Endoderm, gut, yolk, is not. (f) Chorion, chorion, extraembryonic coelom.

8 (a) Endo. (b) Ecto. (c) Meso. (d) Meso. (e) Meso. (f) Ecto. (g) Endo.

9 (a) Amnion. (b) Chorion. (c) Yolk sac. (d) Allantois.

10 (a) Flattened cake (or thick pancake), chorion, chorionic villi, brought into close proximity but do not mix, intervillous. (b) Decidua basalis, endo. (c) It is discarded from mother at birth (as the "afterbirth"); separation of the decidua from the remainder of endometrium accounts for bleeding during and following birth.

11 3, 2, 1.

12 (a) 1. (b) 3. (c) 1. (d) 3. (e) 6. (f) 1. (g) 2. (h) 2. (i) 3. (j) 5. (k) 9. (l) 7. (m) 8. (n) 8.

13 (a) High, corpus luteum, ovary, HCG. (b) First, third, urine. (c) Placenta, GnRF, HCG. (d) HCG, human chorionic somatomammotropin (HCS), stimulates development of breast tissue and alters metabolism during pregnancy. (e) Second, birth. (f) Relaxin, placenta, ovaries.

14 280, the beginning of the last menstrual period.

15 (a) In. (b) In, 45. (c) In. (d) De, constipation. (e) In, varicose veins or hemorrhoids.

16 (a) Amniotic, 16–20, does; to determine fetal maturity and diagnose genetic or chromosomal defects, such as Down's syndrome. (b) Chorionic villi sampling, 8–10, does not, chorion, are, genetic and chromosomal defects.

17 (a) Second. (b) Third. (c) First.

18 (a) T. (b) T, FAS. (c) T, C.

19 (a) Prolactin (PRL), anterior pituitary, PIF, high. (b) Increases. (c) Maintains prolactin level by inhibiting PIF and stimulates release of PRF, PRL, oxytocin. (d) Oxytocin.

20 (a) P, normal. (b) Genotype, homozygous dominant, normal. (c) Pp, normal but a carrier of PKU gene. (d) 25 percent PP, 50 percent, Pp, and 25 percent pp.

21 Dominant, see recessive traits in Exhibit 29-3 (page 761) of the text.

22 (a) Ova, X, Y. (b) XX, XY. (c) Father, as shown in the Punnett square since only the male can contribute the Y chromosome.

23 (a) X-linked, differentiate, X, (b) X^cX^c; X^CX^c; X^CX^C. (c) X^cY; X^CY. (d) All females are carriers and all males are normal. (e) Of females, 50 percent are normal and 50 percent are carriers; of males, 50 percent are normal and 50 percent are color blind. (f) Males.

MASTERY TEST: Chapter 29

Questions 1–2: Arrange the answers in correct sequence.

____ ____ ____ ____ 1. From most superficial to deepest (closest to embryo):

    A. Amnion
    B. Amniotic cavity
    C. Chorion
    D. Decidua

____ ____ ____ ____ ____ 2. Stages in development:

    A. Morula
    B. Blastocyst
    C. Zygote
    D. Fetus
    E. Embryo

Questions 3–10: Choose the one best answer to each question.

____ 3. All of the following are hormones made by the placenta EXCEPT:

    A. HCS   B. HCG   C. GnRF   D. pLRF
    E. Estrogen   F. Progesterone   G. Relaxin

____ 4. All of the following statements are true of males EXCEPT:

    A. Their genotype is XY.
    B. They determine sex of their offspring.
    C. They are more likely than females to be color blind.
    D. They usually produce more lubricating fluid during sexual intercourse than females do.

____ 5. All of the following structures are developed from mesoderm EXCEPT:

    A. Aorta   B. Biceps muscle   C. Heart
    D. Humerus   E. Sciatic nerve   F. Neutrophil

____ 6. Which part of the structures surrounding the fetus is developed from maternal cells (not from fetal)?

    A. Allantois   B. Yolk sac   C. Chorion   D. Decidua
    E. Amnion

____ 7. All of the following events occur during the first month of embryonic development EXCEPT:

    A. Endoderm, mesoderm, and ectoderm are formed.
    B. Amnion and chorion are formed.
    C. Limb buds develop.
    D. Ossification begins.
    E. The heart begins to beat.

_____ 8. Implantation of a developing individual (blastocyst stage) usually occurs about _____ after fertilization.

    A. Three weeks   B. One week   C. One day

    D. Seven hours   E. Seven minutes

_____ 9. The high level of estrogens present during pregnancy is responsible for all of the following EXCEPT:

    A. Stimulates release of prolactin   B. Prevents ovulation

    C. Maintains endometrium   D. Prevents menstruation

    E. Stimulates uterine contractions

_____ 10. Which of the following is a term that refers to discharge from the birth canal during the days following birth?

    A. Autosome   B. Cautery   C. Culdoscopy

    D. Lochia   E. PKU

Questions 11–20: Circle T (true) or F (false). If the statement is false, change the underlined word or phrase so that the statement is correct.

T   F   11. Oxytocin and progesterone both stimulate uterine contractions.

T   F   12. Of the primary germ layers, only the ectoderm forms part of a fetal membrane.

T   F   13. Maternal blood mixes with fetal blood in the placenta.

T   F   14. Amniocentesis is a procedure in which amniotic fluid is withdrawn for examination at about 8–10 weeks during the gestation period.

T   F   15. The normal gestation period is about 266 days following the beginning of the last menstrual period (LMP).

T   F   16. Stage 2 of labor refers to the period during which the "afterbirth" is expelled.

T   F   17. Maternal pulse rate, blood volume, and tidal volume all normally increase during pregnancy.

T   F   18. The decidua is part of the endoderm of the fetus.

T   F   19. Inner cell mass cells of the blastocyst form the chorion of the placenta.

T   F   20. A phenotype is a chemical or other agent that causes physical defects in a developing embryo.

Questions 21–25: fill-ins. Complete each sentence with the word or phrase that best fits.

——————————— 21. The newborn infant generally has a respiratory rate of ——, pulse of ——, and a WBC count that may reach ——, all values that are —— than those of an adult.

——————————— 22. As a prerequisite for fertilization, sperm must remain in the female reproductive tract for at least 4–6 hours so that —— can occur.

——————————— 23. HCG is a hormone made by the ——; its level peaks at about the end of the —— month of pregnancy, when it can be used to detect ——.

——————————— 24. Determine the probable genotypes of children of a couple in which the man has hemophilia and the woman is normal. ——

——————————— 25. This is the —— mastery test question in this book.

Answers to Mastery Tests

CHAPTER 1

Multiple Choice

1. A
2. A
3. C
4. D
5. B
6. B
7. D
8. D
9. E
10. B

True-False

11. T
12. F. Stand erect facing observer, arms at sides, palms forward
13. F. Right iliac
14. F. Physiology
15. T
16. F. Cell (or cellular)
17. F. At the intersection of four regions of the abdomen: right hypochondriac, right lumbar, epigastric, and umbilical
18. F. Two-dimensional
19. T
20. T

Fill-ins

21. Mediastinum
22. Muscular
23. Intracellular
24. Differentiation
25. Liver

CHAPTER 2

Multiple Choice

1. B
2. D
3. B
4. D
5. B
6. A
7. C
8. A
9. B
10. E

True-False

11. F. Inorganic
12. F. Two bonds since it requires two electrons
13. F. Water and NaCl (Oxygen is not a compound; glucose is organic.)
14. F. 20
15. F. Electrons
16. F. K^+ is a cation
17. T
18. T
19. T
20. F. Prostaglandins, steroids, and fats

Fill-ins

21. Amino acid
22. Monosaccharide or hexose or glucose
23. Polysaccharide or starch or glycogen
24. Adenosine triphosphate (ATP)
25. Fat or lipid or triglyceride (unsaturated)

CHAPTER 3

Multiple Choice

1. D
2. E
3. A
4. E
5. C
6. B
7. E
8. D
9. A
10. B

True-False

11. F. Phospholipid molecules
12. F. Some of the characteristics of

13. T
14. F. Flagella
15. T
16. T
17. T
18. F. Do have or possess
19. T
20. F. Nuclear division (Cytokinesis is cytoplasmic division.)

Fill-ins

21. Reduction
22. Peripheral
23. Phagocytosis
24. S (synthesis)
25. U-A-A-G-U-G

CHAPTER 4

Multiple Choice

1. C
2. D
3. B
4. A
5. C

Matching

6. Cartilage
7. Dense connective tissue
8. Stratified squamous epithelium
9. Loose connective tissue
10. Simple squamous epithelium

True-False

11. T
12. F. Good nutrition, good blood supply, and is younger
13. F. Endocrine
14. T
15. F. Poor conductor of heat and therefore reduces heat loss (provides insulation)
16. F. Extracellular or interstitial
17. T
18. F. Stratified
19. T
20. T

Fill-ins

21. Histology
22. Periosteum
23. Simple squamous epithelium
24. Avascular
25. Epithelium

CHAPTER 5

Multiple Choice

1. A
2. B
3. C
4. E
5. A

Arrange

6. C B A
7. A C B
8. C B A
9. B A C
10. B A C

True-False

11. F. Composed of different kinds of cells
12. F. Melanin
13. T
14. T
15. T
16. F. Positive
17. F. External root sheath
18. F. Only the dermis
19. F. Epidermis
20. T

Fill-ins

21. D or D_3
22. Corneum
23. Epidermal ridges or grooves
24. Sebum
25. Stimulating sweat glands to secrete

CHAPTER 6

True-False

1. F. Children
2. T
3. F. Mineral salts, and one third is due to collagenous fibers (The very few cells contribute little to bone weight.)
4. T
5. T
6. F. End of the bone (often bulbous)
7. F. Protrudes through the skin
8. T
9. F. Fluid from blood vessels in Haversian canals, but not blood itself
10. F. Only in origin
11. T
12. T
13. F. Osteoclasts
14. F. Hyaline cartilage; a few start out as fibrous membranes (originally all bones begin as mesenchyme).
15. T
16. F. Lacunae

Arrange

17. B C A
18. A C B
19. A C B
20. A B C

Fill-ins

21. Diaphysis
22. Callus
23. Perichondrium
24. Skull; also the clavicles
25. D (or D_3)

CHAPTER 7

Multiple Choice

1. C
2. B
3. B
4. B
5. D
6. A
7. B
8. C
9. C
10. C

Arrange

11. A B C
12. B A C
13. C B A
14. C A B
15. A C B

True-False

16. F. Intercostal space
17. T
18. F. IX, X, and XI
19. T
20. T. Kyphosis

Fill-ins

21. Axis (second cervical vertebra)
22. Mandible
23. Temporal
24. Ethmoid
25. It contains red bone marrow and it is readily accessible.

CHAPTER 8

Multiple Choice

1. E
2. A
3. D
4. C
5. B

Arrange

6. A C B
7. C A B
8. C B A
9. B C A
10. B C A

True-False

11. F. False
12. F. Lesser
13. F. Do not
14. T
15. F. Shallower and more oval
16. T
17. T
18. F. Only tibia and talus

19. F. No other bone (It is used for muscular attachment only.)
20. F. 30

Fill-ins

21. Scaphoid
22. Pott's
23. Fibula
24. Acetabulum
25. 56 (14 $\times$ 4)

CHAPTER 9

True-False

1. T
2. F. Sutures and syndesmoses
3. F. More common and usually less damaging than
4. T
5. F. Either synarthrotic or amphiarthrotic, and the same is true of cartilaginous
6. F. Do not cover surfaces of articular cartilages
7. T
8. F. Less
9. T
10. F. Flexion
11. T
12. T
13. T
14. F. Less

Arrange

15. B A C
16. C B A

Multiple Choice

17. D
18. A
19. C
20. B

Fill-ins

21. Bursae
22. Sprain
23. Retraction
24. Diarthrotic
25. Pivot (or synovial or diarthrotic)

CHAPTER 10

True-False

1. T
2. F. A bands but not I bands
3. F. Dense connective tissue (which may surround skeletal muscle)
4. F. Facilitate or enhance
5. F. An autoimmune response in which antibodies are produced which bind onto receptors on the sarcolemma
6. T
7. T
8. T

9. T
10. T
11. T

Arrange

12. C A B
13. C A B

Multiple Choice

14. B
15. C
16. A
17. C
18. B
19. E
20. D

Fill-ins

21. Sarcolemma
22. Passage of calcium ions from the extracellular fluid through sarcolemma
23. Latent
24. Acetylcholine
25. ATP + creatine

CHAPTER 11

Arrange

1. B C A
2. A C B

True-False

3. T
4. T
5. F. Antagonists
6. F. Shape (trapezoid)
7. T
8. T
9. T
10. F. The latissimus dorsi extends, but the pectoralis major flexes
11. T
12. F. Two heads of origin; but origins of biceps brachii are on the scapula, and origins of biceps femoris are on ischium and femur

Multiple Choice

13. D
14. C
15. D
16. C
17. B
18. D
19. A
20. D

Fill-ins

21. Flexion
22. Extension

688

23. Posterior
24. Figure LG 10-2: *T* (anterior fibers) *U V*; Figure LG 10-3: none.
25. Figure LG 10-2: *AA BB CC DD*; Figure LG 10-3: *I*.

CHAPTER 12

Arrange

1 A C B
2 A C B

Multiple Choice

3. B
4. C
5. A
6. C
7. C
8. E
9. B

True-False

10. T
11. F. May be myelinated since neuroglia myelinate CNS fibers
12. T
13. F. A number of presynaptic knobs
14. T
15. F. Diverging
16. T
17. T
18. F. Amino acids
19. T
20. F. Spinal nerves (as well as some other structures; but not the brain)

Fill-ins

21. Integrator
22. Cooling of neurons slows down the speed of nerve transmission, for example, of pain impulses
23. End bulbs of axons
24. Peripheral
25. Cell body (soma)

CHAPTER 13

Arrange

1. B C A
2. B C A
3. C B A D
4. D B C A E

Multiple Choice

5. B
6. D
7. C
8. A
9. D
10. B

True-False

11. F. Intact cell body and a neurilemma
12. F. Reflex center, conduction site
13. T
14. T
15. F. Are not
16. T
17. F. Less often than somatic ones since it is difficult to stimulate most visceral receptors
18. T
19. Axons of motor
20. F. L3-L4, L1-L2.

Fill-ins

21. Ganglia
22. Meningitis
23. Poly, ipsi
24. Thoracic, upper lumbar, sacral (further explanation in Chapter 16)
25. Pia

CHAPTER 14

Multiple Choice

1. E
2. B
3. A
4. A
5. D
6. C
7. B
8. B

Arrange

9. B A C
10. C A B
11. A C B
12. A B D C

True-False

13. F. Serotonin and acetylcholine
14. T
15. F. Endorphins and enkephalins
16. F. Cerebral cortex
17. T
18. T
19. T
20. T

Fill-ins

21. Association
22. Midbrain
23. Lysosomes
24. Sweating, dilation of blood vessels of skin
25. Cerebrum, cerebellum

CHAPTER 15

Multiple Choice

1. A
2. D
3. D
4. C

Arrange

5. A B C
6. D A C B
7. D B A C
8. B C D E A

True-False

9. T
10. F. Do not obey
11. F. Lower motor
12. F. Events that occur at approximately 24-hour intervals, such as sleeping and waking
13. F. Sight, hearing, and smell (not pressure)
14. T
15. T
16. F. Randomly distributed, that is, of varying density
17. F. Pain and temperature sensations in the right
18. T
19. F. Referred pain
20. T

Fill-ins

21. Substance P
22. Reticular activating system
23. Extrapyramidal
24. Loss of sensation of left foot
25. Activated

CHAPTER 16

Multiple Answers

1. B, C, D, E
2. A, E, F
3. A, C, E, G
4. A, B, C, D
5. A, B, E, F
6. B

Multiple Choice

7. E
8. A
9. C
10. D

True-False

11. F. Stressed
12. T
13. T
14. F. Dependent on
15. T
16. F. Two neurons are
17. T
18. T
19. F. Some (or most)
20. T

Fill-ins

21. Cardiac muscle, smooth muscle, and glandular epithelium
22. Splanchnic
23. Vagus
24. NE (and also epinephrine)
25. De

CHAPTER 17

True-False

1. F. Decreased, hyperpolarization
2. F. Divergent
3. T
4. F. Aqueous humor, but not vitreous humor, is
5. F. More
6. F. Middle
7. F. Accommodation, but not convergence, is a result

Arrange

8. A C B
9. C D A B
10. C D A B
11. A E C B D
12. A D E B C
13. B C A D E
14. A B E C D

Multiple Choice

15. B
16. C
17. E
18. D
19. C
20. A

Fill-ins

21. Cataract
22. Cis-
23. Ophthalmology
24. Malleus, incus and stapes
25. Production of aqueous humor and alteration of lens shape for accommodation

CHAPTER 18

Arrange

1. B A C D
2. D B A C

Multiple Choice

3. C
4. C
5. A
6. D
7. E
8. B
9. B
10. E

True-False

11. F. Polyphagia and polyuria are both
12. F. Hypothalamus and affect the anterior pituitary
13. F. Sympathetic
14. T
15. F. Cortisol is
16. T
17. F. Does not permit
18. T
19. T
20. T

Fill-ins

21. Oxytocin (OT or pitocin)
22. Thyroid hormones (T_3 and T_4), epinephrine or norepinephrine
23. MSH, ACTH, enkephalin, endorphin, and β-LPH
24. II (maturity onset)
25. Decreased extracellular water, pain, stress, trauma, anxiety, nicotine, morphine, tranquilizers

CHAPTER 19

Arrange

1. C A B
2. A B C

Multiple Choice

3. A
4. E
5. A
6. D
7. D
8. B
9. C
10. A
11. C

True-False

12. T
13. F. Dissolution (or fibrinolysis)
14. T
15. F. Stage 2
16. F. Negative mother and her Rh positive babies
17. F. Extrinsic; intrinsic
18. F. Lower
19. T
20. F. 47 ml (males) to 42 ml (females)

Fill-ins

21. Oxygen, carbon dioxide, nutrients, wastes, sweat, hormones, enzymes
22. Albumin, globulins, and fibrinogen
23. Lymphocyte
24. 5,000–9,000
25. Hemocytoblasts

CHAPTER 20

Arrange

1. C A B D
2. C E D A B

3. D E A B C
4. C D E A B

Multiple Choice

5. D
6. C
7. B
8. D
9. B
10. B
11. C
12. D
13. A
14. A

True-False

15. T
16. F. Diastole (AV valves are open.)
17. F. Close
18. F. Veins carry
19. T
20. F. No part

Fill-ins

21. Right to left
22. Coronary artery disease
23. Electrical changes or currents that precede myocardial contractions
24. Arrhythmia or dysrhythmia
25. In

CHAPTER 21

Multiple Answers

1. A C E F G
2. A D G

Multiple Choice

3. D
4. A
5. B
6. C
7. D
8. B
9. D
10. C
11. D
12. B
13. E
14. B

Arrange

15. A B C D
16. A B C D
17. C D B A
18. A B D C
19. A B C D E
20. B C A

Fill-ins

21. Aneurysm
22. Great Saphenous
23. External jugular
24. Systolic
25. Aorta, anterior communicating cerebral, basilar, brachiocephalic, celiac, gastric, hepatic, splenic, superior and inferior mesenteric, middle sacral

CHAPTER 22

Arrange

1. A C B D
2. D C B A
3. A C D B E

Multiple Choice

4. E
5. D
6. A
7. B
8. C
9. C
10. A
11. C
12. C
13. B
14. C
15. C
16. C
17. D
18. A

True-False

19. F. Stimulate formation of
20. F. Antibodies against the individual's own tissues

Fill-ins

21. B, T
22. Lymphokines
23. Suppressor T
24. E
25. Skin, mucosa, cilia, epiglottis. Also flushing by tears, saliva, and urine.

CHAPTER 23

Arrange

1. A B C
2. A C B
3. D C E B A
4. A E C B D
5. D C E B A F

Multiple Choice

6. C
7. A
8. C
9. A

True-False

10. F. HCO_3^-
11. T
12. F. Dissociate from
13. T
14. T
15. F. Higher
16. T
17. T
18. F. Intraalveolar
19. T
20. T

Fill-ins

21. External respiration (or diffusion)
22. Functional residual capacity (FRC), 2,400
23. Oxygen
24. Larynx
25. Surfactant

CHAPTER 24
Multiple Choice

1. B
2. C
3. C
4. D
5. B
6. A
7. A
8. D
9. A
10. D

Arrange

11. A C B
12. A D B C
13. E D B A C
14. A D C B E
15. A B D C E

True-False

16. T
17. T
18. F. No digestive enzymes, but it does produce mucus
19. F. Villi, microvilli, and goblet cells (Rugae are in wall of the stomach.)
20. T

Fill-ins

21. De, in
22. Parotid
23. Small intestine; water, electrolytes, alcohol, and some drugs such as aspirin
24. Chewing
25. Endo

CHAPTER 25
Multiple Choice

1. C
2. E
3. D
4. B
5. E
6. A
7. E
8. C
9. D
10. C
11. D
12. D
13. A

Arrange

14. A B D C F E

True-False

15. F. Removal of hydrogens
16. F. Twice
17. F. Are necessary for synthesis of body protein, but cannot be synthesized by the body
18. F. 38
19. F. Synthetic reactions that require energy
20. T

Fill-ins

21. 4, 9
22. Carbohydrate, protein
23. Sodium (Na^+)
24. Prostaglandins (PG), hypothalamus
25. Acetoacetic (keto-), fats

CHAPTER 26
Arrange

1. A B C
2. A C B D
3. E B A D C
4. C E A B D
5. A C E B D
6. D B C A E F

Multiple Choice

7. B
8. D
9. B
10. A

True-False

11. F. Increases
12. F. A kidney to the bladder
13. T
14. T
15. T
16. F. More
17. F. Pull substances from filtrate into blood
18. T
19. T
20. F. Incontinence

Fill-ins

21. Renal corpuscle
22. Nephron
23. 1.008 (or other low value; not high values like 1.030)
24. Afferent, de, de
25. H^+, erythropoietic factor, vitamin D

CHAPTER 27

True-False

1. T
2. F. Stronger
3. F. Equals
4. T
5. T
6. T
7. T
8. T
9. F. HCO_3^-, Na^+, and K^+ (Starch and glucose are nonelectrolytes.)
10. F. K^+
11. T
12. T
13. T
14. T
15. T
16. F. Move substances in opposite directions: BHP pushes substances out of blood and IFHP pushes substances into blood.

Multiple Choice

17. D
18. D
19. D
20. E

Fill-ins

21. Extra
22. 67, 33
23. Electrolytes
24. H^+, HCO_3^-
25. 137–150, 3.5–5.0

CHAPTER 28

Arrange

1. B C A
2. A B C
3. A B C
4. B A C
5. A D B C
6. C B A D
7. C D B A
8. B E D A C
9. B C D A E
10. F C B D A E

Multiple Choice

11. B
12. C

Multiple Answers

13. A, B, C
14. B, E, F
15. B, C, D

True-False

16. F. Spermatogonia; primary spermatocytes (Oogonia do not divide after birth of female baby.)
17. T
18. T
19. F. One
20. T

Fill-ins

21. Cryptorchidism
22. Progesterone and estrogens
23. Sustantacular
24. De, progesterone, and estrogens
25. Reproductive organs, fluid and electrolytes, ana

CHAPTER 29

Arrange

1. D C A B
2. C A B E D

Multiple Choice

3. C
4. D
5. E
6. D
7. D
8. B
9. A
10. D

True-False

11. F. Oxytocin but not progesterone stimulates
12. F. All three layers form parts of fetal membranes: ectoderm (trophectoderm) forms part of chorion; mesoderm forms parts of the chorion and the allantois; endoderm lines the yolk sac.
13. F. Does not mix but comes close to fetal blood
14. F. Amniocentesis, 16–20
15. F. 280
16. F. Fetus
17. T
18. F. Endometrium of the uterus
19. F. Trophectoderm
20. F. Teratogen

Fill-ins

21. 45, 160, 45,000, higher
22. Capacitation (dissolving of the covering over the ovum by secretion of acrosomal enzymes of sperm)
23. Chorion of the placenta, second, pregnancy
24. All male children normal; all female children carriers
25. Last or final. Congratulations! Hope you learned a great deal.